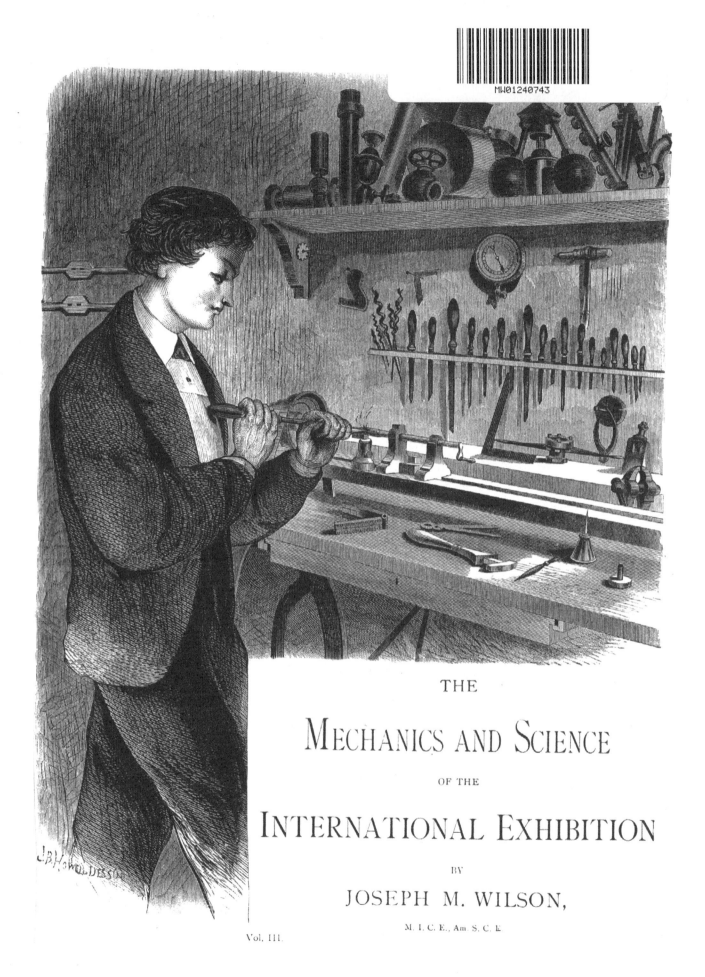

THE

Mechanics and Science

OF THE

International Exhibition

BY

JOSEPH M. WILSON,

M. I. C. E., Am. S. C. E.

Vol. III.

Masterpieces of the Centennial Exposition 1876
Volume 3: Mechanics, Science and History Illustrated
Special Edition

by Joseph Wilson

Copyright © 2020 Inecom, LLC.
All Rights Reserved

No parts of this book may be reproduced or broadcast in any
way without written permission from Inecom, LLC.

www.CGRpublishing.com

More books by CGR Publishing:

1939 New York World's Fair: The World of Tomorrow in Photographs

1904 St. Louis World's Fair: The Louisiana Purchase Exposition in Photographs

Masterpieces of the Centennial Exposition 1876: Volume 1 Fine Art

The International Exhibition, 1876.

MUCH of the inventive genius for which the people of America have always been noted has been developed by the necessities of the case. When first discovered by the inhabitants of Europe, she was in reality a new world, covered by immense forests, possessing vast mineral resources, with a great extent of territory from temperate to semi-tropical, having a variety and richness of soil adapted to the cultivation of wheat, corn, sugar, cotton, and all the productions most useful to man, destined therefore, to be the future producing market of the world, having great natural facilities for the manufacture of these products, and with all this, lying in a state of nature, inhabited by a savage people, and entirely undeveloped.

The great scarcity of human labor in this new country even to the present day, has stimulated and sharpened the inventive faculties of her citizens towards the designing of machinery which should accomplish that for which hand-labor could not be found, until the talent has become a second nature, and stamped its impress upon the nation. Under these circumstances, it is but natural that we should expect to find in the great Centennial Exhibition of this nation an unprecedented display of machinery. The history of previous exhibitions leads us to anticipate this. Notwithstanding her great distance from foreign countries, requiring her representation at these exhibitions to be always small, yet her exhibits have universally made a marked impression and have been noted for those unexpected labor-saving novelties which so cheapen the cost of manufacture and increase productiveness.

The Corliss Engine; Side Elevation.

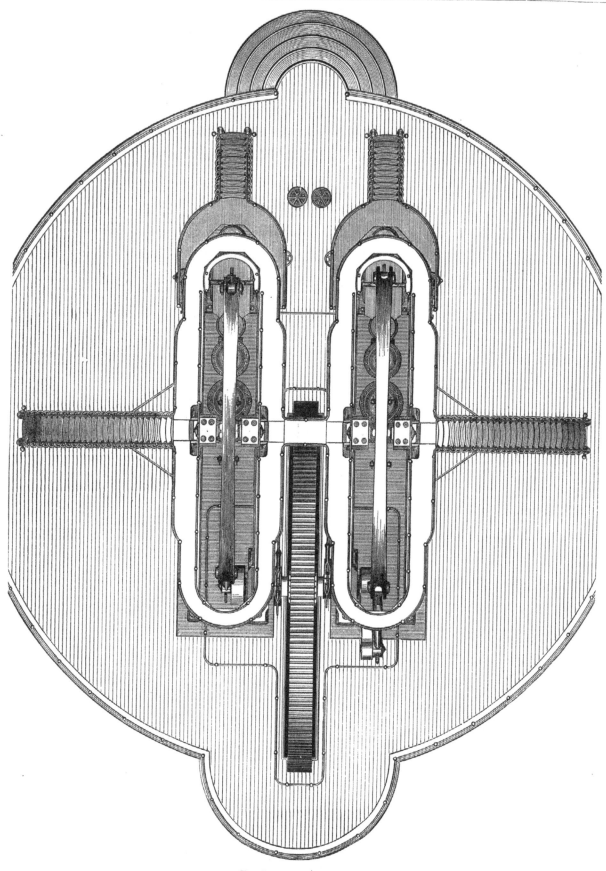

The Corliss Engine; Plan.

Greater prominence has been given to the Machinery Department of this exhibition than has ever been the case before, and a building has been erected which has been pronounced by those most competent to judge as the best ever provided for the purposes intended. In the centre of this building stands that wonder of the modern era, that thing which needs but the breath of life—what no human being has yet been able to invent, but must always supply by himself—to be a creation; that invention upon which now depends the daily bread of hundreds of thousands of people; a grand and noble specimen of the "Steam Engine." There it stands, holding its place as a veritable king among machinery, so powerful and yet so gentle, capable of producing the most ponderous blows upon the anvil, or of weaving the most delicate fabrics; that to which all other machines must be subservient, and without whose labor our efforts would be small indeed; the breathing pulse, the soul of the machinery exhibition. This engine, which may be seen from all parts of the building towering up above everything else, comes from little Rhode Island, from the city of Providence, and for its existence the Exhibition is indebted to the energy and perseverance of Mr. George H. Corliss, proprietor of the Corliss Steam Engine Company.

Early in 1875, when the question came up of the power that would be required in the building to run the fourteen acres of machinery which it was expected would be on exhibition, Mr. George H. Corliss, Centennial Commissioner from Rhode Island, conceived the idea of providing a single engine that should furnish this power—estimated at 1400 horses—placing it in the centre of the Machinery Hall, and he made an offer to the Commission to this effect. The Commission feeling that the honor of supplying this power should be distributed among different establishments, did not accept the offer, but invited proposals from prominent firms for what would be necessary to operate either the whole of the shafting in the building, or each line of shafting separately, so that it might be distributed among as many different parties as there were lines of shafting. When the proposals were received, it was discovered that they were not sufficient altogether to cover the requirements of the exhibition, that none of the bidders would agree to furnish the whole of the power, and that none would provide the boilers and connections necessary for the complete execution of the work. It was also found that the cost to the Commission would be much greater, and that the care and attention required in giving out a number of varied contracts to different establishments, with the additional trouble—

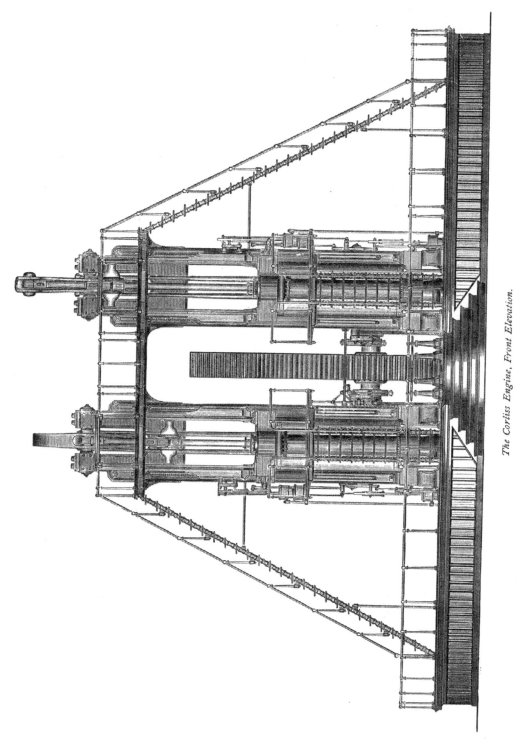

The Corliss Engine, Front Elevation.

while the Exhibition would be in working order—of looking to those contractors for proper supply of power, would, in all human probability, be more than by Mr. Corliss' proposition. After considerable delay, Mr. Corliss was then unanimously requested to renew his offer, and in June, 1875, eleven months before the opening, a contract was closed with him for the work, giving him only about ten months to construct that for which there had not as yet been even the first sketch prepared. How promptly the work has been performed, will be apparent when we state that on the day fixed by the contract—April 17, 1876—the engine was up in place, steam was turned on, and it was run for some time with perfect success.

The special characteristics of the Corliss Engine as compared with other steam engines may be said to consist in the valve gear, the form of valve and the peculiar method adopted by which steam is freely admitted at the full boiler pressure and discharged after use, without presenting any resistance to the piston. Independent parts are used for admitting and exhausting the steam with four separate valves, the steam being cut off from the cylinder entirely by the main steam valves without the employment of any supplementary valves. The steam valves are opened against the resistance of springs, and a liberating gear is called into play, disconnecting the valves and leaving them free to be closed by the springs. These springs are brought to rest without shock after closing by means of an air cushion formed by a small cylinder with a closed bottom, in which a piston is fitted to work easily, a certain amount of air being imprisoned just as this piston approaches the bottom, acting as a cushion and preventing any shock. The valves of admission are regulated by direct connection with the governor, thus controlling the speed of the engine without the use of a throttle-valve in the main supply-pipe.

The principal inventions which distinguish the Corliss Engine were devised in 1848 and patented in 1849, a beam engine of 260 horse-power having been constructed with the new improvements for use in Providence. The success of this engine was so great as to induce the parties with whom Mr. Corliss was associated to erect new works in the summer of 1848 for the purpose of extending their business. The machine-shop then built, covering an acre of ground, is yet standing intact. About two years after this time, a course of patent litigation commenced, extending over a period from 1850 to 1865, involving the firm and very much embarrassing its operations. Mr. Corliss from being the defendant in the first place, was afterwards obliged to take the place of plaintiff to maintain his position. The various cases incurred an expense of over $106,000 cash, taking up about one-third of Mr. Corliss' personal time, but in the end he was fully vindicated, and the final judgments of the court rendered in his favor.

In 1857, the Corliss Steam Engine Company was organized under the laws of the State to take charge of the business of the concern, and for more than five years past Mr. Corliss has been the sole owner of the whole of the stock. The engine which we have before us is the result of one individual's effort.

In reference to the special points which distinguish this particular engine, it may be said to be a strictly engineering design, the material of the framework being arranged directly on the constructive lines so as to best resist the action of the forces which come upon it at the least waste and expense. The forms of the lines of curvature of the walking-beams are important in considering the effects of cooling in the castings, and the shape is that best adapted to resist the stresses to which they are exposed. The arrangements by which the keys of the connecting-rods are accessible in any position of the beams, and the cutting away of the lower portions of the beams, making the lower lines different from the upper, are all noticeable. In fact, every detail of the design has been made with a view to the strength, symmetry and accessibility of the parts. The valve-gearing is a very novel and characteristic feature, all the peculiarities having been emphasized and carried to greater perfection than in any previous engine manufactured by the company. The crank-pins are covered with a coating of the best steel, hardened and ground as smooth as glass, and all work on the machine has been carried out with equal thoroughness. The engine is really a double engine formed of two large beam engines of 700 horse-power each, set upon a raised platform fifty-five feet in diameter, and having between them a single fly-wheel (a gear), the cranks of both connecting with the same crank-shaft. Although possessing nominally 1400 horse-power, it may be increased, if found necessary, to 2500 actual horse-power. The cylinders are 40 inches in diameter, and the stroke is 10 feet, the intention being to work with from twenty-five to eighty pounds of steam, according to the requirements of the Exhibition. An engine at the Wamsutta Mills, New Bedford, has a larger cylinder, but it is not so heavy an engine, and will not stand so high a pressure or do as much work as this one. The air-pump and condensing apparatus are fully provided as required. The gear fly-wheel, which is 30 feet in diameter and cast in sections, weighs about 56 tons, and is believed to be the heaviest cut wheel ever made. It has two hundred and sixteen teeth, finished with the greatest accuracy, moving without noise at the required rate of running, viz., thirty-six revolutions per minute. The crank-shaft is 19 inches in diameter and 12 feet long, made of the best hammered iron, and the bearings are 18 inches in diameter and 27 inches long. The cranks are of gun-metal, weighing over 3 tons each. The walking-beams—which weigh about 11 tons each—are cast in one piece, and are 9 feet wide in the centre and 27 feet long. The connecting-rods are of horse-shoe scrap-iron, and the piston-rods, 6¼ inches in diameter and of steel, their velocity at the regular rate of speed being 720 feet per minute. The large gear-wheel, connecting with the gear fly-wheel, is 10 feet in diameter, and is a single casting of 17,000 pounds weight. It is placed on a main shaft 252 feet in length, running crosswise of the building, and connecting at the ends and at two intermediate points by nests of bevelled gear 6 feet in diameter to shafts 108 feet in length, running at right-angles to the main shaft, and extending to points directly under the lines of overhead machinery shafting. These four connecting shafts have at their ends the main pulleys—eight in all—seven of them 8 and one 9 feet in diameter, and each 32 inches across the face. They are connected to the main machinery shafting overhead by double belting 30 inches in width, making an aggregate width for all together of 20 feet, required for the transmission of the whole power of the engine. Each belt drives a line of shafting of over 600 feet in length, with a separate section of the machinery, and where the belts rise from the floor, they are enclosed by glass partitions so that they may be out of the way and yet be visible to the visitor as exhibits.

The engine as a whole is 39 feet in height from the main floor of building to the top of walking-beam at its highest pitch, and every part is easily accessible by means of balconies and stairways. Its total weight, including everything connected with it, is about 680 tons. The general proportions are exceedingly harmonious and graceful, and the details simple and in excellent taste, the framework, walking-beams, balconies, etc., being painted of a quiet, uniform tint, relieved only by the polished work of the cylinders and moving parts.

The boiler-house—just south of the Machinery Hall—contains twenty Corliss upright boilers of 70 horse-power each, of a simple, vertical tubular type, entirely accessible inside and out. The water rising very rapidly around the tubes in the central part of the boiler is provided with a large return space next to the outside where the current moving more slowly allows an opportunity for the deposit of sediment, which may readily be removed. At the top, each tube is easily reached for cleaning when required. Horizontal flues, lined with fire-brick connect with two brick chimneys, and the steam is conveyed to the engine by means of a double-riveted wrought-iron pipe 18 inches in diameter and 320 feet long, well protected with felting to retain the heat of the steam, and carried through an underground passage, lighted by gas, and sufficiently large for a man to easily walk its whole length.

Among the firms represented at the Centennial Exhibition none occupy greater prominence than MESSRS. BURNHAM, PARRY, WILLIAMS & CO., of the BALDWIN LOCOMOTIVE WORKS, PHILADELPHIA.

This establishment owes its origin to Matthias W. Baldwin, who commenced business in Philadelphia in 1817 as a jeweller, and eight years later entered into partnership with Mr. David Mason, a machinist, for the manufacture of bookbinders' tools and cylinders for calico-printing. Requiring a stationary engine for carrying on the work, and not finding a satisfactory one in the market, Mr. Baldwin determined to make one for himself. His efforts were successful, and an engine was built of such excellent design, workmanship and efficiency as readily to procure for the firm orders for similar engines. Thus Mr. Baldwin's attention was turned to steam-engineering and the way prepared for taking up the problem of the locomotive when the time should arrive. The original engine is still in use at the works of the present firm, quietly performing its allotted duty from day to day, and preserved as the germ of the mammoth industry now being carried on by the largest exclusively locomotive-building firm in the world.

Mr. Mason soon leaving the firm, Mr. Baldwin continued the business alone. His first locomotive was in miniature, constructed for Peale's Museum, of Philadelphia, in 1831, and its success was such that he received an order for a locomotive for the Philadelphia, Germantown and Norristown Railroad Company, —then being worked by horses—which he completed and delivered in November, 1832. The "Old Ironsides," as it was called, although constructed under many difficulties, was, compared with others of the day, a marked and gratifying triumph, and it did effective duty for many years afterwards.

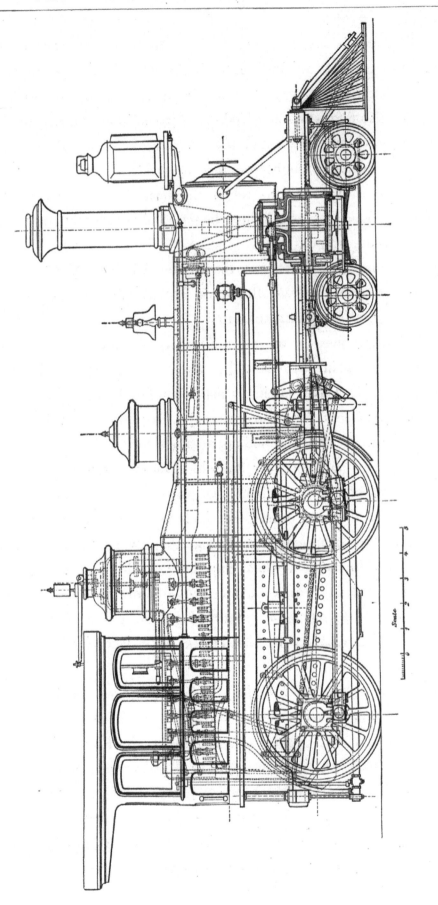

Passenger Locomotive: Burnham, Parry, Williams & Co.

Van Ingen & Snyder, Engs.

From this the business has progressed up to the present time, passing through many changes but always with Mr. Baldwin at the head until his death in 1866, when a new organization was effected and afterwards modified to the present firm. Its history corresponds with that of many others of our prominent business organizations, filling an important page in the history of the progress of machine manufacture in this country. The works now have a capacity of five hundred locomotives per year, the total number that have been constructed in all being about five thousand.

Among the exhibits of this firm is a Passenger Locomotive built for the Central Railroad of New Jersey, and represented by the engraving on the preceding page. It has a gauge of road of 4 feet 8½ inches and a total wheel-base of 44 feet 2 inches, including tender, or of locomotive alone, 22 feet 5 inches. The driving-wheels are 8 feet 6 inches to centres and 5 feet 2 inches in diameter, having centres of cast-iron with hollow spokes and rims. The truck-wheels are 2 feet 4 inches in diameter. The Washburn steel tires are used. The total weight of the locomotive in working order is 75,000 pounds, and the weight on driving-wheels 51,500 pounds. The cylinders are placed horizontally, each cylinder cast in one piece with half saddle, right- and left-hand cylinders, reversible and interchangeable, the diameter of cylinder being 1 foot 5 inches, and length of stroke 1 foot 10 inches. The oil-valves to cylinders are placed in the cab and connected to steam-chests by pipes running under jacket of boiler. The boiler is of the wagon-top type, furnished with one dome and made of best homogeneous cast-steel three-eighths of an inch thick, manufactured by Hussey, Wells & Co., the outside diameter at the smallest ring being 4 feet. The fire-box is of cast-steel of the same manufacture, 8 feet 6 inches long by 2 feet 9¾ inches wide, the side sheets being one-fourth and the back sheet five-sixteenths of an inch thick. The tubes are of iron, lap-welded, made by W. C. Allison & Sons, with copper ferrules on the fire-box ends. They are 2 inches in diameter, 11 feet 3 inches long and one hundred and sixty-three in number. The heating surface comprises 1065 square feet, including grate, fire-box and tubes.

The tender has eight wheels of 2 feet 6 inches diameter, furnished with Taylor's steel tires. The capacity of tank is 2200 gallons, the tank-iron being manufactured by the Catasauqua Manufacturing Company. The engine throughout is finished according to the high standard for which this firm is so celebrated—solid, substantial and neat, with no useless ornamentation, every part fitted to gauges and thoroughly interchangeable, the whole being an excellent specimen of American manufacture.

In connection with the motive-power exhibits we may very appropriately mention Lonergan's Patent Oil-Cups and Automatic Lubricators. The principles upon which they work have been beautifully carried out, resulting in most excellent forms of apparatus for the requirements. The oil-cups are of several varieties to suit different purposes. Our first engraving, on page 15, shows the usual construction for stationary motion, being partly in section and partly an exterior view.

It consists of a metallic cup or casing, A, A, pierced by diamond-shaped openings in the cylindrical part, with a tube, B, to be connected with, and passing to, the part to be lubricated. Inside of this casing is a glass cylinder, C, with cork rings, D, D, at top and bottom. The cap, E, screws down tightly on to the cork, making an oil-tight joint. A plug, F, with ground joint, and held in place by a spiral spring, G, effectually closes the tube B and prevents the passage of oil unless desired otherwise. This plug connects with the handle, H, H, on top, the connection being movable through the cap, E, of the casing and hollow in the upper portion as shown at I, I, there being openings, K, at the lower end of this hollow space, and a cap, L, screwed on at the top, the latter having an air-hole, O, pierced through it. A set screw, M, passes through the rim of the handle, H, with a rest, N, for the same in the cap, E. When it is desired to fill the cup, the cap, L, is unscrewed and the oil poured in, the handle, H, being turned around until the set-screw, M, is off of its rest, N, the plug, F, then tightly closing the entrance to the tube, B. After filling and replacing the cap, L, then by turning the handle, H, and placing the set-screw, M, on its rest, we can, by adjusting this screw, regulate exactly the required amount of opening necessary at F for the proper oiling of the machine. When the machine is at rest and no oiling needed, it is only requisite to raise the handle, H, and turn it so as to move the set-screw from its rest, and the spring, G, at once closes the plug, F, into the opening of B and stops the consumption of oil.

Figure 2 shows a modification adapted to movable parts under rotary motion. The spring, G, is dispensed with and the loose plug, F, has a little stop, P, in it, the set-screw, M, being differently arranged as shown. At each rotation the loose plug,

F, is thrown up, the distance of its throw being regulated by the set-screw, and a certain amount of oil finds its way down the tube, B, B.

The Automatic Lubricator, as shown by Fig. 3, has a cup composed of the best quality steam metal and made extra heavy, with a regulating arm on top allowing adjustment to any feed desired, there being small holes or rests for the end of the jam-screw of this arm at short intervals all around the circumference of the cap of cup. In arranging it for use, the valve, B, is closed and the cup filled with lubricant through the stem, E. The top, D, which has a lignum-vitæ handle to prevent heating, is then screwed down tight, and the valve, B, opened by turning the handle until the indicator or arm is half-way around the cup. After a moment it is moved partly back to within say six or eight holes from the starting-point. By actual experience, the engineer can adjust this to the exact requirements of his engine. Steam passes into the cup by the valve, B, and condensing into water sinks to the bottom of the oil, lifting an equal amount of the latter to the top, which flows down the pipe to the parts where lubrication is desired. The large opening in the pipe to the top of the feed-valve allows a circulation of steam, keeping the lubricant in a liquid state independent of outside temperature and securing thereby a uniform feed. A waste-cock is provided for drawing off the condensed water and impurities which collect in the bottom of the cup.

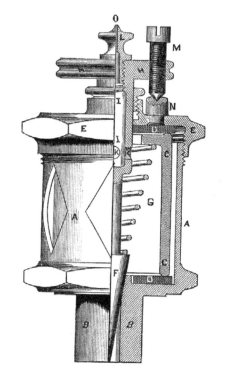

Lonergan's Oil-Cup; Fig. 1.

Especial attention may be called to the ease and precision with which these cups may be regulated to any desired feed and the evident saving in oil thus effected, and in the case of the last form described, the admission of steam into the cup at all times, whether feeding or not, is an important point, keeping the lubricator always in a condition for use without waste or trouble. These oil-cups have been introduced on the engine "Dom Pedro II," of "Baldwin Locomotive Works," on exhibition in the Machinery Hall.

Machine tools, as distinguished from hand tools, are those designed for planing, shaping, drilling, or boring metal, wood, or stone, in which mechanical

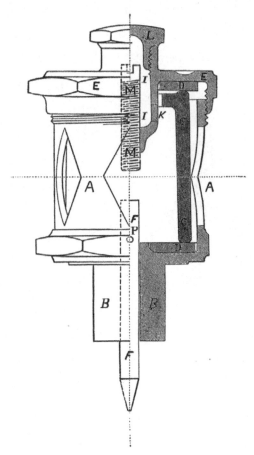

Lonergan's Oil-Cup; Fig. 2.

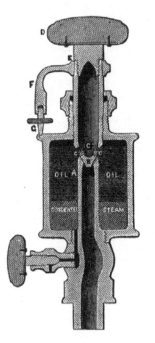

Lonergan's Oil-Cup; Fig. 3.

appliances take the place of manual skill in guiding the tool or cutting edge in its determined path. The term comprises not only turning-lathes and drill-presses, but also hydraulic forging-machines, steam-hammers, riveting-machines, punching and shearing-machines, and, in fact, all those for working metal or other material in which the above condition is fulfilled.

The manufacture of machine tools of late years has become a specialty. The requirements of modern engineering demand an extensive use of these tools, and exact that they shall be made in the greatest attainable perfection, and as far as possible self-acting or automatic, capable when once started at work of doing perfectly what is given them to do, without depending upon the attendant for the result.

We have already mentioned in the "History of Previous Exhibitions" the exceedingly favorable impression produced by the machine tools of Messrs. William Sellers & Co. as exhibited abroad. This firm has taken a prominent position in the present exhibition, occupying a large extent of space in the Machinery Hall with their characteristic and novel machines. From among the number we may mention a patent self-acting planing-machine for horizontal, vertical and angular planing of any required length. This machine differs essentially from ordinary planers, possessing peculiarities that impress the beholder at once with the amount of master-thought that has been expended upon its design. Attention is first attracted to the method by which motion is given to the table holding the object to be planed. A special form of spiral pinion being used, placed upon a driving-shaft which crosses the bed diagonally, passing out in the rear of the upright on the side next to the operator and connecting with the pulley-shaft by means of a bevel-wheel and pinion. The location of this pulley-shaft, as may be seen from the engravings, brings the driving-belts within easy reach of the workman, and the fact of its axis being parallel to the line of motion of the table permits the machine to be placed side by side with lathes, thus economizing space in the shop. There are four teeth on the pinion, arranged like the threads of a coarse screw of steep pitch, and working into a rack on the table, the teeth of which are straight and are placed at an angle of five degrees to its line of motion, to counterbalance any tendency of the pinion to move the table sideways.

We understand that this arrangement for moving the table has been found to be very durable, the operation of the teeth being more of a rolling action than a rubbing or sliding one. A strong box-shaped connection between the sides of the bed, just at the uprights, holds them very firmly together, an advantage not attainable in most other forms of planers where the methods adopted for giving motion to the table do not allow the required space. The plan of diagonal shaft adopted has another superiority in throwing the bevel-wheel and pinion driving it, out from under the table and allowing the former to be made of any necessary size compared to the latter as may be required to give the requisite reduction in speed and transmission of power from a high-speed belt without the interposition of other gearing, whereas, in the ordinary screw-planer, the projection of the table over the ends of the bed limits the size of the gearing at the end of the screw. The method adopted for shifting the belts constitutes

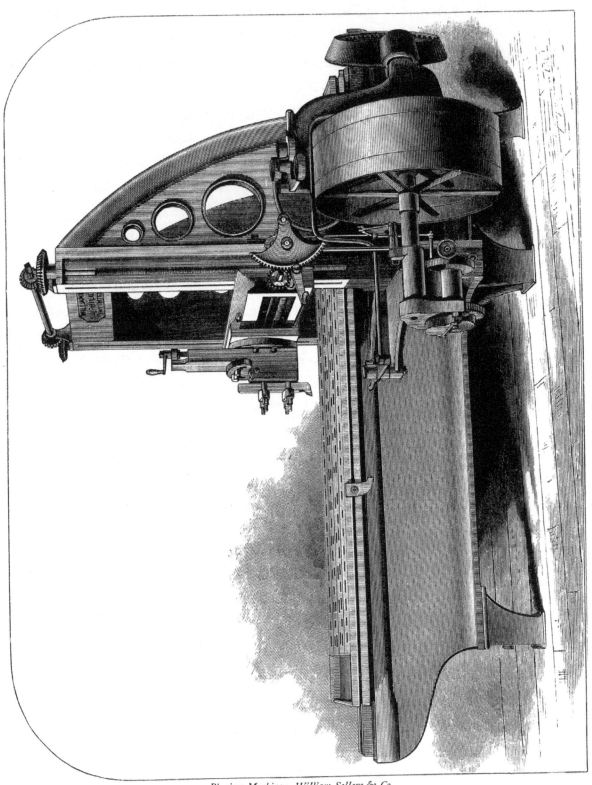

Planing-Machine: William Sellers & Co.

another novelty. An arm rises from the rear bearing of the pulley-shaft extending over the pulleys and supporting three fulcrum-pins. On the centre one is a peculiarly shaped lever, swinging horizontally between the other two upon which are placed the two belt-shifters. The shifters are operated from the middle lever by teeth or projections on each, the arrangement being such that one shifter is always moved before the motion of the other is commenced, one belt always leaving the driving-pulley before the other begins to take hold and reverse the motion, requiring but little power, allowing the least possible lateral

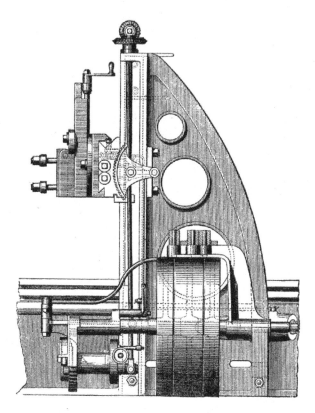

Planing-Machine. Fig. 1. Side Elevation.

motion of the belts and avoiding all undue straining and shrieking. The usual adjustable stops are provided on the sides of the table, operating the belt-shifting apparatus by means of a double-armed lever and link connection. The position of this apparatus is exceedingly convenient for the workman if he desires to change the belts without reference to the stops, allowing him to easily control and reverse the motion of the table by hand or even to stop it entirely by shifting both belts on to the loose pulleys without arresting the motion of the counter-shaft. This is a great advantage at times when planing surfaces of irregular shape.

The machine has positive geared feeds, self-acting in all directions with tool-lifter operating at all angles. The feed motion of the cutting-tool is obtained in nearly all planing-machines from the belt-shifter, entailing upon the stops on the table an undue amount of work and really resulting in quite limited variations of feed. The usual screw and central feed shafts are provided in this

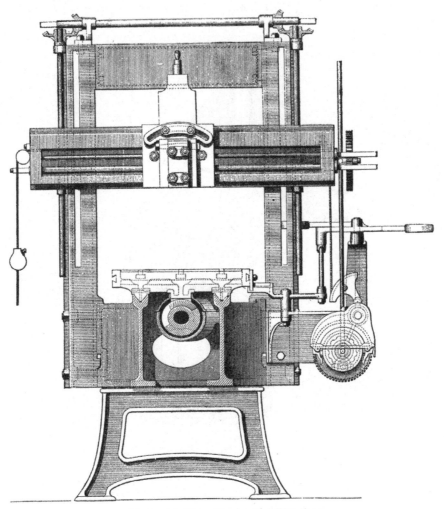

Planing-Machine. Fig. 2. End Elevation.

machine in the cross-head for horizontal or vertical motion, receiving the variable motion for any required amount of feed through a ratchet-wheel fitted interchangeably to their squared end projections. This ratchet-wheel is operated by a toothed segment, which receives at each end of the stroke of the tool the required alternate movements in opposite directions by means of a light vertical feed-rod from a crank-disk below, on which the crank-pin is so arranged as to allow any variation and adjustment of throw and amount of feed that may be desired to be made during the cutting stroke of the machine. The crank-plate

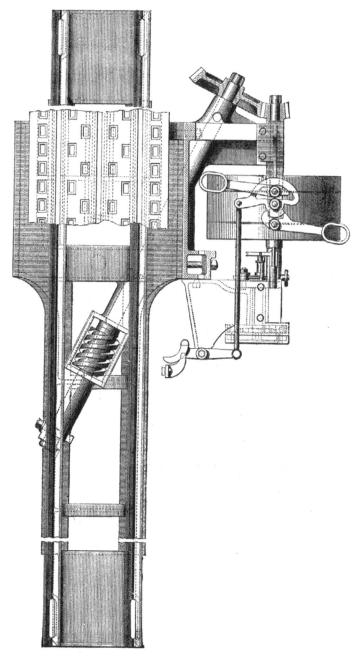

Planing-Machine. Fig. 3. Plan.

is alternately moved a half revolution and disengaged in either direction at each reversion of the stroke of the tool by means of an ingeniously contrived double pawl and ratchet-wheel, receiving motion from a pinion on the front end of the pulley-shaft. At each change of motion the pawl is thrown into gear by friction, keeping up a positive motion of the crank-disk by the ratchet-wheel until the pawl is disengaged from its teeth by a positive stop.

Messrs. Sellers' method of lifting the tool-point on back motion in this machine merits attention as another improvement on the usual plan, which, in most planing-machines consists in hanging the cutting-tool in what is called an apron, so adjusted as to allow it to swing loose on the back stroke, but to be held rigidly when cutting. This arrangement is very objectionable in all fine planing, and especially in large planers where the tool is quite heavy. Various ideas have been put into practice for actually lifting the tool-point clear of the work on the back stroke and dropping it into place again ready for action on the return, but the method here shown possesses especial ingenuity, lifting the tool in every position of the slide-rest, and doing so from within the cross-head without interfering in any way with the automatic feed motion, the machinery for working the feeds occupying the centre about which the adjustable part of the saddle rotates. This lifting apparatus is operated by a cord attached to a grooved segment which is connected with the crank-plate of the feed motion by a link, a reciprocating motion being imparted to the cord corresponding with the motions of the table, and occurring only at the end of each table movement, beginning with the reversion of each stroke. The cord is guided over sheaves at the ends of the cross-head and passes around a cord-wheel in the saddle, having at its other end a weight to keep it in tension. The cord-wheel, by means of a pinion at the other end of its shaft, operates on a light annular plate-wheel recessed into the saddle, around the central part containing the small feed bevels. In a spiral groove on the face of this plate-wheel slides a block which is attached to the end of a pipe surrounding the vertical feed-screw, and extending upward through the casting, with a pair of elastic clamps at its upper end. These clamps operate by friction on a flat rod which passes the whole length of the vertical slide on its side next to the saddle, and has at its lower end, which is thickened up, a hole. The long arm of a bell-crank lever fits loosely into this hole, and the short arm extends down directly behind the tool-apron. The action of the cord imparts motion to this bell-crank and affects the tool apron, pushing it forward and letting it fall back again into place as required. The action is perfect and beautiful, without interference with any of the functions of the machine in the least. When the vertical slide is turned into any new position upon the horizontal axis of the saddle, the pin in the spiral slot drags the plate-wheel around, the cord-wheel slipping within its encircling cord, and as soon as the machine is started adjustment to the new position takes place at once among the parts, the lifting apparatus operating as before.

The mechanical engineers of this country, as a rule, adopt original methods of design, taking the problem presented to them with certain given conditions, such and such results to be accomplished, and working the whole out from a new basis, without any blind adherence to old-established forms or precedents, and it is this which distinguishes our American practice and is productive of such satisfactory results. No firm has had a greater share in work of this kind than Messrs. William Sellers & Co., and their machinery throughout bears evidence of it. We hope to be able to refer to them again at a future time.

Numerous efforts have been made in reference to the manufacture of bricks by means of machinery which should be equal in quality and finish to those made by hand, but great difficulty has been experienced, especially in the case of face or front brick, in that it has been found practically impossible to supply an equal amount of clay to each of the mould-boxes, thereby producing brick of an unequal size and density, and also that the pressure being imparted to the clay from one side only—as has been the general custom—the bricks are often defective in strength, particularly at the corners and edges, and consequently not of first quality. It is claimed that these serious difficulties have been overcome in "Gregg's Triple Pressure Brick-Machine," of which a working specimen is shown in the Exhibition. The principle of this machine is such that the heavy developing pressures take place while the mould-table is at rest, an advantage not attained in any other machine, to our knowledge. There is, accordingly, but a nominal amount of power required to operate it, and a large amount of wear and tear, strain and breakage usual in other machines is avoided.

We may designate Brick-Machines under three classes: Dry Clay, Slush, and Crude or Moist Clay Machines. In the first class, the clay after being dried and granulated, is filled into the mould-boxes through "filler-boxes" or graduating measures. When a number of moulds are grouped together it seems to be a practical impossibility to fill them all alike, and the bricks in some are turned out imperfect. The extraction of the moisture in the clay before moulding also destroys its cohesive power, preventing complete fusion in the burning, and producing bricks unable to withstand the action of the weather. In the manufacture of slush-brick, the great amount of water that is used while exceedingly favorable to the production of good brick, is otherwise objectionable on account of the great length of time expended in the slow, out-door process

Triple Pressure Brick-Making Machine: W. L. Gregg.

of drying, and the risk attendant from unfavorable weather, as at least twenty-five per cent. of water must be evaporated from slush-brick before it is safe to burn them. In works producing say thirty thousand of this kind of brick per day, it is stated that upwards of twenty-three tons of water must be evaporated every twenty-four hours.

The present machine occupies a position between the "Dry Clay" and "Slush" machines, and may be designated as a Crude or Moist Clay machine, manufacturing to advantage with crude clay in a state so stiff as to require an evaporation of only about one-eighth that necessary with slush-brick before burning, and yet retaining all the cohesive qualities of the material. The brick, when

The Great Clock in Machinery Hall: Seth Thomas Clock Co.

burned, are of closer grain, less porous and, therefore, stronger and more durable than those manufactured by the other methods.

The engraving given on page 24 shows the general form of the machine. It is provided with a circular mould-table having an intermittent motion and containing eight sets of moulds, with four to a set, thus making in all thirty-two moulds. There are three distinct places for producing pressure on the clay in the moulds. The first is produced by a pressure-wheel from above, the second by a toggle-joint actuated by cams, and the third from above and below by toggle-joints and cams. The brick are delivered by a sweep-motion on to an endless belt, or carrier. The moulds are of hardened steel and the balance of the machine is of iron worked up in a very solid and substantial manner. A ten horse-power engine, with one of these machines, will produce forty-thousand brick in ten hours, including the preparation of the clay, which is performed by the same apparatus, although not shown in the engraving.

The United States has been noted for many years for the manufacture of cheap, serviceable varieties of clocks, the production of late amounting to over a million annually, and immense exportations being made to all parts of Europe, Asia, Africa, China, Japan, South America, etc., until the "Yankee Clock" has become a household word in every quarter of the known world. It is only within the past twenty years, however, that the construction of Tower Clocks has been undertaken in this country, the supply having hitherto been procured from abroad. Lately, a number of firms have been engaged in this business, with remarkable success, obtaining in a short space of time quite a celebrity, and notably among them may be mentioned the "Seth Thomas Clock Company," one of whose clocks has been placed on exhibition, in working condition, over the east entrance of the Machinery Hall. Its mechanism is shown by the engraving given on the preceding page, and the details of its construction have been carried out in great perfection, bronze metal being used for the wheels, except in the case of the winding-gear, which is of iron, It strikes the hours and quarters upon two bells, the power being sufficient for these to be extra heavy.

An idea may be gained of the size of the clock when it is stated that the main frame is ten feet long by three and a half feet wide and seven feet high from the floor, and that the total weight is seven thousand pounds. In the striking apparatus the main wheels are forty-one inches in diameter and the drums for cords twenty-three inches. The main time-wheel has a diameter of twenty-four inches, and the drum for cord twelve inches. The pendulum has a zinc compensation-rod fourteen and a half feet long and beats once in two seconds, the weight, including pendulum-bob, being five hundred pounds. Dennison's Gravity Escapement is used. Arrangements have been made to run twenty-six electrical clocks from the main clock, to be located in different parts of the building, and to make connection every twenty seconds. The clocks manufactured by this Company are remarkable for their accuracy and the per-

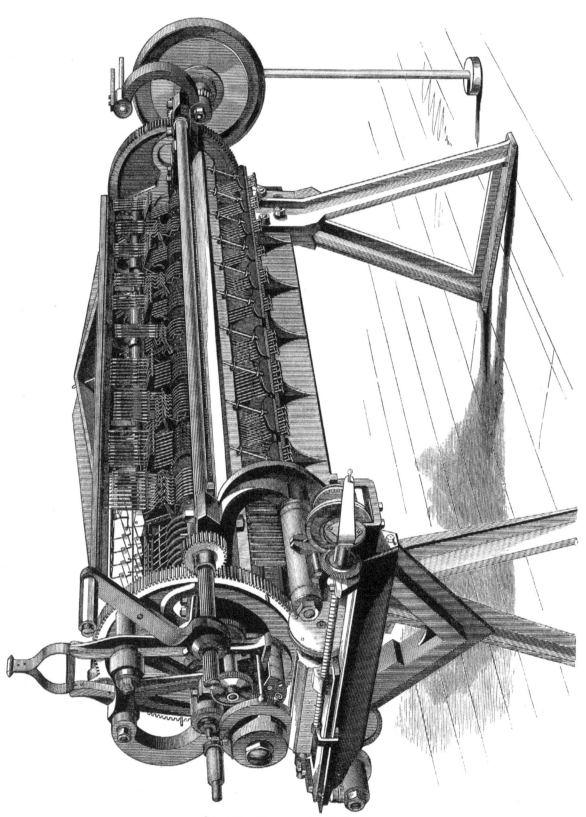

Calculating Machine: Geo. B. Grant.

fection of their mechanism, and have obtained a reputation as first-class timekeepers in every respect.

In a quiet corner of the Machinery Hall, just back of the exhibit of locomotive engines, stands a machine which at first sight would probably puzzle some of our best mechanics to give an opinion as to its use, and even then, would require a master-mind to analyze its mode of action. We refer to the Difference Engine of George B. Grant, designed for the construction of large Mathematical Tables, such as Tables of Logarithms, Sines, Tangents, Reciprocals, Square and Cube Roots, etc., and built for the University of Pennsylvania. All those interested in such subjects are familiar with the Difference Engine of the late Charles Babbage and its failure. Following him came George Schentz, a printer of Stockholm, whose machine, however, never came up to the full requirements of a difference engine, being of slow speed, sensitive and delicate in its details, containing radical defects in the theory of its mechanism and never reaching beyond the entrance to the goal for which its inventor contended. The subject was first taken up by Mr. Grant in 1869, when he was as yet entirely ignorant of the labors of Babbage and Schentz, and the year after, he prepared full drawings of his machine, but met with so much discouragement from those he consulted in the matter that the work was given up. In 1871, however, Professor Wollcot Gibbs—now of Harvard College—heard of his labors on the problem, and after a thorough examination into the subject, approved of the plans and gave so much encouragement to the inventor by his deep interest and constant efforts of support as to contribute largely to the final success that has been attained. After several failures to procure the necessary funds for expenses, a liberal subscription was made by the Boston Thursday Club in 1874, and the same year the means requisite for the construction of a large engine were furnished by Mr. Fairman Rogers, of Philadelphia, to whose munificence science is indebted for the machine now before us, which was finished only a few days before the opening of the Exhibition. When it is remembered how important numerical tables are in practical applications of mathematics, and the great labor and time necessarily occupied in their calculation and publication by the usual methods, involving errors which it seems impossible to prevent even with the greatest care and the most watchful proof-reading, the value of such a machine may readily be seen.

The accompanying engravings—shown on pages 27 and 29—give a very fair idea of the apparatus. It occupies a space of about five feet in height by eight feet in length, and weighs about two thousand pounds, containing, when in full

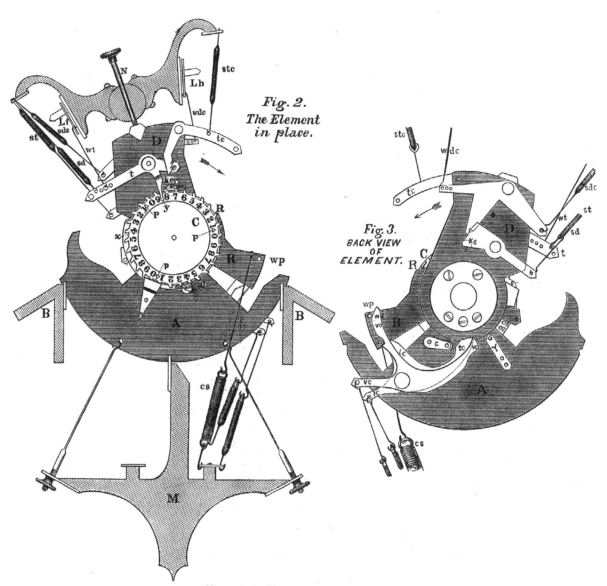

Views of the Element; Calculating Machine.

working order, from twelve thousand to fifteen thousand pieces. The long body contains the calculating mechanism, while at the front end is an apparatus for printing a wax mould of the results, from which an electrotype may be made directly for printing, requiring no setting of type and no risks of error. The machine is driven either by hand, by a crank at the front end, or by a power appliance at the rear end. The calculating portion of the machine consists of a number of elements placed in the long frame side by side, each element representing one decimal place of the work, and there are from twenty-five to one hundred of these, according to the particular requirements of the problem in hand.

Figures 2 and 3—presented on page 29—show front and rear views of an element constructed of pieces of sheet metal, all in the machine being alike and interchangeable. They are placed in the frame one-half inch apart from each other and arranged in groups, each group representing an order of differences. Referring to the figures, the portion A of the element is fixed and the portion D is a rocking arm revolving upon a bearing at the centre of the plate, C, and and having its upper part—in the case of all the elements—fixed to a long frame, shown in cross-section in the figure at N, which is oscillated back and forth by the driving-gear of the machine. When this frame is moved forward in the direction indicated by the arrow, the arm of each element moves with it and adds to itself the figure that is upon the element of the same decimal place in the group or order below. The pin, k, e, in the figure moves over the teeth of the wheel, C, until a proper point is reached, when it is released by means of mechanism attached to the corresponding element in the order below and falls on to the cogs of the wheel, C, carrying it with it in its forward movement until lifted out from the cogs by the riser at R. As it goes back it sends its figure up to be added to the element of the same decimal place in the next group or order above. The trip, t, striking one of the pins, p, draws down upon the wire, w t, and by means of a longitudinal lever, L f, acts upon a wire that releases the driver, k e, of the element added to in the order above. (See L b and w d c in the figures.) The action throughout is the same, each motion forward adding all the odd orders to the even ones, and each motion back adding the even orders to the odd ones and at the same time printing the tabular number at the front end of the machine, stamping it into the wax plate there for that purpose.

This machine possesses a very great advantage over any previous invention of the kind, in that any wheel may be connected so as to add its number to any wheel of the next higher order. In the machines of Babbage and Scheutz each element was arranged to add its figure to a given fixed wheel in the order above. With a given amount of mechanism, by means of this arrangement, this machine can accomplish work of three times the complexity of any former machine.

Many of the figures used in an operation are constant. Thus, the last order is constant at various values and many of the other elements are fixed at zero or at nine. To provide for this, each element carries a constant-wheel, consisting of a thin disk of brass, x, which turns on the bearing of the calculating-wheel, and being set at any figure by the spring and pin, y,—Fig. 2—allows the driver, k, e, to drop on the wheel at the same tooth each time and add a constant figure. Thus—see Fig. 2—if set at 8, there are 8 units added at each movement.

The operation of carriage, while simple as to mechanism, is exceedingly complex in theory of action, the apparatus for the same, although by far the most important part of any calculating-machine, being, nevertheless, the most difficult to contrive, and it was to this that Babbage's machine owed its failure, and Scheutz's machine its slow working speed. The riser, R, is hung on the centre-pin of the wheel, C, and it falls as the wheel passes from nine to zero, throwing the driver, k e, out one space further on than otherwise would be done and carrying a unit to its wheel. This riser, R, is held up by two catches which are operated by the pins, p, p, p. On the next wheel below, one catch is drawn aside as the wheel passes from eight to nine and the other as the zero is reached. As the riser drops, a pin, w p, upon it strikes and draws the upper catch of the next wheel above. The arrangement makes a perfect and simultaneous carrying apparatus, acting under any possible combination of requirements.

The construction of the printing apparatus is rather complex as many conditions must be satisfied. Each of the upper ten calculating-wheels is connected by gearing with a die-plate, in the edge of which common printing type are set. While the machine is in motion, these plates are separated slightly and work easily without interference with each other. When an impression is to be taken, they are brought closely and firmly together and a pair of plungers at the same time straightens the line of figures and presents it ready for the plate of wax below, which rises and receives the impression.

The terms of the table can be arranged in almost any way desired for printing, either each under the preceding, or before it, or they may be run across the page, as is generally done, and either forward or backward. It is also possible to adjust the distance between the lines and vary it from line to line as required.

When the machine is worked by hand, a speed may be made of ten to twelve terms per minute, and from twenty to thirty when by power by the attachment at the rear end. All that limits the speed is the imperfection of the mechanism, and in the case of the present machine—the first ever constructed of so complex a character—imperfections are to be expected which will not exist in future machines. Thirty of the elements of this machine were placed in a light wooden frame and worked successfully at a speed of over one hundred terms per minute, and if the whole machine were used and sufficient power applied, this speed would be perfectly practicable, provided that the mechanism of the driving-gear and printing apparatus were in accurate working order and made sufficiently strong to stand the wear and tear resulting from the same.

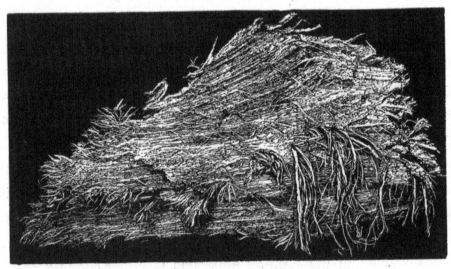

Asbestos in its Natural State: H. W. Johns.

The mineral Asbestos, although familiar to the ancients and employed by them in the manufacture of a fire-proof cremation-cloth and for some other purposes, has, in modern times—until within the last few years—been classed among those substances more curious than useful. The silky, fibrous nature which it possesses, and its well-known fire-proof and non-conducting qualities and resistance to the action of acids have, however, at last attracted attention, and we are indebted to Mr. H. W. Johns, of New York for its adaptation to some very important purposes in the useful arts. We present above an engraving of an exceedingly characteristic specimen of this mineral which has been placed among the exhibits.

Asbestos exists in vast quantities in the United States and numerous other parts of the world. It is obtained from the mines either in bundles of soft, silky fibre or in hard blocks which are capable of separation into fibres. These fibres vary in length from two to forty inches, are of a greasy nature and exceedingly flexible, possessing great strength in the direction of their length, and are therefore capable of being woven into cloth as used by the ancients. These properties possessed by asbestos render it an excellent substance to incorporate into cements—as hair is put into plaster—to bind the parts together and at the same time to give body to the material, and it was this use that Mr. Johns first made of it.

The facilities for obtaining the mineral were very poor—there never having been any demand for it in the market—but as the want was created and it really existed in nature in great quantities, these facilities soon increased and with abundance of material Mr. Johns was enabled to utilize it for other and more important purposes. It was found to make an excellent roofing material. Sometimes it is applied in the form of an asbestos concrete and spread over the roof by a trowel, but more generally a peculiar roofing felt—into the composition of which asbestos largely enters—is first nailed down on the sheathing boards, and this is then covered by means of a brush with a preparation of flocculent asbestos, silica paint, etc., making an entirely water-, fire- and weather-proof surface, a good non-conductor of heat, well adapted to all climates and costing a very reasonable price.

Its non-conducting qualities render asbestos peculiarly applicable as a covering for steam-boilers, pipes, etc., and it has largely been used for this purpose. One of its most recent applications has been for steam-packing. The elevated temperature, moisture and friction to which steam-packing is subjected requires a material possessing just the qualifications existing in asbestos, and experience has shown its great adaptability to this use. As a body for paints, being mixed with linseed-oil and colors, it has succeeded remarkably well; an asbestos paper is made incombustible and very useful for filtering acids, and every day new applications are discovered for this material, so few years ago supposed to be worthless.

Mr. Morris L. Orum, of Philadelphia, exhibits a Flexible Mandrel for Bending Metal Pipe that has attracted considerable attention, being an exceedingly ingenious and novel arrangement of great value in its particular department. By the usual method of bending pipes they are first filled with melted rosin, some fusible metal as lead, or sand, so as to preserve their shape, requiring considerable trouble and care in cleaning out afterwards and making it a tedious, expensive and imperfect job at the best, requiring almost universally—unless with very small pipe—the use of the hammer and file to straighten the irregular crimps formed on the interior curve of the bend. By this method, a mandrel is used, being a strong, closely wound steel helix, formed of square or rectangular wire, and having a uniform external diameter such that it may fit easily in the pipe to be bent. A nut is fastened into one end of the mandrel into which a stem is screwed, which may be made of any length, for removing the mandrel after the work is accomplished. This is done by simply revolving the stem in the direction to wind up the helix, when its diameter is reduced, allowing it to be easily withdrawn, and the spring in the metal restores it to its original size after removal. If the mandrel is not sufficiently long for the whole bend of the pipe, it may be moved from place to place as required, or reverse bends introduced, if desired, with the greatest ease. If the proper sized mandrel is not on hand to bend a given pipe, the next size smaller may be used without any appreciable error.

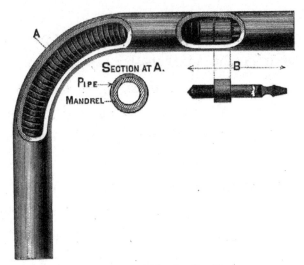

Oram's Mandrel, for Bending Pipe.

Large pipe can be bent as accurately and readily as small pipe, a matter of great difficulty by the old methods, and impossible when the pipe became quite large, requiring previous softening by heat, and resulting in elliptical, unequal and irregular shapes of cross-section that were exceedingly undesirable, the material at the outer diameter becoming very thin and weak. By this method, the pipe after bending has a practically uniform internal section without appreciable variation in diameters, a qualification that will be fully appreciated by the manufacturers of pneumatic despatch-tubes, where this requirement is so necessary.

Mr. Oram has bent ¾-inch butted zinc pipe to a curve of 1¼ inches radius, 1-inch pipe to 2 inches radius, and 2½-inch pipe to 5 inches radius. A 2-inch pipe has been bent to 4 inches radius, cold, without any difficulty. Square pipe may also be bent as readily as round, and the process seems to supply a want long felt.

Messrs. Frederick Hurd & Co., of Wakeford, England, exhibit a Coal-Cutting Machine, which appears to present some novelties in design and to be a very

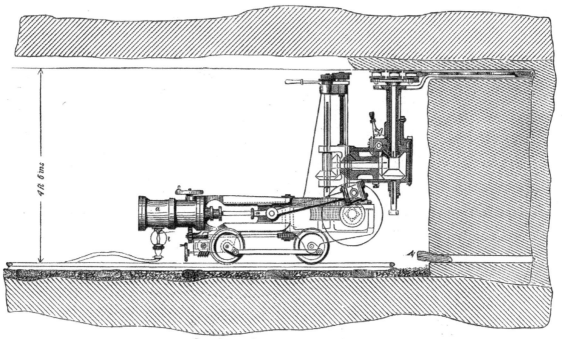

Coal Cutting Machine: F. Hurd & Co.

desirable form of apparatus for our mines, cutting gangway as well as room work. We refer more particularly to the pattern for four feet to four and a half feet seams, smaller seams not being, as a rule, worked with profit in this country. The illustrations accompanying—presented on pages 36 and 37—will give the reader an idea of the construction and mode of action of this machine. The value and importance of reliable coal cutting machinery is increasing every day, tending to dispense with the most exhausting and dangerous part of a miner's work, to lessen by a very large amount the waste or slack always obtained in getting out coal, and to obviate the expense, trouble and unreliability of hand-labor.

The machine is provided with two cylinders of six inches diameter and twelve inches stroke, and works by the action of compressed air. This has been found the most satisfactory motive power that can be used for mining machinery, being easy of application and, at the same time, improving the ventilation and reducing liability to fire. Motion is given to the cutters by bevel gearing, and the shaft driving the cutters, by a simple arrangement, is made capable of being revolved in a vertical plane about the horizontal shaft, thus providing for cutting out all four faces of the drift, quite an advantage over the usual machines which make only the under-cut. In the cutting-wheel the periphery in which the cutters are fixed is placed eccentric to the fulcrum on which they revolve, and the pressure is resisted by anti-friction bowls, which also act as drivers, thus dispensing with guides and slides. The cutters are put in or taken

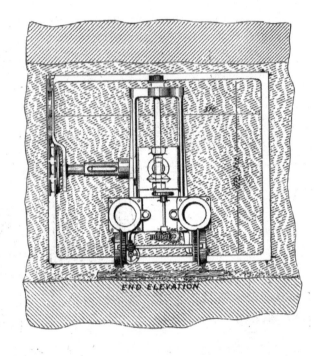

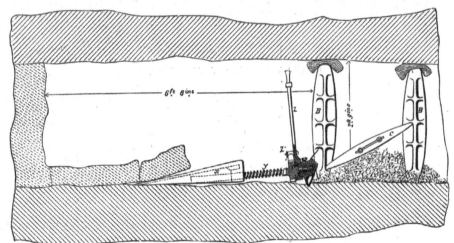

Coal Cutting Machine, Figs. 2 and 3: F. Hurd & Co.

out of cut by a swivel-nut and screw acting on the lever or radial arm in which they revolve, or by a pinion and quadrant. They are made of plain, square, Titanic steel, manufactured by S. Osborne & Co., of Sheffield, and are set sideways, above and below, allowing for the clearance of the disk, being readily adjusted radially to vary the depth of the cut according to the quality of the coal or mineral.

The leading wheels of the machine are kept in position on the rails when at work, by a bowl mounted in a differential lever, with self-acting adjustment to adapt itself to irregularities on the coal face without the possibility of getting off the rails while at work.

After removing the coal included within the cuts, a sort of wedge-shovel,

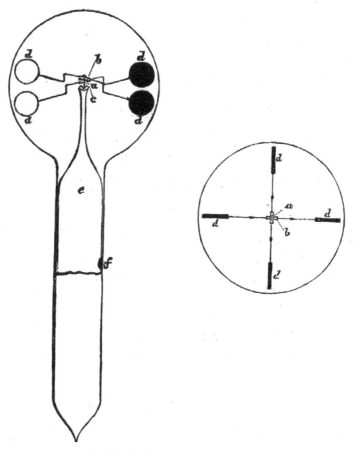

Crookes' Radiometer: Jas. J. Hicks.

as shown by Fig. 3, is used to raise up and remove the lower portion of the seam. The form of post shown in this figure is not nearly so good as the French post, where the bottom is placed in a ring containing sand, so that if the roof presses down and holds it, it can be relieved by letting some of the sand run out, being on the same principle as the method of supporting the centres for arch bridges by sand-tubes.

Mr. James J. Hicks, of London, England, exhibits a curious apparatus in the form of a Radiometer for demonstrating the mechanical action of light and the conversion of radiation into motive power, constructed according to the design of Mr. William Crookes, F.R.S., to whom is due the discovery of this force. Mr. Crookes' attention was first drawn to the matter from noticing that in weighing heavy pieces of glass apparatus in a vacuum, there appeared to be a variation in weight, corresponding to variations in temperature of the material weighed from that of the surrounding air and weights of balance. This led him to institute a series of experiments with very delicate forms of apparatus, and he discovered that there was a force depending upon the action of the light. In the case of an exceedingly fine and light arm suspended in a glass tube with balls of various materials at the ends of the arm, the whole being thoroughly exhausted from air, he found that on the approach of a heated or luminous body to one of the balls a very decided repulsion took place and an attraction if a cold body, such as ice, was used. He also ascertained that when the different rays from the spectrum were thrown on white and black surfaces there was a decided difference between the action of light and of radiant heat, dark heat having no perceptible difference of action on white or black substances, but luminous rays repelling black surfaces much more energetically than white. Acting upon these facts he designed and constructed an instrument, completely and beautifully exemplifying the principles of his discovery, which he called the "Radiometer."

This instrument is shown on the preceding page in section and plan by Figs. 1 and 2, and consists of four arms of some light material, to the ends of which are fixed thin disks of pith with one side black and the other white, the black sides for the four disks all facing the same way. These arms cross each other at right angles and are balanced at their centre points on a hard steel point, a, resting on a jewel-cup, c, so that they may freely revolve in a horizontal plane. A thin glass globe, drawn out to a tube at the lower part so as to form a support, encloses the whole and is exhausted to the greatest attainable vacuum and hermetically sealed.

When this instrument is placed subject to the influence of light, the arms rotate with greater or less velocity directly in proportion to the intensity of the incident rays, and in the case of very intense light, like that from the sun or burning magnesium, the rapidity of rotation becomes so great that the separate disks are lost in a circle of light. Experiments made by varying the distance from the source of light show that the mechanical action is inversely proportional to the square of the distance. Dark heat produces no rotation.

This new apparatus may be applied practically to a number of uses. A standard candle may be defined as one which at a given fixed distance causes a certain number of revolutions to the apparatus per minute, and comparison can readily be made between this standard and various other kinds of light, or these various kinds may be compared among themselves. The effects of light through different media can also be ascertained and the photographer may use this instrument to great advantage in his so-called dark-room to ascertain whether the light he is using is likely to injure his sensitive preparations or not. He can also measure the intensity of the light in his operating-room, and by means of one of these instruments, instead of a watch, regulate the amount of exposure necessary for a subject with the greatest accuracy, working according to so many revolutions of the instrument independent of the time.

The discoverer has also very lately invented a torsion-balance by which he is enabled to weigh the force of radiation from a lighted body—as a candle—the principle being similar to that of Ritchie's torsion-balance, and fully confirming the previously ascertained law of inverse squares. By calculations deduced from an experiment with sunlight, it was demonstrated that the pressure of sunshine amounted to two and three-tenths tons per square mile. Mr. Crookes' discovery is one of great value, and promises to be the means of solving in future many problems, as yet unexplained, in the action of forces in our vast universe.

Messrs. Aveling & Porter, of Rochester, England, well known as occupying a high position in the specialty of Locomotives applicable to common roads, to agricultural purposes, to road rollers, etc., make a very creditable exhibit in the British Department of the Machinery Hall. One of their latest improvements, and an exceedingly important one in this class of machines, consists in their method of mounting the principal working parts of the locomotive, the crank-shaft, the counter-shaft and the driving axle, so as to prevent the unequal working of these parts from producing any injurious strain upon the boiler, a defect that has long been a fertile source of trouble in all engines of this kind. This is accomplished by prolonging the side plates of the fire-box upwards, as will readily be seen by the accompanying engraving, thus forming a complete arrangement for carrying the bearings of these working parts without connecting with the boiler directly. Many other improvements have been made in these machines from time to time, and they may be regarded as possessing great simplicity, strength, durability and economy of working.

Steam Road-Roller: Aveling & Porter.

In the "Road Locomotive Engine," a single cylinder is used and it is placed on the forward part of the boiler, preventing priming, dispensing with steam-pipes and resulting in considerable economy of fuel. This cylinder is surrounded by a jacket, with which it is in direct communication, and the steam is taken into it from a dome connected with the jacket. The driving-wheels are of wrought-iron, provided with compensation motion, so as to allow the turning of sharp curves without disconnecting either wheel, and sustain about 85 per cent. of the weight of the engine. The steering is done from the foot-plate.

It is stated that the working expenses vary from 2½ to 6 per cent. per ton per mile, depending upon whether the work is continuous or intermittent and the condition of the roads travelled over. Whenever circumstances render the use of the ordinary rigid tires objectionable, the wheels are fitted with the spring tires of W. Bridges, Adams, consisting of inner and outer tire-frames, having solid blocks of India-rubber between them, and connected together by a "drag-link," to prevent friction on the rubber blocks.

Farm Locomotive Engine: Aveling & Porter.

Cranes are attached to the front of these engines when desired, very much increasing their usefulness in dock-yards, quarries, etc., and those visiting the Exhibition Grounds during the moving and placing of the exhibits have no doubt noticed the effective work performed by one of these machines in use by the British Commission.

The general characteristics of the "Agricultural Locomotive" are very similar to those of the "Road Locomotive," and this apparatus is expressly adapted to the working of machines for steam cultivation, thrashing, sawing, pumping, or other agricultural duty.

Among the motors exhibited in the Machinery Hall, LANGEN & OTTO's Patent ATMOSPHERIC GAS ENGINE, in the German Department, is a remarkably ingenious and exceedingly useful and effective machine, applicable with advantage in many cases in the industrial arts, more than three thousand being now in use on the Continent of Europe, working with considerable economy and effect.

In all gas-engines, as previously designed, the motive power has been obtained by the direct action of the force of the explosion of a mixture of gases on the surface of the piston, in a similar manner to that in which steam acts in the steam-engine. In this engine the main characteristic may be said to be in the use of a "free piston," rising without resistance upon explosion, the motive power being obtained indirectly during its descent by the pressure of the atmosphere acting upon its upper surface, owing to the partial vacuum below, following the explosion. Great advantage obtains from this arrangement in furnishing a power of longer application, less suddenly exerted and more steady in its action. In former machines the sudden, intense force necessarily given out could not be immediately made available, resulting in destructive action on the machinery of the engine and in great loss of power, which expended itself in

Atmospheric Gas Engine: Langen & Otto.

heat, requiring a large application of water to keep the parts cool, and avoid oxidization of the lubricant and destruction of the piston. All this could be accomplished, but the consumption of water was great and the heat taken up by it was carried off without doing work, amounting to just so much power lost.

The present engine—as shown by the perspective—consists of a large vertical piston-cylinder, having on one side, near the bottom, a valve-system for the admission and ignition of the mixed gases, and at the top a fly-wheel and the necessary mechanism for utilizing the force from the piston and for working the valves below. Fig. 1 gives a cross-section at the base, showing the valve arrangement, and Fig. 2 a partly sectional detail of the mechanism at the top.

The most novel part of the design consists in the method adopted by which the piston is allowed to move freely upward, but must connect with the fly-wheel as it returns downward. The piston-rod is a rack, and gears into a toothed wheel, shown in Fig. 2, which consists of a central pulley, a, keyed on to the shaft of the fly-wheel and surrounded by a toothed ring, b, on the interior face of which three inclined surfaces are cut. Between each of these and a corresponding curved wedge, c, faced on the side next to the pulley with leather, a set of live rollers—made of rubber—travels freely. On the upward motion of the piston, the toothed ring is free and revolves backward upon its central pulley; but upon the return stroke, the live rollers wedge in between the ring and pulley, giving a firm hold on the shaft and allowing the piston to impart its motion to it. One might suppose that a ratchet-wheel and pawl would accomplish this result better, but it was found too sudden and rigid in its action and on trial abandoned.

To start the engine the piston is lifted up about one-eleventh of its stroke, causing a proper mixture of gas and air to be drawn in under it by the passage x, x, Fig. 1, and at the same time, by the motion of the valve, the chamber, y, filled with gas and air, goes first downward, igniting its contents by a flame, w, continually burning outside, and then upward to x, where it fires the charge under the piston. The resulting expansion drives the piston up very rapidly, reducing the temperature of the gaseous contents of the cylinder, consuming the heat in work, and the result is a partial vacuum under the piston which, by the pressure of the atmosphere, then makes a return stroke. The atmosphere is prevented from entering the lower part of the cylinder from the outside and destroying the vacuum; but a valve is provided that allows the products of combustion to be expelled by the piston as it reaches the bottom. The pawl, *d*, Fig. 2, in gear with the ratchet-wheel, *e*, on the main shaft gives the proper valve-motion. This pawl is controlled by an ordinary governor, worked by bevel-gearing from the fly-wheel shaft, and when the engine is running at full speed, a lever connected with this governor holds the pawl away from the rat-

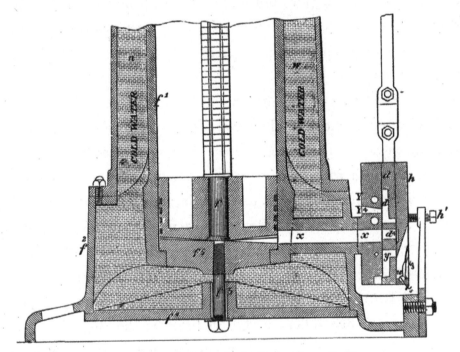

Fig. 1. Atmospheric Gas Engine (base): Langen & Otto.

chet-wheel, *e*, and no motion of the piston takes place. As soon as the speed decreases in the least, however, the action of the governor releases the pawl, which makes connection with the ratchet, and, causing one turn, operates the valve-rod, and, at the same moment, by means of the mechanism, *i, k, l, m, n*, raises the piston-rod far enough from the bottom to receive the charge of mixed gases; the explosion follows and the piston rises, repeating its operation as before. It will be seen by this that the piston does not necessarily act at every revolution of the driver, unless the full capacity of the engine is used, but only works when required to keep up the power. The driving-wheel may make even forty or more revolutions without any motion of the piston, unless the falling of the governor brings it into action. The economical effect of this is at once apparent, as the consumption of gas becomes directly proportional to the amount of work done.

The driving pressure on the return stroke varies from eleven pounds per square inch at first, to nothing, at about four-fifths of the stroke; or, say, a mean effective pressure of about seven pounds per square inch through the entire

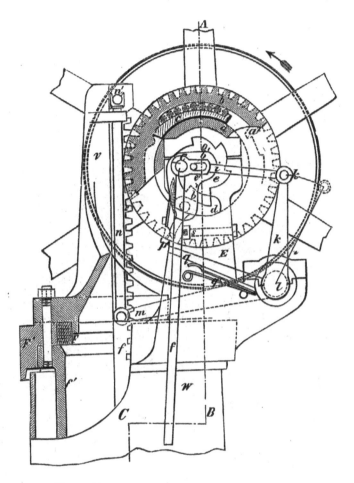

Fig. 2. Atmospheric Gas Engine (top): Langen & Otto.

movement. The parties claim a consumption of gas of 26.5 cubic feet per horse-power per hour, and about 12 per cent. effective power on the theoretical amount supplied by the fuel.

The engines make a slight noise in working,—an inherent feature in machines of this kind,—but there are undoubtedly decided advantages in economy of fuel, cleanliness, absolute safety from accident by explosion, and capability of starting at full power at a moment's notice, making these machines formidable rivals of steam-engines—in the limited sphere for which they are adapted. Messrs. SCHLEICHER BROS., of Philadelphia, represent the exhibitors of the GAS-ENGINE, and have made arrangements to manufacture it in the United States.

Messrs. POOLE & HUNT, of Baltimore Md., exhibit several sizes of "Leffel's Patent, AMERICAN DOUBLE TURBINE, WATER-WHEELS;" and we present an engra-

Double Turbine Water-Wheel: Poole & Hunt, Baltimore.

ving of one having a wheel thirty and a-half inches in diameter, the largest shown, although the manufacturers make up to a diameter of ninety-six inches.

These wheels are exceedingly popular in the United States, there being over seven thousand now in use, and the owners of the patent-right claim great advantages from the peculiarities of construction. The unsteady motion, variable speed, and irregular quantity of water always obtaining in practical manufacturing operations, necessitate requirements in the construction of water-wheels that do not always show themselves in the usual test-trials, where everything is arranged for the purpose, and it is claimed that the Leffel Double-Turbine meets just these points and possesses remarkable efficiency and durability under actual long-continued use. This Turbine belongs to that class in which the water enters at the circumference and discharges at the centre, and it has two sets of buckets on one wheel, one over the other, and each constructed upon a different principle. The upper ones curve only slightly downward and run in towards the centre of the wheel, having the faces at quite an angle to a radial line, while the lower ones curve down almost immediately and bend sideways in the direction of the circumference very considerably, before reaching the lower face. The upper buckets, therefore, receive a great side-pressure and the lower ones almost a vertical pressure. By this arrangement it is claimed that there is admitted to a wheel of any given size, the greatest possible volume of water, consistent with its economical use, and at the same time the greatest attainable area is provided for its discharge.

Movable guides around the outer circumference direct the water in to both sets of buckets, and these guides may be adjusted to any position at pleasure, even to shutting off the water entirely if desired; working by guide-rods and a segment of a toothed wheel running in a pinion regulated by a vertical rod convenient of access. The wheel is surrounded by an ample spherical cast-iron flume or penstock, seven feet six inches in diameter in the present case, furnished with a movable cap large enough to allow the passage of the entire wheel, if necessary, at any time for repairs, and also provided with man-holes. The penstock is attached at one side by a flange connection to a cast-iron supply-pipe, three feet eleven and a-half inches in diameter. The wheel is supported at the foot of its axle by a hard-wood rest, set on end, and working with excellent effect.

It is claimed for these wheels that they give a maximum discharge of water with a minimum friction, and produce an actual work of eighty-five per cent. on an average of the absolute work of the fall. The cast-iron penstock presents great difficulties of construction, and has been executed in superior style, reflecting great credit on the makers. In fact all the parts of the machine have been manufactured with great care and precision, this alone adding materially to its economy and efficiency; and the whole combination, in proportion and general mechanical arrangement, is first-class, well adapted to its intended purpose, and securing the best results, both in efficiency of action and durability under use.

We also engrave on this page a PATENT FEED-WATER HEATER and DOUBLE-ACTING FORCE-PUMP, exhibited by MESSRS. POOLE & HUNT. In feed-water heaters as usually constructed, the exhaust steam from the engine is discharged into the water and there permitted to escape; whereas in the present example the heater is simply connected by a branch-pipe with the exhaust in such a way that a sufficient volume of the exhaust steam is attracted into the heater to raise the feed-water to a temperature of 200 degrees without in any way impeding the free escape of the exhaust, or causing back pressure in the cylinder. The supply of cold water is admitted into the heater in a small but continuous stream, and flows over and through a series of disks set inside, but shown separately in the engraving.

Feed-Water Heater and Double-Acting Force-Pump: Poole & Hunt, Baltimore.

By the arrangement here adopted the condensation, refuse grease and other matter from the cylinder does not enter the heater, and is therefore prevented from access to the boiler. In the ordinary heaters, where the steam discharges directly into the water, these impurities are carried on into the boiler, forming injurious combinations, causing foaming, etc., and raising serious objections to the use of "open heaters" as usually constructed.

MESSRS. FERRIS & MILES, of Philadelphia, exhibit a large variety of steam hammers, one of which, known as the DOUBLE FRAME STEAM HAMMER WITH PARALLEL RAM, is shown by the accompanying engravings. This hammer follows very closely in general arrangements the well-known steam hammers as introduced by James Nasmyth, of England, and consists of a steam cylinder mounted upon two frames, which form convenient guides for the hammer ram playing between them. The piston-rod connects the hammer ram with the piston, and operates it directly by the steam pressure in the cylinder. The ram in the present case is set parallel with the frames, although Messrs. Ferris & Miles, under one of their patents, frequently set the ram diagonally, for the purpose of enabling the operator to utilize the whole extent of the lower end of the ram for die surface, and for placing this surface in proper

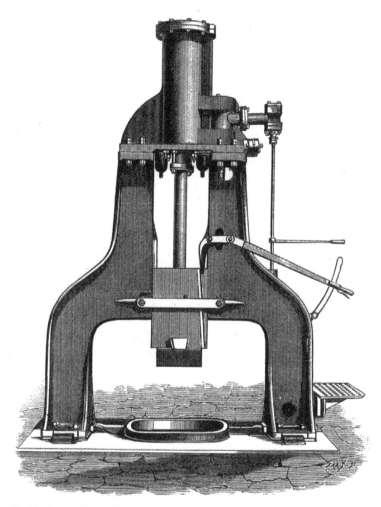

Double Frame Steam Hammer with Parallel Ram; Ferris & Miles, Philadelphia.

direction for convenient working. To afford additional stiffness to the frames, which are box castings, flanges are placed on their exterior edges, and the ram is made rather thicker than the frames, projecting slightly on each side, so as to give a neat appearance and allow the workman to readily see every motion it makes.

The main steam valve is a cast-iron hollow cylinder with enlarged ends, which work in cylindrical seats, where are situated the steam ports. The steam supply enters the valve chamber between the valve seats, circulating freely around the valve; and the exhaust chamber, fitted with proper drain and exhaust pipes, lies below the floor of the cylinder, allowing any condensation which may occur to drain into it. The exhaust steam from above the piston passes down into this chamber through the hollow cylindrical main valve. The valve stem making its connection with the valve in this exhaust chamber, has only atmospheric pressure on it, and requires no stuffing-box. This method of connection also prevents any disturbance of the perfect balance of the valve, as would be the case if attached on the steam side, and reduces the amount of power required to work the valve, as well as the friction of the parts, to a minimum.

Motion is given to the valve automatically by means of an inclined plane on the ram, which operates a cam, rocker or bell crank, connected by a link to the valve stem, and a hand lever pivoted on to this rocker places the whole apparatus directly under the control of the operator. The mechanism is characterized by extreme simplicity, consisting of only three pieces, and yet by means of it every possible gradation of stroke can be given, continuous or intermittent, light or heavy, dead or elastic, long or short, fast or slow, with as much ease and certainty as if the enormous machine, sometimes as much as ten tons in weight, and having a stroke of eight feet, were a mere tack-hammer in the hands of the operator.

Spring buffers below the cylinder protect it from injury, and by removing these the piston packing can be examined or replaced in a short time without disconnecting any of the other parts. Stuffing-boxes are provided at the connections of steam and exhaust pipes to prevent leakage from vibration due to action of hammer.

Drop-treadles, worked by the hammer-man's foot, are frequently fitted to the smaller sizes of hammers, which generally have only single frames, and they work exceedingly well, stopping and starting the hammer "on the blow," the ram always stopping "up," ready to strike again. The guides are made adjustable, so that any wear can easily be taken up and the dies kept in accurate position for stamping work into moulds. Much thought has been bestowed by the makers upon the arrangement of the anvil and foundation, so as to provide the most perfect attainable base for the machine at the least possible expense consistent with the requirements.

The provinces of the Netherlands from time immemorial have been obliged to contend with great difficulties from encroachments of the sea and from inland floods of the rivers. The waters of the North Sea, under storms in certain directions, are driven with great violence on the coast, and although in many places a natural barrier exists in the sand thrown up from the ocean, forming rows of hillocks, which give protection to the country back of them, yet at other places, and necessarily at the mouths of the rivers, this does not occur, and sea-walls or dikes of great strength have to be constructed to resist the action of the waves. On the other hand the nature of the country, consisting of vast areas of low bottom-land like that on the Mississippi, sometimes even twenty feet or more below the beds of the rivers, subjects it to great risk of overflow, especially in the spring of the year, when the upper waters, coming from more temperate climes, break up before a free passage has been opened to the sea, and accumulating, completely fill the channels of the rivers and spread over the country, carrying destruction in all directions.

The proper protection of Holland against these encroachments demanded the mutual coöperation of all the inhabitants, and therefore became a national undertaking, and resulted, after much trouble from want of union among the different provinces, in the establishment, in 1798, of what is called the Waterstaat, an organization clothed with almost absolute power, with authority to compel service from all sources if any sudden demand should occur for it, and into whose hands was intrusted the construction and maintenance of all hydrographical undertakings in the kingdom of the Netherlands.

The causes previously given, and the fact that the soil is generally composed of a soft alluvium over sand of unknown depth, and also that the great sedimentary deposits brought down by the rivers are continually acting to elevate them above their surroundings, have together resulted in a strong development of the principles of hydraulic engineering, and the actual execution of great operations in the reclamation of flooded lands, the opening of rivers to navigation, the building of sluices, locks, canals, the construction of difficult bridge foundations, etc.

In 1842 a Royal Academy for Engineers was founded at Delft, where

Bridge over the Hollandsch Diep, and Bridge at Bommel, with details: Department of the Netherlands.

MECHANICS AND SCIENCE.

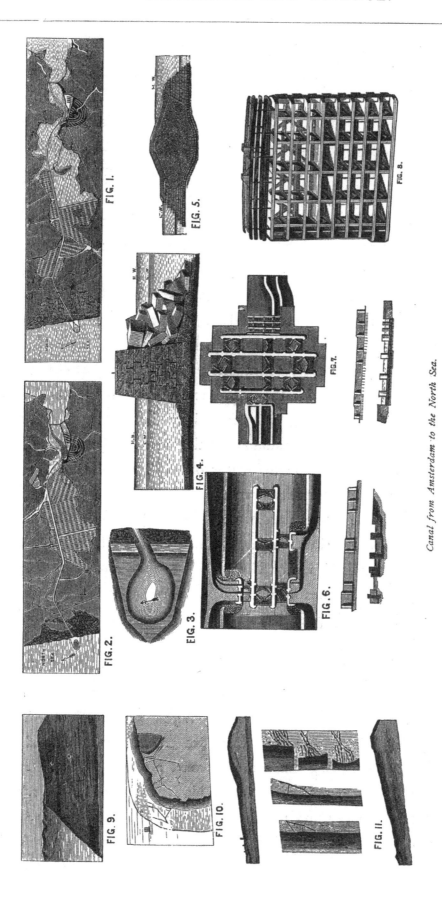

Canal from Amsterdam to the North Sea.

Fig. 1. Plan showing state of the work in 1872. Fig. 2. Plan showing completed work. Fig. 3. Jetty at North Sea entrance. Fig. 4. Section of jetty-walls at North Sea entrance. Fig. 5. Section of dam across the Y below Amsterdam. Fig. 6. North Sea locks. Fig. 7. Zuyder-Zee locks. Fig. 8. Lock-gate. Figs. 9, 10, 11 Exemplify the denudation and protection of the sea-coast of North Holland.

scientific training was provided not only for engineers intended for the Waterstaat, but also for those proposing to engage in kindred pursuits of industry and trade. In 1860 the construction of all railroads came under the charge of the Government, and the quarter century just passed has been more prolific in bringing forward engineers of distinguished ability and in the completion of grand engineering works of most important character, than any previous period in the history of the country.

The Department of Public Works of the Netherlands makes an exceedingly handsome display at the Exhibition, illustrating, by means of maps, plans and models, its principal great works.

Among the first of these was the drainage of the lake of Haarlem, ordered on the 27th of March, 1839, and completed in 1852, at a cost of over four and a half millions of dollars. This lake had been formed by an overflow very many years before, and the present operation recovered some 42,481 acres of valuable land, by the sale of which the entire expenditure was returned. All that section of Holland surrounding and covered by the Zuyder Zee has been entirely changed from its original condition by the inroads of the ocean, which have occurred from time to time on a greater or smaller scale, and on record from the fourth century. The gradual formation of the Zuyder Zee itself has been due to this cause, the whole area having been at one time a fertile, well-cultivated country. A comparison of the two maps on exhibition, of Holland in 1576 and 1876, illustrates this very clearly. The question of the drainage of the vast area of this sea has also been taken up, and the investigations show that the work is entirely practicable and can be carried out in from eight to twelve years' time.

Another important work has been the construction of the North Sea Ship Canal from Amsterdam, saving thirty-six miles in distance, and restoring a large amount of land from the waste waters of the Wijker-Meer and the Ij. The building of the piers or jetties for the North Sea entrance was effected under great difficulties, the treacherous sands of the coast rendering unsuccessful the method first adopted, of using heavy concrete blocks as previously carried out on the English coast at Dover, and requiring considerable modifications of plan. The construction of the great embankment or dam across the Y below Amsterdam, from Schellingwoude to Paardenhock, a distance of three-fourths of a mile, rendered necessary in carrying out the project, and involving the building of a system of locks sufficiently large to pass the immense shipping business without allowing the barrier to be even *felt* as an obstruction, was one of the most important features connected with this undertaking, and it was here that the use of mattresses of fascine work, weighted with ballast stone, one of the peculiarities of dike construction in Holland, came so extensively into play.

Next in order came the project for the improvement of navigation from Rotterdam to the sea, by the selection of the most favorable of the numerous channels by which the Rhine, the Meuse and the Scheldt communicate with each other and discharge into the ocean, and by the introduction of lateral dikes, contracting the width where too great, or widening it where too narrow, so improve it as to allow the most effective action of the tides in removing deposits; also the formation of a new outlet to the sea by a cut through the Hoek van Holland and the building of jetties into the sea for the proper protection of its entrance from storms. The whole of this work involved great practical difficulties, and required engineering skill of the very highest order. It has been carried out with the greatest success, and rests, it is to be hoped, an enduring monument to those by whom it was undertaken.

The particular feature in the construction of the jetties has been in the use of the fascine mattress, a thing not before attempted in the open sea, although very extensively used in dams and dikes obstructing or confining inland waters. The mattresses are made on the sea-shore between high and low water, so that they can be floated off to their final locations by the tide. Fascines are made from willow branches, osiers, etc., which are grown for the purpose as a regular business on the low lands of the estuaries of the rivers, the crop being cut every third or fourth year. The largest sticks are used for cask hoops, and the smaller ones tied up into bundles, each bundle being made to contain at least three sticks, from ten to eleven and a half feet long, and two of six and a half to eight feet, the twigs and smaller branches remaining attached, and more being added if necessary to make up the standard size. The whole is bound together with two osiers, making a bundle about seventeen inches in circumference at the thick end, and fourteen at the other. Long ropes, called "wiepen," are made up with a series of fascines, each bundle bending well into the next, all firmly tied together, and making a rope of seventeen inches circumference. To build the mattress a series of these "wiepen" are first laid parallel to each other, at distances apart of about three feet, and on top of these, at right angles to them, a second series, the same distances apart. These are then tied together at each crossing-point with two twigs, except those on the exterior lines and the alternate ones on the interior, which are fastened with tarred rope from old ships' cordage. The ropes are twisted upon small stakes, so that the ends may be used for tying upper wiepen to be mentioned hereafter.

A continuous layer of bundles of twigs is now laid upon the lower lines of wiepen, at right angles to them, covering the whole of the open spaces, the top ends of the twigs being upward, and the different bundles lapping partly over each other. A second layer is then placed over the whole at right angles to the first, and on top of this a third layer of the same kind, rectangular to the second. Wiepen are then placed at right angles in both directions directly over the lower wiepen, the small stakes at the corners fixing the locations of the cross points, and they are tied at the alternate crossings with the ends of the tarred ropes used for tying the lower wiepen, the whole being firmly pressed and held together. The intermediate crossings are tied with twigs, as in the case of the lower wiepen, and the

Bridge at Kuilenburg: Department of the Netherlands.

small stakes at the alternate corners are pulled out. The upper surface of the mattress is finally provided with divisions or cells to contain the ballast or broken stone necessary for sinking the mattress and holding it in position in the work. These are formed by driving stakes into the upper lines of wiepen, and weaving basket-work in between them with willow twigs up to a height of seven to nine inches, the stakes being then driven down well into the mattress, leaving the ends about six inches above the willow-work. At the third and fourth crossings of the wiepen from the corners, and also at distances of about fifty feet from each other over the surface of the mattress, a large stake is driven with some six or seven others in a sloping direction from the centre one, these being used to fasten the cables by which the mattress is conveyed to its destination and held in place while being sunk. Iron rings are

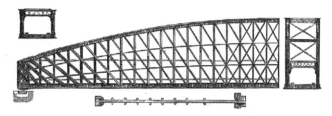

Details of Bridge at Kuilenburg.

also provided, fixed to cross points of the wiepen, so as to attach vessels by lines when the mattress is placed in position. The mattress is now towed to its position, and anchored as far as may be necessary, dependent upon the velocity of the sea; and small vessels of ten to fifteen tons are secured all around, filled with good supplies of stone. The pockets of the mattress are gradually loaded from the centre outward, and then more rapidly as it begins to sink, care being taken to let it settle evenly. At the proper time all ropes are let go simultaneously, the stone being thrown out as rapidly as possible, and the mattress finally sinks to the bottom. The ballasting proceeds until about three and a half to seven hundredweight of stone per square yard are deposited over the whole surface.

The work is built up in this manner, one layer of mattresses over the other, until the required height is reached. The head of the southern jetty has a width at the top of eighty-two and a half feet, the centre being three and three-tenths feet above mean low-water line. At the height of ten feet above that level is a platform of timber, supported on rows of square piles driven down through the whole construction, the outside rows being inclined and bolted at the tops to the adjoining rows. The platform has a width of about twenty-five feet, and carries a line of railroad track. The sides of the jetty are filled with large blocks of basalt stone set to a slope of one to one on the south side, and one and one-fourth to one on the north, the lower mattresses projecting from twenty-seven to thirty-three feet beyond the body of the work. The head is protected by two ranks of piles, bolted together at the top, and two other ranks of shorter piles are driven in the stone-work as a protection around the outside of the whole.

The Utrecht Boxtel line of State railways in the Netherlands crosses three large rivers, the Lek, the Waal and the Maas, within a distance of ten miles, the bridges at these points being known by the respective names of the Kuilenburg, the Bommel and the Crèvecœur viaducts. The great lengths of these bridges, the nature of the streams that they cross, and the local circumstances necessitated engineering skill of a high class. The conditions of the foundations were such as to require piling. The piles varied from twenty-three to fifty three feet in length, being driven in some cases by the ordinary pile-driving engine, and in others by a steam ram. After the piles were cut off to a level below water, the space between them was filled with beton or concrete, projecting from three to five and a half feet beyond the footings of the masonry above, and varying from eleven to twenty-one feet in thickness. The tops of the piles were completely floored over, and masonry built up, well bonded on to the floors to prevent sliding by longitudinal and cross walings of oak, and the faces of piers and ice-breakers were finished in Belgian ashlar. The footings of the piers were thoroughly protected by a close row of long piles to each, and heavy rip-rapping of rough stone.

The superstructure of the Kuilenburg bridge was built by the well-known Dutch firm of HARCORT & Co., under the superintendence of MR. N. T. MICHAËLIS, Engineer-in-chief. It consists of nine spans, entirely of wrought-iron construction, there being one span of 492 feet clear opening, one of 262 feet 6 inches, and seven of 187 feet each, making, with the widths of piers, a total length between the faces of abutments of 2181 feet. The bridge consists of two open trusses, built of riveted plates and angles, the upper and lower flanges being formed in the shape of double T's, side by side, the inclined ties of thin rectangular bars, except toward the centres of spans, where they require stiffening for compression under variable load, and the vertical struts of I-shape, some of the largest being strengthened by the introduction of two series of channel-bars between the verticals. The trusses are placed so as to give a clear width of roadway of 27 feet, and height of 16 feet 5 inches, the structure being a *through* bridge. Cross-girders 2 feet 11¼ inches deep connect the main trusses, and the whole is well stiffened by a thorough system of lateral and diagonal bracing. The span of 492 feet has a parabolic upper member, the depth of truss in centre being 35.6 feet, and at the ends 26.24 feet. The other spans have rectangular trusses of the same depth as the ends of the parabolic truss.

All holes for riveting were drilled, no punching being allowed in the work. The bridge is built for double track, there being only a single track placed on at present. Two foot-paths are provided for the service of administration. The total weight of material in the structure is as follows:—

Wrought iron,	$4394\frac{6}{10}$ tons.
Bessemer steel,	$610\frac{5}{10}$ "
Cast iron,	30 "
Lead,	$3\frac{2}{10}$ "

There were also 8000 cubic feet of oak, 9500 cubic feet of fir timber used, and 350 tons of plates placed between the rails to form the floor of the bridge. The total cost for masonry and superstructure was $1,187,100.

The Bommel bridge consists of eight land openings of 187 feet each, and three openings of 393 feet, making a total length between abutments of 2839 feet 7½ inches. The Crèvecœur bridge has ten openings of 187 feet, and one of 328 feet, making a total span of 2346 feet. The superstructure of these two bridges is of the same character as that at Kuilenburg, except that the masonry is made for two separate single-track bridges, only one being erected at present, and an additional line of superstructure must be put up when double track becomes necessary. Curved upper members are used in the longer spans of both bridges. The weights of material at Bommel are—

Wrought iron, ... 3468$\frac{2}{10}$ tons.
Bessemer steel, ... 227$\frac{1}{10}$ "
Lead, ... 2$\frac{3}{10}$ "

the total cost being $1,358,125 for masonry and superstructure.

The Crèvecœur bridge contained—

Wrought iron, . . 2106$\frac{8}{10}$ tons.
Bessemer steel, . 84$\frac{9}{10}$ "
Lead, ... $\frac{5}{10}$ "

and cost $465,000.

The whole of the iron and steel work in these bridges received six coats of best lead-oil paint after being cleaned in a bath of muriatic acid. The limiting strains for tension and compression were taken at 6½ and 4½ tons per square inch respectively.

Among the other noteworthy exhibits may be mentioned the bridge over the Hollandsch Diep, the steel bridge at Dordrecht, the swing bridge across the North Holland Canal, etc., and a handsome model of the Blanken lock-gates, used at a number of places in Holland. They are arranged by means of communicating ducts between the chambers of the locks and a recess into which the gates open, that the gates may be opened and closed simply by regulating the passage of the water through these ducts

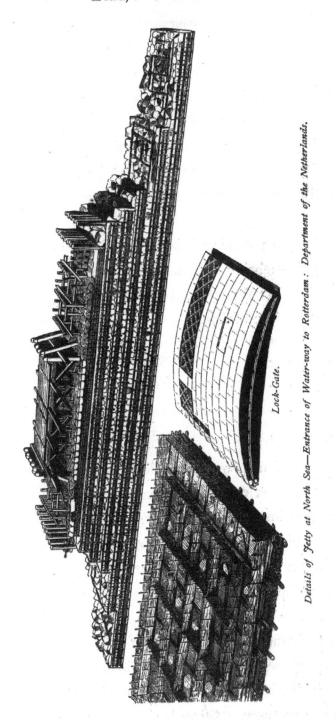

Details of Jetty at North Sea—Entrance of Water-way to Rotterdam: Department of the Netherlands.

Lock-Gate.

During the early years of railway management, when traffic was light and the number of trains few, a very simple system of signals was quite sufficient for the proper regulation of these trains. Hand signals, with flags by day and lamps by night—different colors being employed, red generally for danger, blue or green for caution, and white for safety—answered all purposes. As business developed, however, and as the turnouts and crossings at stations increased in complication, it became evident that something better than these primitive methods must be adopted. Stationary signals were then introduced, elevated at some height above the level of the rails, so as to be seen from a considerable distance, and they were placed at safety or danger as required to correspond with the clearing or blocking of the line. That known as the Semaphore signal, consisting of a vertical post with a movable arm attached near the top by a pivot, and capable of hanging vertically or of being moved out at right angles to the post, was the form of signal most universally adopted, proving so superior to all other kinds as to rapidly replace them. When the arm was thrown out at right angles to the post, it signified danger; when hanging vertically, it denoted safety; and when inclined at an angle of forty-five degrees, it expressed caution. The movable arm was counterweighted, so that in case of derangement of the apparatus or breakage of connections, it always flew out to "Danger," stopping all traffic, and at the worst only causing delay. As the system of tracks became still more complicated, arms were introduced on both sides of the post, and occasionally two or more tiers of arms, one above the other, the arms on one side always referring to trains in one direc-

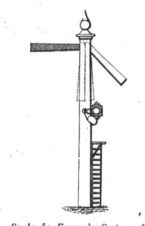

Saxby & Farmer's System of Railway Signals. Fig. 1.

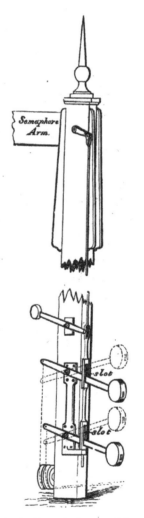

Saxby & Farmer's System of Railway Signals. Fig. 2.

tion, and those on the other side to trains in the other direction. The arms were painted red on the side next to the approaching trains they were intended to govern, and white on the other side, so as to appear less prominent and avoid confusion. Lamps were attached to the post for night use, the movement of red or green glass over the white glass indicating the desired signals. Where several arms were in use it became necessary to number them or mark them with symbols so as to distinguish them apart and signify to which set of tracks each signal belonged.

The Semaphore signal was found eminently satisfactory, continuing in use to the present day, but the method of operating it was quite inefficient. By errors of the signal-men, signals were sometimes given for wrong tracks, or switches were opened before the danger signal was turned on, or sometimes the danger signal changed to safety before the switches were set right, resulting often in cases of serious accident. The obvious remedy for this was to make the movement of the signal automatic with the movement of the switch, or, better still, to make the movement of the danger signals obligatory before the track could be blocked, or the clearing of the track obligatory before the safety signals could be set. The solution of this problem, step by step, has resulted in a system of apparatus so complete as to leave little if anything to be desired.

MESSRS. SAXBY & FARMER, of LONDON and BRUSSELS, have been early and strongly identified with this system, taking out patents in England as long ago as 1856, and in this country in 1868. They are represented by an exceedingly handsome and complete working model at the Exhibition, and we cannot explain better the latest and most novel improvements in this direction than by a description of their apparatus.

On the preceding page we present an engraving, Fig. 1, showing the Semaphore signal, and also one of the Semaphore with slotted rod, Fig. 2. It is exceedingly desirable in certain cases that two signal-men should control one signal, so that the consent of both should be necessary to its use. This is arranged in a simple manner by means of two slots on the signal-rod, in which pivoted levers move up and down, each being operated by a separate signal-man. It is evident that both levers must be moved the same way before any change can be effected in the signal, and that such change must be in concordance with the intentions of both operators. This principle, which is capable of almost unlimited extension, and allows any number of slots, in which a pointed lever or pin may work, was the germ of the system introduced by Mr. Saxby, in 1856, in his invention of combined interlocking signals, by which it was rendered mechanically impossible to make the position of the switches contradictory to the position of the signals, or to allow irreconcilable signals to be given, no matter how complicated the system of tracks or switches.

MESSRS. SAXBY & FARMER claim, with their system of apparatus, perfect safety in working switches, signaling junctions, stations, &c.; great facility and rapidity in manœuvring trains; and great economy of working, one signal-man being able to operate all but the very largest stations.

In arranging a Saxby & Farmer apparatus at any particular station or junction, a convenient site is selected, on which is erected a signal-tower, or building with a second story, having large windows, and well overlooking the arrangement of tracks. In this is placed a set of levers, arranged in a cast-iron frame side by side, and by which the whole system of signals and switches is operated. Some work the switches and others the signals, the former by rod connections, and the latter usually by wire rope.

Fig. 3 shows this arrangement of levers for some particular station, the name generally given to it being the "Locking Apparatus," for the reason that the levers are so interlocked that the switches must be properly set and locked in the right direction before it is possible to move the signal-lever corresponding, and the signals themselves are so interlocked, so as to protect the path of a signaled train, throughout its whole length, from all crossing lines. It will be noticed that each lever is numbered, and that some have one or more secondary numbers under the principal number. These numbers are to guide the operator, and the secondary numbers specify the levers which control the principal lever, and which must be moved before any movement of the principal can be made. For instance, suppose lever No. 11, having under it secondary No. 7, operates a certain switch. Before it is possible to use it, lever No. 7 must be moved, acting on the danger signal, and covering the opening of this switch by No. 11. It is evident that with this system it is impossible to make a mistake causing any further inconvenience to an approaching train than possible delay.

In describing the method by which this locking is accomplished, we will refer to Fig. 4, and we would state that an important advantage belongs to

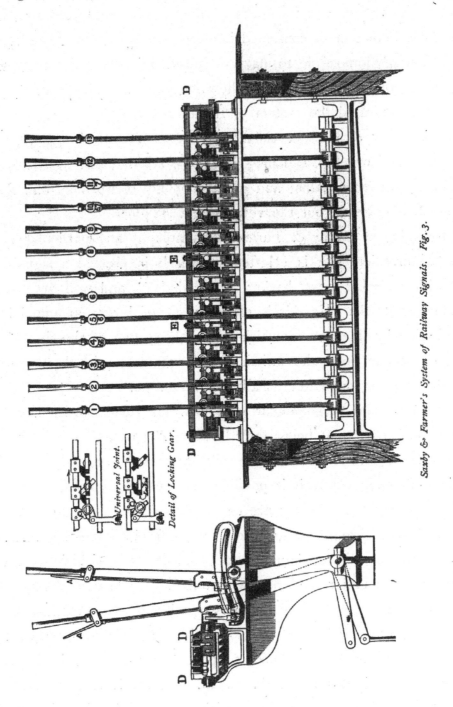

this apparatus in the fact that the interlocking gear is actuated solely by the movement of the spring catch-rod in front of the lever and attached to it. This spring catch-rod carries a stud, upon which is a small block, B, which travels in the curved slot of the rocker, D, a segmental plate, movable on the centre, A. When the lever is thrown forward or backward to its full position, the spring catch fits into a notch in the fixed quadrant on which it moves. When it is in its forward or normal position, to the front of the frame, with the spring catch-rod down, the left-hand end of the rocker is depressed, and the right-hand end raised, as shown by the dotted lines. When the spring catch-rod is raised, the rocker moves into the position shown by full lines, and keeps this position until the spring catch falls into the notch at the rear of the frame, when it assumes a third position, elevated on the left and depressed on the right. A jaw at the left-hand end of the rocker carries a universal jointed vertical link, E, giving motion to a small crank at the end of a spindle, the bearings of which are shown at G, G, there being a spindle for each lever. These spindles lie directly under a series of horizontal rectangular bars, shown at D, D, in Fig. 3, called locking-bars, which slide to and fro in the guides E, E, and to these bars are attached locks L, L, Fig. 4.

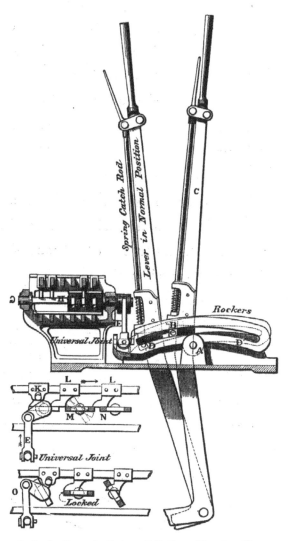

Saxby & Farmer's System of Railway Signals. Fig. 4.

The spindles are flat in their central portion, as shown at M and N, Fig. 4, and when they stand in their normal position they are horizontal, and the locking-bars and locks are free to move forward and back over them. When turned up, however, out of the horizontal, as shown at I, and in dotted lines at M, they catch on the locks and stop their movement and that of the bars to which they are attached. Some of the spindles are required to work locking-bars, and are provided with a short vertical crank, the stud of which works between two horns on the locking-bar, as shown at K, giving a horizontal motion to the bar at any movement of the spindle.

It will be seen that the locks and crank attachments may be fixed at any locations desired on the locking-bars, by means of set screws or a similar arrangement, some being made to allow free movement of one spindle, as at M, while at the same time another spindle, as at N, is locked.

Whenever the spring catch of any lever is raised, its rocker is lifted and the corresponding spindle turned. If the spindle is locked it will be impossible to move the spring catch. A very small movement of the spring catch, if the spindle is free, will cant it up sufficiently to lock the locking-bar upon which it works and prevent any movement of it by other levers.

The spindle can occupy three positions, as shown at M: first, the horizontal, when the lever is in its normal position; second, slightly inclined, as shown by dotted lines, when the lever is being moved; and third, a more inclined position, also shown dotted, when the lever has been pulled to its full open position and the spring catch released.

The third position is a very important one, some of the locks not being released until this position is attained. Thus the spindle N is not released until that at O is in the third position, and the lock over it has moved sufficiently to be able to enter a hole in it. The spindle then by its new position prevents a return movement of the lock or any change of the spindle O until brought back to the horizontal again, corresponding to the lowering of the spring catch into the forward notch of the quadrant.

By this interlocking apparatus it is possible to absolutely prevent any signal-man from even commencing to make a movement of either switches or signals until all switches or signals which have any relation to the movement intended to be made have been effectually locked in their proper positions, a matter of the utmost importance.

By the system here explained it may readily be understood that any combination or arrangement of interlocking different levers may be made as desired, that changes can be made to accommodate alterations to tracks or switches by merely moving the positions of the locks, and also that extensions of the apparatus can be economically effected to suit extensions of business without throwing away the portion already in use.

Messrs. Saxby & Farmer also provide a "Patent Facing Point Lock," which insures that the switches shall be properly closed and locked before a train can be signaled over them, and also prevents the possibility of the switchman moving the switches while the train is passing. This apparatus, shown by Fig. 5, is worked by a lever in the "locking apparatus," and the interlocking of the levers forces the switchman to use this lock before giving the signal. A cross-rod connects the two switch-rails near their movable ends, and in this are two holes, so that the switch may be locked for either position. A taper bolt shoots into one or the other of these holes, bringing the apparatus up home, and keeping it snug and tight. If the switches are not properly closed, the bolt will not enter the hole, and the facing switch lock lever in the apparatus will not work, and will consequently prevent the safety signal being given.

In order to prevent any danger of a signal-man carelessly moving the switch-rails while a train is passing over them, a switch-locking bar is provided, consisting of a bar at least as long as the greatest distance between any two pairs of wheels of a car, placed on the inner side of one of the fixed rails immediately adjoining the switch-rail, and connected with the system by which the lock-bolt just described is worked. It is hinged on short links lying in a vertical plane, so that it cannot be moved lengthwise without at the same time being raised. When the lock-bolt is at either end of its stroke, the bar is at its lowest level, and just allows the flanges of the wheels to pass over its length. When an attempt is made to move the lock-bolt for either of its extreme positions, the switch-man is unable to do so without raising the bar, which cannot be done so long as the cars are passing over it.

We have before mentioned the great advantages resulting from the use of slotted signal-rods, rendering it impossible for disagreement to take place between signal-men who by mechanical means jointly control a signal, although they may be a considerable distance apart. Messrs. Saxby & Farmer exhibit in this connection Farmer & Tyer's patent "Electric Slot Apparatus," by means of which the same result can be accomplished for an unlimited distance between the signal-men. This may be applied to their ordinary locking apparatus without

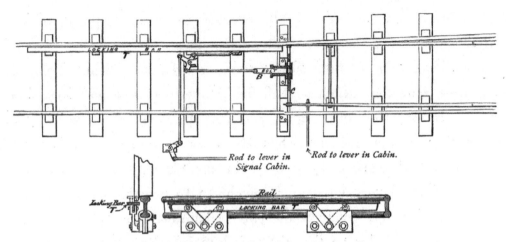

Saxby & Farmer's System of Railway Signals. Fig. 5.

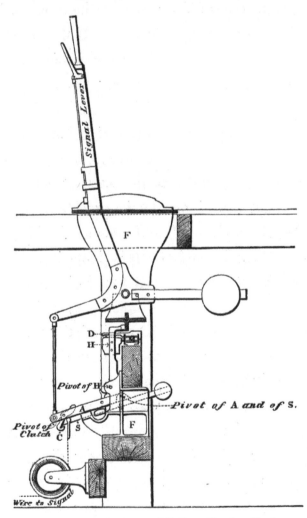

Saxby & Farmer's System of Railway Signals. Fig. 6.

difficulty, and carries out, by means of an electric current, all the advantages of the mechanical slot. Fig. 6 represents one of the signal levers of the locking apparatus to which this arrangement has been attached. A is a lever consisting of a pair of wrought-iron plates placed side by side, with a space of about three inches between them. This lever is worked by a connecting-rod from the signal lever, and the clutch, C, pivoted to it, connects it with the lever, S, acting on the same pivot as A, and working the signal. H is a hammer, so pivoted that a small upward movement of A will raise H to the nearly vertical position in which it is shown. M is an electro-magnet, and D a detent. When the lever, S, is free, and not held by the clutch, C, the signal always flies to "Danger."

The electro-magnet, M, when under the action of an electric current, holds the hammer, H, in place, being further assisted by the detent, D, also under its influence. If the current be broken, the hammer, H, falls on to the clutch, C, releasing the bar, S, and placing the signal at danger. When the current is broken, therefore, it is impossible for the signal-man at the lever to move the signal from danger, no matter how often he moves the lever back and forth, and by putting the control of the current around the electro-magnet, M, under the charge of the other signal-man, he can control the signal as he desires.

This invention gives each signal-man on a block system, in addition to the usual block telegraph instruments, the actual mechanical control over the signals at the next station, controlling the coming train even before it enters his section of road. Messrs. Saxby & Farmer also show an admirable arrangement of gates for level crossings, so arranged that danger-signals are displayed whenever the gates are shut across the railway, and cannot be lowered until the gates are opened again and shut across the public roadway. The gates are connected to a lever similar to a switch-lever, and by a rack and pinion movement all gates are shut and opened simultaneously, the lever being made to interlock with the signals.

There are two general methods employed for the reduction of the metal copper from its ores—one the old system of repeated fusions, or the smelting process, always used in the case of rich ores, and the other designated the wet process, by which the material containing the metal is subjected to the action of a solvent which dissolves the copper out, leaving the other ingredients behind as a residue. This wet process has been very extensively used in the extraction of copper from lean or poor ores, which could not be profitably worked by the old method. Rich ores are easily smelted, and the proportion of fuel used compared with the metal obtained is small; but as the percentage of copper decreases, the difficulty of separation increases, and the amount of fuel required becomes greater, making the operation too expensive for practical use. With the wet process, however, only one furnace operation is generally required, and but little fuel. To reduce a ton of copper from a five per cent. ore by the smelting process would consume about fifteen tons of coal, while only three tons would be necessary in obtaining the same amount by the wet process. There are, however, expenses in the latter method which do not obtain in the former, the principal one being the precipitant necessary to throw down the copper from its solution. The amount of this varies directly with the amount of copper obtained; and while its use as compared with the consumption of fuel by the other process makes the wet method the most economical for a low-grade ore, it does not do so for a rich ore. The manner of procedure must therefore vary with the kind of ore available.

The precipitant usually employed in the wet process is metallic iron, depositing the copper in a pure state, and it depends on the form in which the copper exists in solution how much iron is required. With the proto-salts one ton of copper will require nearly a ton of iron, the atomic weight of the latter being a little less than the former, while with the sub-salts less than half a ton will be necessary.

In England copper is reduced from some six hundred thousand tons of Spanish pyrites annually, most of the sulphur having been first extracted for the manufacture of sulphuric acid by roasting the already partially roasted ore with salt. By this means the copper is converted into proto-chloride, becoming soluble in water; and to render the process still more certain, the water is acidulated with muriatic acid, which at the English extraction-works is almost a waste product. The solution is then brought into contact with metallic iron, which precipitates the copper, and the liquid, containing chloride of iron, is thrown away. At some works in England and on the Continent, where acid is a waste product and carbonates of copper are available, the copper is extracted simply by acids. Both these methods are exceedingly effective, but are applicable only where salt and acids are very cheap, which is nowhere the case in this country.

In order to utilize a secondary product of sulphuric acid manufacture, Mr. Monnier patented in the United States, some years ago, a process which consisted in sulphatizing, and thus rendering soluble in water, the copper in sulphuretted ore, by roasting with sulphate of soda, and he obtained, when his method was worked with skill, very satisfactory results. The limited supply of the reagent, however, and its cost elsewhere than at the chemical works, interfered with the general adoption of the process.

Copper is very generally distributed, almost every one of the United States claiming copper-mines, but the ores are seldom abundant enough in one place to justify the erection of works at the mines, or sufficiently rich to bear transportation to a market. Displays are contributed to the Exhibition from quite a number of localities, both of ores and furnace products. The CHEMICAL COPPER COMPANY, of PHŒNIXVILLE, PENNSYLVANIA, exhibits in the Government Building a series of specimens illustrative of a wet method called the HUNT & DOUGLAS PROCESS, showing the ores treated, the different steps of the method employed, and the marketable products obtained. In the exhibit of the American Society of Civil Engineers, in the west gallery of the Main Exhibition Building, is shown a drawing of the works, from which we have taken our engraving. MESSRS. HUNT & DOUGLAS claim, as the chief merit of their process, the use, as an efficient solvent, of the waste liquors containing chloride of iron which are run off as worthless after the precipitation of the copper in the salt and acid methods. They find that a solution of chloride of iron, when used in connection with a strong brine, dissolves readily the copper from either, naturally or artificially, oxidized ores. They therefore make such a solution by dissolving, in the proper proportions, sulphate of iron (copperas) and salt, or in any other convenient way, and in this solution, heated to about 150° Fahr., they digest the ore, which if massive must first be ground to such fineness that its copper contents will be exposed. If the ore is already oxidized by nature, and consists of a native carbonate or oxide, or a silicate like chrysocolla, it will yield up its copper to the solvent without other treatment; but if the copper be combined with sulphur, it must first be roasted to drive off the sulphur and oxidize the metal.

Ores that are very slimy and form mud on settling must be agitated with the chloride of iron solution in vats provided with a stirring apparatus, as

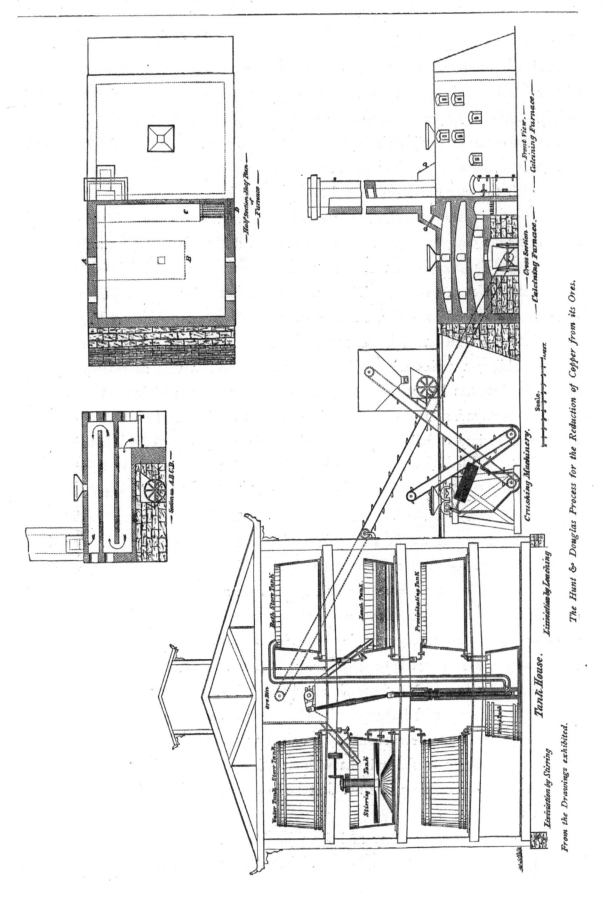

The Hunt & Douglas Process for the Reduction of Copper from its Ores.

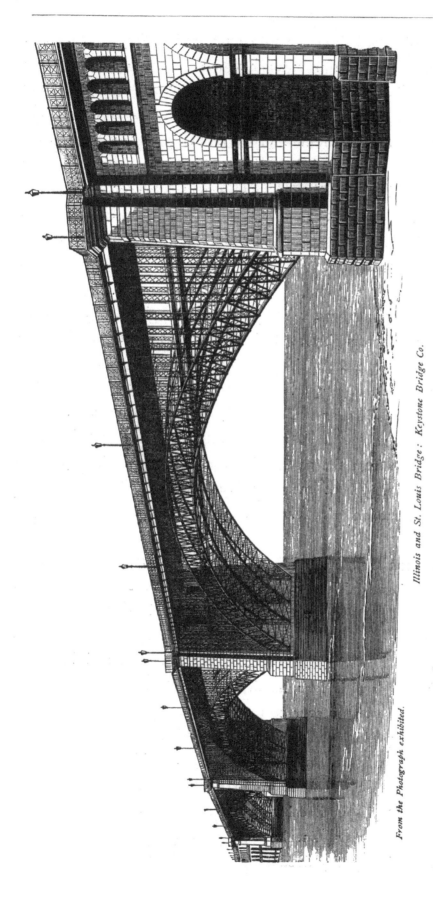

Illinois and St. Louis Bridge: Keystone Bridge Co.
From the Photograph exhibited.

shown in the engraving, while if the ores be coarse and gritty, they may be laid a foot or more in depth on the false bottom of a leach-tub, and the solvent solution be allowed to filter slowly through them. This method by filtration is slower but quite as thorough as by agitation. The chloride of iron in passing through the ore reacts with the oxide of copper, and there results an insoluble peroxide of iron and a mixture of soluble chloride and dichloride of copper which flows into the precipitating tanks, where it comes into contact with metallic iron, and the copper is thrown down in crystalline metallic grains. The liquor becomes recharged with chloride of iron, and ready therefore to be pumped back into a storage tank and heated so that it may be used over again and passed through the ore to dissolve a fresh charge of copper. The solvent is thus constantly regenerated, and the same liquor circulates indefinitely, alternately charged with chloride of copper and with chloride of iron. The only reagents used are a little salt, which must be added from time to time to supply inevitable loss, and the iron consumed in precipitation. This item, however, is less than in the ordinary methods, as about two-thirds of the copper is dissolved as a dichloride (that which is kept in solution by the brine), and therefore only from sixty to seventy-five parts of iron are consumed in yielding one hundred parts of copper.

Messrs. Hunt & Douglas do not claim that the neutral solution as used will dissolve the copper as thoroughly as acids would; but if the ore be suitable and at all skillfully treated, the residue from a four per cent. ore will not retain over one-half per cent. of copper.

In the exhibits of the Chemical Copper Company may be seen some remarkably perfect and large crystals of metallic copper obtained by slow precipitation, and ingots made from the cement by a single fusion. This method of reduction is also employed at the Ore Knob Copper Mine in North Carolina, and for the extraction of silver and copper at the Stewart Works, Georgetown, Colorado.

The Keystone Bridge Company, of Pittsburgh and Philadelphia, exhibit in the Main Building a beautiful model, on a scale of one twenty-fourth the full size, of the swing-bridge which it has recently erected over the Raritan Bay, New Jersey, for the New York and Long Branch Railroad Company. The bridge was designed by J. H. Linville, C. E., the President and Engineer of the Keystone Company, and is four hundred and seventy-two feet in total length, being the longest pivot-span ever built in the United States, and believed to be the longest in the world. It is what is denominated a through bridge for a single track, and consists of two trusses, forty feet high at the centre, and thirty feet at the extremities, calculated for a maximum live load of two thousand five hundred pounds per lineal foot, and a constant dead load of four thousand five hundred pounds per lineal foot, the parts being proportioned so as to resist the stresses due to the dead weight of the bridge when open, as well as the dead and live load when closed, and operating as two spans of ordinary bridge. The upper and lower chords are consequently calculated for both tension and compression. The whole construction of the bridge is in wrought iron, the posts being hollow and readily accessible for painting on the interior surface. The structure is supported on the rotating pier by a central drum, the load being transferred either to a central anti-friction cone-bearing pivot or to a series of thirty bearing-wheels under the drum, two feet in diameter, twelve inches face, and traversing between steel-faced tracks. By means of centre suspension bolts in the drum, the weight may be adjusted either partly or entirely on the central pivot or on the bearing-wheels, as found to be most desirable.

The lower chords of the bridge are parted at the centre, so that by means of wedges the ends of the trusses may be adjusted to any required elevation. Before swinging the bridge open, the entire structure is lifted about four inches by four hydraulic rams placed in transverse girders of the central drum under the four central posts of the trusses, and operated by a double engine having two cylinders of eight inches diameter and ten inches stroke, the same engine being also employed to turn the bridge. The turning-gear is brought into action by a friction-clutch, there being two pinions which are made to act equally on opposite parts of the rack by an equalizing attachment in a large mitre-wheel placed between the shafts that drive them.

Automatic locks fasten the bridge at both ends when it is rotated into the line of track, and it is then lowered on to solid bearings on the drum and at the extremities by simply turning a valve, the projecting rails of the track fitting into the same shoes that receive the rails of the permanent spans. About one-half of the dead weight of the structure is made to bear upon the solid adjustable supports of the rest piers, the bridge by this arrangement being rendered firm and steady under the load of passing trains, the trusses acting as two independent spans. Therefore when one arm is loaded no effect takes place in the other arm, and no provision is required for holding this arm down, or otherwise providing as necessary in the case of continuous girder swing-bridges. The bridge has been designed as a cantilever under dead load when swung, and as two independent spans under dead and live load when closed, this being considered by the engineer as the most satisfactory plan; and by means of a slotted link connection at the centres of upper chords, the continuity may easily be destroyed when the bridge acts as two spans. The operation of lifting the bridge by the rams as previously mentioned, before swinging, renders the practical execution of this principle entirely feasible. The weight of the entire structure above the drum is six hundred tons, and the engine, turning-gear and hydraulic machinery operate the whole with ease.

The KEYSTONE BRIDGE COMPANY in this connection also make an exhibit illustrating the method adopted for forming and uniting the steel tubes used in the arches of the great bridge over the Mississippi River at St. Louis, and the

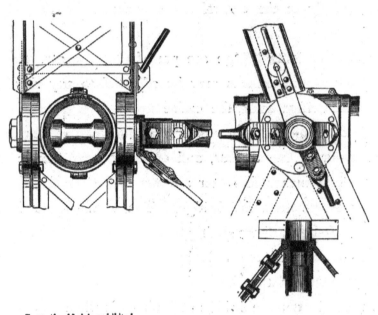

From the Models exhibited.

Details of the Illinois and St. Louis Bridge: Keystone Bridge Co.

accompanying engraving will serve to explain more clearly our description of the same. This bridge, the largest arched bridge in the world, with its spans of five hundred and twenty and five hundred and fifteen feet, was designed by Captain James B. Eads as Chief Engineer, and the superstructure was manufactured from his designs and erected by the KEYSTONE BRIDGE COMPANY. The tubes of the arches are composed of six rolled cast-steel staves forced into a cylindrical envelope of steel, the lengths of sections between the joints being about twelve feet, and the depth of the arched rib between the centres of two concentric tubes about the same. The two lines of tubes are braced together, and the ends of contiguous sections are united by couplings made in two parts with projections turned on the inner surface to fit into corresponding grooves on the ends of the tubes. The connecting-pin for lateral struts, diagonals and lateral bracing between the several arches is tapered and driven tightly into the joint, the whole connection being made water-tight.

We present an engraving of the bridge which will give the reader an excellent idea of its general design and magnitude. The arches were built outward simultaneously from the abutments and from each side of the piers, being supported by means of direct guys, composed of two lines of main cables of forty-two square inches section, passing over towers to anchorages on the shore, and by guys balanced over towers on the piers. The towers stood on hydraulic rams, which were caused by automatic gauges to rise and fall, to compensate for changes of temperature in the arches and cables.

The KEYSTONE BRIDGE COMPANY also make a handsome display of photographs of large span bridges constructed by them, the Parkersburg and Bellaire bridges of three hundred and forty-eight feet span, the Newport and Cincinnati bridge of four hundred and twenty feet, etc., etc., and a fine perspective of the High River bridge of the Cincinnati Southern Railway over the Ohio, with spans up to five hundred and twenty feet.

Swing-bridge over the Raritan Bay: Keystone Bridge Co.
From Model Exhibited.

Among the various articles shown in the Machinery Hall, the SPOOL-WINDING MACHINE exhibited by the CLARK THREAD COMPANY, of NEWARK, NEW JERSEY, and PAISLEY, SCOTLAND, merits particular attention, as an automatic apparatus of considerable beauty from a mechanical point of view, and one in which the combinations and details necessary to carry out the requirements have been designed with great skill, and evince that study and forethought on the part of the inventor which never fail to show themselves in the results.

Our engraving gives a front view of the machine, which may be considered as composed of two parts, the portion on the left, really consisting of eight little duplicate machines arranged in a row, and that on the right, which forms a single piece of apparatus and operates the whole. Each of these little machines winds a spool of thread, and hence eight spools may be wound at a time. Back of these may be seen a trough, which is to contain empty spools, and back of this again, in the rear of the apparatus, is a shelf which is intended to hold the bobbins of thread roughly wound as they come from the mill. From these bobbins, threads are taken up and passed through a tension apparatus above them, which keeps them at the proper stretch for winding, and then carried on, each to its little machine. The machines are held rigidly together by longitudinal rods so as to oblige them to work firmly as one piece, and there are three longitudinal shafts or rods passing through the whole set from the machinery at the right-hand end; the upper one, which we will call the guide-rod, moving back and forth and giving side motion to the thread; the main shaft below and behind, with cog-wheel attachments at each machine for revolving the spools; and a rod in front, which carries a steel finger for moving the thread as will be explained presently. The spools are held horizontally and longitudinally in position just back of the front finger-rod by clamping-pins like axles, which pass into the holes at the ends and revolve with the spools very rapidly. Just back of each spool is a swinging curved hopper, its upper end reaching almost to the spool-trough previously mentioned, and its lower end open and curving up just under the position of the revolving spool. In front is a small receiving-trough to each machine for wound spools. All that the attendant has to do is to keep the hoppers filled with empty spools, remove the full spools from the lower troughs as they accumulate, and see that the thread is regularly supplied by the bobbins behind, adding new ones when the thread winds off of those attached. The machine otherwise is entirely self-acting.

A piece of highly-tempered steel called a thread-guide, in the front edge of which is a groove the size of the thread to be wound, is fastened on the upper sliding or guide-rod at each machine. The thread passes down from the tension apparatus over this guide to the spool. As the spool revolves, the longitudinal motion of the guide-rod back and forth moves the thread to and fro over the spool, which winds it up layer by layer; and all those who have noticed the beautiful regularity with which a spool is wound, each layer of thread, one over the other, and each thread in the same layer next to its brother, close up, but no lapping, no confusion, will bear witness to the nicety requisite in machinery which shall accomplish this end.

There is a measuring-gauge attached to the machine. Just as two hundred yards are wound, the spool ceases to revolve; a little chisel moves up and nicks its edge; the sliding-rod in front with its steel finger moves longitudinally and draws the thread over; a hook passes up and pulls it down tightly into the nick; another chisel cuts it off, and the spool drops down into the receptacle provided for it in front. The swinging hopper then flies up with an empty spool in its curved lower end, which is taken up by the axle-clamp and starts into revolution. At the same time the thread, the cut end of which has been held down by the apparatus for the purpose, is pulled over and started on the new spool, and the operation proceeds as before.

It will be noticed that the part of a spool on which the thread is wound always has a variable length, increasing as the winding proceeds outward from the centre. Provision must therefore be made to give this variable motion to the guide-rod carrying the thread-guides. This is effected in its feed at the right end by giving a variable motion to the stops changing its direction. There are attached to this guide-rod two segmental nuts which are made to come alternately into contact with a revolving shaft having reverse screws contiguous to each other, one screw working in each half nut, causing the nuts to travel first in one direction and then in the other. These nuts connect with an arm with a forked end, which works on a fulcrum and operates over a pair of stops or jaws, pressing on to them and moving above them for one motion, and below them for the other, two heavy springs operating to

Spool-winding Machine: The Clark Thread Co., Newark, N. J.

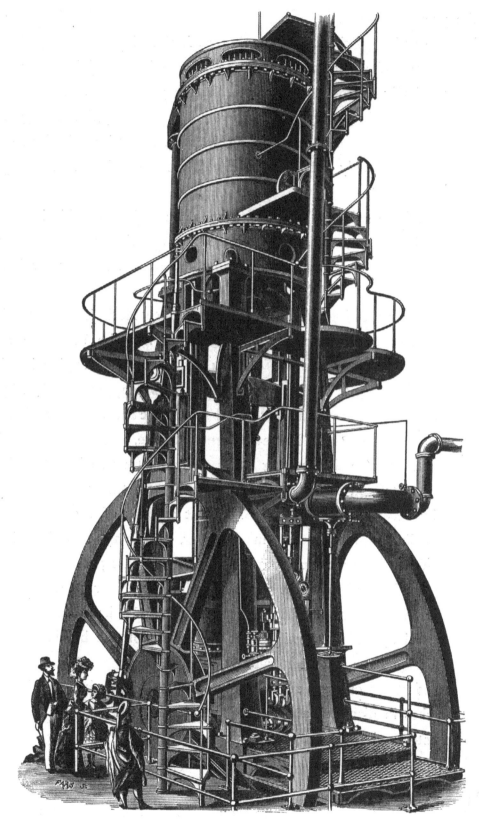

Blast Engine: I. P. Morris Co., Philadelphia.

produce the pressure and change the motion, alternately forcing it down and up, the alternate action of the nuts changing each time in accordance with this motion. By means of a cam and an arrangement of toggle-joint the pair of jaws opens gradually as the thread winds, keeping at a certain distance to correspond to each particular layer, thereby regulating exactly the sliding movement of the guide-rod. When the winding is finished and ready for another spool, the jaws are suddenly closed to their smallest dimensions and the operation is repeated.

We have endeavored thus to bring the main points of this interesting machine before the reader, our limited space not allowing us to pass on to further details, but we believe we have sufficiently explained its peculiar features to make its operations intelligible.

Prominent among the exhibits in the Machinery Hall, just west of the Corliss Engine, and towering almost to the roof, may be seen the BLAST ENGINE of the I. P. MORRIS COMPANY, of PHILADELPHIA. This engine has been designed to meet the wants of American Furnace Managers, certain requirements having been laid down as a standard which the firm have endeavored to follow as closely as possible. These requirements are, "completeness without sacrifice of accessibility to the moving parts, self-adjustment of parts liable to irregularities of wear, and steadiness of the whole structure and preservation of alignment by being self-contained." The first engines of this class—a pair having steam cylinders forty inches in diameter, and blast fifty-eight inches, with a stroke of four feet six inches, and producing a blast pressure of twenty-five pounds—were built about eight years ago for Bessemer steel production. Since that time twenty-four, including the present engine, have been built and put into successful operation, showing that the efforts of the builders towards perfection of design have not been without their reward.

The firm construct engines on this plan with blast cylinders varying from seventy-five inches in diameter and six feet stroke to one hundred and eight inches in diameter and nine feet stroke, and nearly all of them are provided with condensing apparatus sufficient for initial steam pressure of forty pounds per square inch, admitted during three-fourths of the stroke, and producing a vacuum of twenty-four and one-half to twenty-six inches. All parts are proportioned to the work of supplying steadily a blast of forty pounds pressure if required; and although this is beyond the ordinary working of anthracite coal-burning furnaces, it has been exceeded in one case, a pressure of thirteen and one-half pounds having been blown for a considerable time by one of these engines without causing it any injury.

The engines are fitted with the Wanich equilibrium valve, designed by Mr. A. Wanich, foreman of the machine-shop of the Company. The essential feature of this valve consists in the use of a ring cast on the back of the main valve, extending upward and bored out so as to envelope and slide freely upon the outside of another ring cast on the steam-chest bonnet above, extending downward and turned off evenly on the outer circumference. These rings are of course concentric, and the annular space between them is quite small, very much less than the aggregate area of the holes for the passage of steam below the pilot-valve, consequently any steam passing this annular opening when the pilot is raised, goes freely through into the cylinder, exerting no appreciable pressure on the back of the main valve, and permitting it to rise easily. This has been confirmed by connecting an ordinary steam-gauge with the space enclosed by the rings, showing the pressure, when the pilot was seated, to be say thirty-five pounds, and dropping suddenly almost to zero when the pilot was raised, until the main valve opened, when it rose again to thirty-five pounds. This valve has been in use for about four years with highly satisfactory results, saving steam and proving easily manageable.

The blast-valves are of selected thick sole-leather, backed with plate-iron, and the blast-piston is fitted for either metal, wood or bag-packing. The steam-piston is provided with metal double rings held out by springs. The valves are lifted by cams operating directly against rollers fitted into the bottom ends of the lifting-rods, and these cams are adjustable but not variable, giving facilities for experimenting so as to determine the best distribution of steam without interference with each other. The cam-shaft is driven by spur-gears fitted to the main shaft. The rim of the fly-wheels on the side in line with the crank-pin is cored out, so that the excess of weight on the other side will counterbalance the weights of piston-rods, cross-heads, etc. The shaft is of wrought-iron, and the cross-head swivels in the yoke connecting the two piston-rods, so that it may accommodate itself to any irregularities of wear in the main shaft or crank-pins.

This particular engine has a height of thirty-six and one-half feet, weighs two hundred and fourteen thousand seven hundred and ninety-four pounds, and exerts seven hundred and fifty horse-power, delivering ten thousand cubic feet of air per minute. The bed-plate upon which the whole construction rests is eight feet wide and thirteen feet long, weighs seventeen thousand pounds, and is laid on a foundation of hard brick or good stone at least ten feet in depth and well anchored to it so as to insure stability. The steam-cylinder is fifty inches in diameter, and the blast-cylinder ninety inches, the stroke being seven feet. The fly-wheels weigh forty thousand pounds each.

The height of the engine is principally due to the length of stroke, and this has been done so that a given quantity of air can be supplied by a less number of revolutions and with fewer beats of the blast-valves than is generally adopted in other engines. The direct loss in delivery due to piston clearance and space in the passage being a quantity depending on the diameter of the blast-cylinder, then if we take a fixed diameter of cylinder, it is clear that the percentage of loss of useful effect will diminish as the stroke increases.

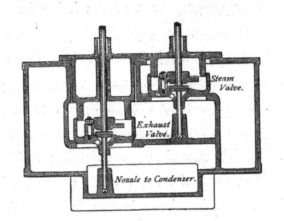

The Wanich Equilibrium Valve.

The engine is provided with a condensing apparatus situated just back of the main working parts, and in the entire construction everything has been carried out with a view to proper economy both in first construction and in future use. The firm claim for this style of blowing-engine, as compared with others, a reduced cost, not only of the engine itself, but also of the foundations required in setting it up, and the buildings necessary to cover and protect it when placed in working condition. Great advantage also results from the direct action of the engine, the power being transmitted directly from the piston-cylinder to the blast-cylinder without the action of a beam, as in many engines of this kind. The I. P. MORRIS COMPANY have for many years been engaged in the manufacture of heavy machinery for iron blast-furnaces, and their exhibit does them great credit.

Steam Type-Casting Machine: MacKellar, Smiths & Jordan, Philadelphia.

Great advances have been made in the methods of casting type for printing purposes from the time of the wooden blocks and rude types of Laurentius of Haarlem to the improved hand-moulds of Archibald Binny of Philadelphia at the beginning of the present century. By the latter as many as six thousand types per day were produced. The hand-moulds were supplanted in 1845 by the complex and effective American type-casting machines, which have wrought an important revolution in the business.

Among the large type-foundries in the United States, that of MacKellar, Smiths & Jordan, of Philadelphia, occupies the first position, and is well represented by an extensive display in Machinery Hall. This firm exhibits a number of modern type-casting machines, which may be operated by hand or power. These machines are constructed upon the same principle (whether operated by hand- or steam-power), and their average production is about one hundred per minute for the ordinary sizes of printing-type, being far beyond the amount of product of the earlier methods. The advantage in using power is that it enables one man to attend to two machines. Our illustration on page 85 shows a machine with steam attachment.

Type-metal is an amalgam of lead, antimony, copper and tin in such proportions as to produce a material hard but not brittle, ductile yet tough, flowing freely yet hardening quickly. Each letter is first cut in reverse shape on the end of a short strip of steel, the greatest care being taken to insure accuracy of proportion and harmony of appearance in the letters of the entire alphabet. The least variation is inadmissible, as it would destroy the harmonious effect of the types when composed or formed into columns or pages. The steel strips when finished are termed punches; and after criticism and approval, each punch is placed in a stamping-machine, and a deep impression made of it in one side of an oblong piece of copper near its end. These pieces of copper are called matrices. They are dressed and fitted up with delicate skill, so that the types cast from them shall be of uniform height and accurate range. They are then ready for use in the casting-machine.

The machine casts but one type at each revolution. It consists of a furnace, on the top of which is a small reservoir of metal kept in a fluid state. In this reservoir is a pump, the plunger of which operates in a cylinder in the bottom, and projects at each stroke a small quantity of the molten metal out from a small hole in a spout or nipple in the front face. The mould in which the stem or body of the type is formed is of steel and is movable, being set in place in front of the reservoir and worked by the action of the same machinery which operates the pump. The copper matrix, containing any special letter stamped into it with the punch, rests with its face against the bottom opening of the mould, being held in position by a curved steel spring shown in the engraving. The method of operation is as follows: The initial movement of the machine brings the upper opening in the mould opposite to the matrix exactly against the hole in the nipple. A simultaneous action of the pump projects a stream of the liquid metal into the mould with considerable force, at the same time stopping the opening in the nipple by a small plug from behind to prevent the further escape of metal. The next movement draws the mould away from the nipple and opens it, throwing back the matrix, extricating the type and dropping it by a slide into a box below. This operation is repeated over and over again as rapidly as the crank or wheel of the machine is turned, and a type is cast each time. On the rapidity of the motion depends the quantity produced. Such is the modern type-casting machine—turning out one hundred types per minute, or sixty thousand per working-day of ten hours, every one of which is a mite contributed to the spreading of knowledge over the world for good or for evil.

The type as thus formed is passed to boys, who break off the jets or waste ends; then to the dressing-room, where the rough edges are rubbed off on the faces of large circular stones; and finally, they are set up in lines, slipped into a long stick, screwed tight, and the bottom of the type is neatly grooved by a planing-tool. The letters are afterward closely inspected with a magnifying-glass, and all imperfect ones rejected.

The exhibit of this firm is exceedingly well arranged, evincing great taste and a considerate regard for the interests of visitors, by showing them not only the modern type-machines themselves, but the various adjuncts of their establishment, as well as the tools used by this house in the last century. The exceptional excellence of their type is proved by the handsome appearance of our book, which is printed from them. Cases are also displayed containing type—the smallest not thicker than a pin—ancient and modern, plain and highly ornamented, and exquisite borders, crochet and music type, and numerous other essential matters for printers' use. These are all shown in their two magnificent Specimen Books, also on exhibition, which are printed in the highest style of typography: the matter of the lines displaying the types being original and exceedingly quaint, these remarkable volumes have no counterpart in the world. This foundry is the oldest in America, having been established in 1796 by Binny & Ronaldson, and claims to be the most complete in the world.

In the profession of Dental Surgery great progress has been made of late years in the introduction of machinery for the use of the practitioner. The manufacture of instruments, apparatus, furniture, artificial teeth, and dentists' materials generally, has been largely increased and developed, and one may now obtain at the dental depots, ready for use, all of the latest and most improved appliances required in this department of business.

Prominent among these establishments is that of S. S. WHITE, of PHILADELPHIA, represented by an exceedingly elaborate display in the Main Exhibition Building. We desire particularly to draw attention to the DENTAL ENGINE, exhibited by this house, as an exemplification of the modern application of machinery, and well illustrated by the accompanying engravings. By means of this engine all the operations of drilling, filing, polishing, etc., are accomplished with great saving of labor and time to the operator and of pain to the patient, affording better-shaped cavities than by the old methods, and giving great facilities for finishing the fillings and cleaning the teeth. It combines great steadiness of motion with ease and quietness of working, possessing at the same time elegance of construction and simplicity of action. It is operated by foot-power.

Fig. 1 gives a general view of the apparatus. The base is divided into three feet well spread out and making a firm support. To one of these feet, lengthened for the purpose, is attached the foot-pedal, which connects by a flat steel spring, called a pitman, with a steel crank, moving a driving-wheel, which is supported by a post rising from the centre of the base and forked so as to provide bearings for its axle-shaft. Above the driving-wheel is an upright rod, the lower portion formed into a yoke passing over the upper part of the wheel and hinging on to the journal-bosses of the axle-shaft, thus making what may be termed a rocking-arm. Its primary statical condition is assured in an upright position by a prolongation beyond the axle-shaft of that arm of the yoke on the opposite side of the crank and the attachment of a spiral spring with its lower end fastened near the base of the apparatus. A screwed

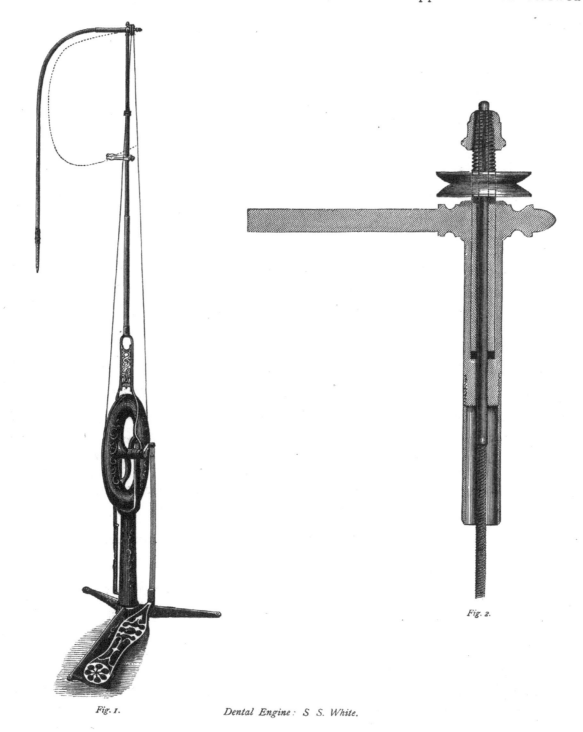

Fig. 1.   Dental Engine: S S. White.   Fig. 2.

extension-joint and jam-nut are provided in the vertical rocking-arm, by which it may be lengthened and the driving-cord tightened if necessary. To the top of the rocking-arm is fixed a right-angled head-piece shown in detail by Fig. 2, the horizontal part of which is drilled to receive a stem, upon which is fastened a pulley by a squeeze-nut on a conical screw. This pulley is driven by a cord passing around the driving-wheel, and revolves the stem, the other end of which connects with the rotating shaft of a flexible arm. The head-piece is pivoted and has free horizontal motion, and the arm is flexible at nearly every point in its length of twenty-six inches, being a rotating spiral within a fixed spiral sheath. The hand-piece is fastened to the end of the flexible arm, and the tool fits in to a tool-holder or chuck, being held by a simple yet perfectly satisfactory arrangement, revolving with the chuck without any vibration, and easily removed in a moment if a change of tool is required.

A large variety of tools is provided for this machine, such as excavating-burrs, drills, burnishers, finishing-burrs, corundum points, boxwood disks, wood-polishing points, etc., and a right-angle attachment is also furnished, which can be fixed to the hand-piece and is of great advantage in certain operations.

By means of an extension-treadle the operator can produce motion from either side of the patient's chair without moving the machine. An air-injector apparatus is also provided, consisting of a rubber bulb or bellows, which is compressed automatically by a simple mechanism connected with and working by the driven pulley. The air is forced from the bulb through a connecting rubber tube to a fixed nozzle at the hand-piece, from which it is thrown into the cavity of the tooth under operation, keeping it clear of burr-dust and cuttings and also keeping the bit cool.

The spring pitman which connects the foot-pedal with the crank is one of the novelties of the machine, giving the crank, when on the "down centre," an upward or live motion, and allowing the performer to operate with perfect ease. It is set at such an angle as always to keep the crank off the dead centre, being adjusted to throw it above its centre and allow greater length of turn in starting from rest.

The pivoted rocking-arm with its return spring always recovering the perpendicular when let free, constitutes another important novelty, affording the operator greatly increased freedom of motion and practically nullifying the tremor which always obtains in rigid machines, and communicates itself to the tool even with the greatest care, raising a fatal objection to their use. The flexible working-arm is also a special feature, bending, curving and yielding to every motion, and allowing the operator a freedom of touch which he could not possibly have with a rigid arm.

For many years no uniform standard existed for screw-threads for ordinary bolt and nut use in the United States. The form of thread as adopted by English engineers and known as the Whitworth standard, while possessing some advantages had many objectionable features. From results of investiga-

Bolt and Nut-screwing Machine: William Sellers & Co., Philadelphia.

tions of Mr. William Sellers, of Philadelphia, presented by him before the Franklin Institute in 1864, that corporation recommended a system of forms and proportions for screw-threads, bolt-heads and nuts for general adoption by American engineers, and urged the same upon the officers of the General Government, requesting their influence towards its selection as an American standard. In 1868 this system was fully indorsed and accepted by the Navy Department of the United States, and afterwards adopted by other departments, at the same time meeting with such general favor throughout the country as to have become in reality the standard of the nation. We show here an engraving of one of Messrs. William Sellers & Co.'s Patent Bolt and Nut-screwing

MACHINES, constructed according to this standard and exhibited in Machinery Hall. It represents their ¾-inch size, cutting screws from ¼ to ¾-inch, other sizes being made up to four inches. On this machine one man has cut three thousand ¾-inch set screws in a day of ten hours, threaded up close to the head, and two inches long on the part threaded, only one set of dies being used and without heating.

A number of important advantages are claimed for these machines over others in use. The dies revolve and the bolt is stationary, thus enabling the workman to put in a fresh bolt without stopping the machine. The motion of the dies is always in one direction, and the bolt is cut at one operation; the

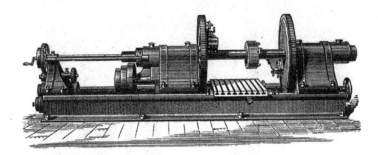

Cylinder Boring and Facing Machine: William Sellers & Co., Philadelphia.

dies open under cut while revolving and remove all trace of the chip made by the cutting tools. They never run backward, the cutting edge therefore lasting much longer. An index on the back of the large driving-wheel is set to numbers given on a card accompanying, and when so set the bolt will fit a nut of corresponding size with the tap sent with the machine. By moving this index one way or the other the bolt may be cut larger or smaller to suit special requirements, and any necessary adjustment may also be made to compensate for wear of the dies. The dies can be changed without taking off any of the die-holding apparatus, and in less time than in a common hand-screwing stock. The bolt-holder always chucks the bolts in the centre of the dies, thereby insuring correct work. A self-acting oil-feeder supplies oil to the back of the dies, thoroughly lubricating the work, preventing heating, and washing out the chips from the die-box. An automatic self-opening attachment is also provided, opening the dies at a given length of thread and insuring uniformity.

Messrs. William Sellers & Co., in classifying their drilling and boring machines, designate all those in which the cutters revolve and the work remains stationary as Drill Presses, while those in which the work revolves and the cutters are stationary they call Boring Machines. Some of their drill presses, however, in the common acceptation of the term, would be called boring machines, such being generally those in which the size of the hole to be made requires the use of independent cutters inserted in a boring-bar.

Among the horizontal drills exhibited by them is a Cylinder Boring and Facing Machine, represented in the accompanying cut, designed to bore locomotive cylinders, and which may be classed as among "the most notable of modern special tools." It has a 6-inch boring-bar driven at both ends of the cylinder, independent slide-rests for facing off both ends, and six changes of boring feed, with quick hand feed. The bar may be taken entirely out of the cylinder by hand or by power, so as to allow shifting of the work. The cutter-heads bore from ten to twenty-two inches. The machine possesses great rapidity of work, and will take one of the largest freight or express passenger-engine cylinders, boring, facing up flanges, and counter-boring for clearance of pistons at end of stroke, in three and a half hours, the quickest time ever previously made on the same work before the construction of this machine being nine hours, and generally on ordinary

Car-wheel Boring Mill: William Sellers & Co., Philadelphia.

boring machines thirteen hours. This rapidity is largely effected by an improvement, by which one cut with a fine feed may be made to take out the greater quantity of the metal, and the machine be then readily and quickly shifted to an exceedingly coarse feed for the finishing cut, resulting in a saving of time, with less wear of cutters and more accurate work, especially with deep holes, than if done in the old way.

Among the boring machines of this same firm we would draw attention to a Car-wheel Boring Mill, shown by the engraving on page 93, and designed for car-wheels up to 36-inch, or general work up to 48-inch diameter. It is provided with a horizontal face-plate and universal chuck for all sizes up to thirty-six inches in diameter, and is arranged with power-feed and quick hand traverse in either direction, thus allowing of rapid work as in the machine previously described and insuring uniformity of hole. The boring-bar is forced down from above into the wheel being bored, and its bearing may be adjusted vertically. The machine possesses quite an advantage in allowing the faces of hubs of locomotive truck-wheels to be turned off, at the same time that they are bored, by an adjustable hub-facing attachment, the slide of which is independent of the boring machinery. A patent safety-crane attachment, as shown in the figure, is made whenever desired. The capacity of this machine is fifty car-wheels per day of ten hours.

Messrs. William Sellers & Co. also exhibit a number of punching and shearing machines of excellent design, and we would mention especially a heavy Plate Shearing Machine for trimming the edges of long plates or for cutting plates of five feet in width or under, off to length. This machine was designed to meet the requirements of modern ship-building or bridge construction. It is provided with a bed for holding the plate and clamping it if necessary, and will shear plates one inch thick with exceeding exactness. The upper blade is guided vertically, and is driven downwards by a pitman as wide as the blade is long, receiving its motion from a long rocking shaft above it, which is operated by an arm or lever in the rear of the machine and not seen in the engraving. This arm has a segmental rack working into the teeth of a spiral pinion driven by a bevel-wheel and pinion, and open and crossed belt similar to the method adopted by this firm for their planing-machines. The driving arrangement is exceedingly efficient, and an automatic adjustment is provided to the belt-shift motion gauging the length of stroke. The blade after making the down stroke immediately ascends again at double its descending speed, and stops up ready for the next cut. It is at all times under the control of the operator, and can be made to cut to any fixed point in its length, and then stopped or raised, the hand-rod in front, operated from either side, being used for shifting the belts and starting or stopping. Curved blades can be placed in the vertical slide if desired, and the bed-plate connected with

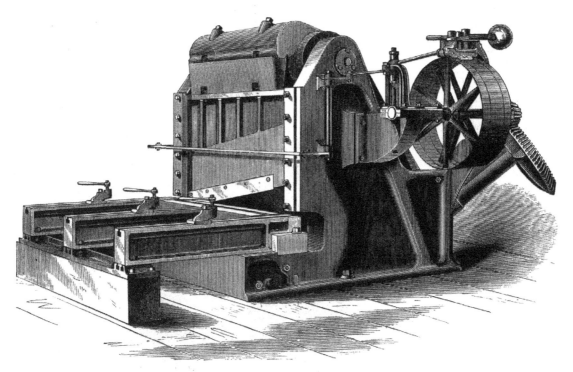

Plate Shearing Machine: William Sellers & Co., Philadelphia.

the lower blade may readily be removed to receive a curved bed-plate, with shear-plate bent to correspond with the curve of the upper blade.

The subject of riveting by power has for some time attracted the attention of mechanical engineers, and steam-riveting machines have been used with considerable success. There are objections, however, to the use of steam which have been most effectually met by the application of hydraulic power, and we are indebted to Mr. Ralph H. Tweddell, of Sunderland, Great Britain, for the invention of a hydraulic riveting machine combining the advantages and avoiding the difficulties of previous systems. MESSRS. WILLIAM SELLERS & CO. as assignees and sole manufacturers in the United States for Mr. Tweddell's patent, exhibit one of their make of these machines, possessing many improvements of their own, and arranged with convenient overhead carriage and hoisting machinery to facilitate its use. The essential point of this invention consists in the use of an accumulator, from which a continuous regular pressure may be obtained as wanted. The adjustable accumulator is arranged with weights suspended below the main casting, and easily released, if required, to adjust the pressure to the kind of work being done, each weight representing two hundred and fifty pounds per square inch on the ram of the riveting machine, and the maximum pressure obtainable being two thousand pounds per square inch. A double-acting pump is connected with it, operated by crank motion, and taking its water from a reservoir in the upright column to which it is attached. The pump is arranged so that when once started for work it is never stopped while the machine is in use. By an improved relief-valve, as soon as the accumulator is full, the direction of the water coming into it from the pump is changed back into the same reservoir from which it was taken, and it continues so to flow until wanted in the accumulator, when the action of the valve directs it back again. The pump is maintained in motion ready for immediate action, and yet relieved from strain when not required for work, avoiding all risk of delay at starting or of loss of water and entrance of air in the chamber while standing.

Portable Riveting Machine William Sellers & Co., Philadelphia.

The portable riveter is suspended from a hoisting machine and overhead-carriage, having both longitudinal and transverse motion. The water under pressure is carried by jointed or flexible pipes from the accumulator to the machine, and passes into a compressing cylinder in which a piston works.

Two levers or jaws of the machine contain dies in the short ends, the long ends being connected by a spiral spring, and one or the other of these levers is attached to and moved by the piston, the dies driving the rivet. The action of the water pressure on the piston is controlled by a valve opened and shut by the operator. When the valve is opened, the piston moves the die to which it is attached until the rivet is headed, acting without blow and with a force positively defined by the pressure on the accumulator. The pressure is continuous and uniform, and may be maintained as long or short a time as desired, entirely independent of the action of the pump. One man can raise and lower the riveter, adjust it to the rivets and operate it. Our three engravings show the machine in three positions, corresponding to work in which the seams are vertical, oblique or horizontal. The rivets are supplied by boys, ahead of the operator, and on straight beam-work ten to sixteen rivets can be driven per minute.

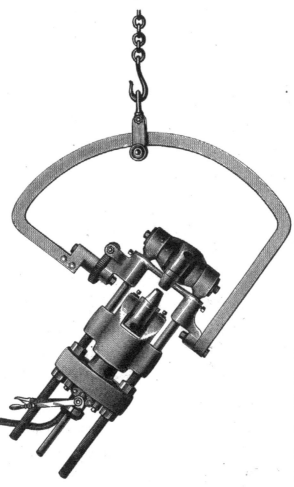

Portable Riveting Machine: William Sellers & Co., Philadelphia.

The GATLING GUN COMPANY, of HARTFORD, CONNECTICUT, exhibits in the Main Building a number of specimens of its famous GATLING GUN, invented by an American, RICHARD JORDAN GATLING, in 1861–62, and after extensive trials, adopted into the service of our own Government as well as by most of the civilized nations of the world.

There are two styles of gun on exhibition, the original type as first constructed and a new design but recently brought forward, which possesses many advantages in arrangement of details over the old gun. Descriptions of the original type have been published and are accessible to the reader. It is of the new gun that we propose to speak.

The gun, as illustrated by the accompanying engraving, consists of five parallel breech-loading rifle barrels, open from end to end, and grouped about a central shaft to which they are rigidly connected by forward and rear disks or plates. The breech of each barrel is chambered to receive a flanged centre-fire metallic-cased cartridge. The shaft extends back some distance in the rear and immediately behind the barrels a cylinder of metal, called a carrier-block, is fastened to it, having on its exterior surface five semi-cylindrical grooves, cut parallel to its axis and forming trough-like extensions to the barrel chambers. These are to take and guide the cartridges into the barrels, and also to receive and discharge the empty cases after they are fired. A prolongation of this cylinder back forms another cylinder, called the lock-cylinder, which carries, in prolongations from the cartridge grooves, five long breech-plugs or locks. A breech-casing, rigidly connected with the gun-carriage by a screw by which the gun may be elevated or depressed, covers the lock-cylinder, and through the centre of the back plate of this breech-casing the rear end of the shaft is journaled. A cylindrical envelope covers the group of gun-barrels from muzzle to breech, and it is attached to the gun-carriage on the lower side by a vertical joint. The front end of the shaft with the front barrel-plate revolves within the end of this cylindrical envelope. A hand-crank is attached directly to the rear end, by which the shaft with its group of barrels, the carrier-block and the lock-cylinder, all rigidly connected with it, may be freely revolved. On the inner face of the breech-casing is arranged a truncated, wedge-shaped, projecting, annular or spiral cam, inclined back both ways from a flat portion, the apex of the truncated wedge pointing towards the barrels, and against this cam the rear ends of the breech-plugs or locks bear, being held in place by a lug from each, working in a groove at the base of the cam. Each lock has in it a firing-pin operated by a spiral main-spring. The firing-pin projects at each end beyond the lock, the front end being a point, and the rear end being finished with a knob which at a certain stage in the revolution of the shaft is drawn back by a groove in which it works, and then suddenly released, causing the front to enter the cartridge and explode it. The breech-casing extends over the carrier-block, covering it, except a portion from near the bottom upwards on the left side, where it is open, so that discharged cartridge-cases as withdrawn from the barrels may drop out on the ground. In the top of this casing is an opening, placed in the correct position and of the proper size, for a single cartridge to fall through into one of the channels of the carrier-block when it revolves underneath. The upper part of this opening is formed into a hopper, to which can be attached a cartridge or feed-case, holding a number of cartridges, lying in single file, one above the other. The cam in the rear of the locks is so arranged that each lock, when it gets in position behind the cartridge-hopper, is drawn back to its full extent so as to admit a cartridge in front. The action is as follows: Turning the crank, the shaft and its appurtenances rapidly revolve. Cartridge after cartridge from the feed-case drops into its respective receptacle in the carrier-block as it comes under the hopper. As each one passes on in revolution, the lock behind it, being pushed by the inclined cam, follows it up, thrusting it into its barrel, and, just before the shaft has reached half a revolution, drives it home and closes the breech. At this moment the firing-pin, which has been drawn back, is released and fires the cartridge, the reaction being resisted by the lock. The lock still revolving onward now begins to withdraw, and a hooked extractor attached to it, which had previously caught over the flange of the cartridge, draws the shell out, dropping it on the ground. By the time a complete revolution is accomplished, the lock is back again all ready for a fresh cartridge in front. The gun thus fires each barrel only once in a revolution, as many shots being fired in one turn as there are barrels. The working is very simple. One man turns the crank, and another supplies the feed-cases, one after another, as rapidly as exhausted, and the operation proceeds indefinitely.

The gun is mounted on wheels in the same way as ordinary field-pieces, or it may be placed on a tripod. In addition to the screw before mentioned for elevating or depressing the breech, there is also an adjustable arrangement at the rear, by which a limited angular movement in a horizontal plane may be given to the gun if desired. This operates very prettily by the centrifugal force from the turning of the handle, making one movement back and forth for each turn, the handle moving in an ellipse instead of a circle.

The details of construction in the new gun have been very much modified from those in the old type, resulting in great simplicity of assemblage and more substantial design, greatly increasing its endurance. The gun is very easily taken to pieces for cleaning or repairs by merely removing the nut at the rear, when the crank can be taken off, and part after part removed, the whole coming to pieces. By this nut, also, which is a set nut, an adjustment

Gatling Gun: Gatling Gun Company, Hartford, Connecticut.

can be made at a moment's notice, in the length of the spaces for the cartridges, to accommodate the breech-chamber to cartridges from different manufacturing establishments, which often differ considerably in thickness of head. In the old type of gun this adjustment was a matter of considerable trouble, and had to be made at the front end. A great improvement has been effected in the new gun in the ejecting of the locks. By opening an aperture in the back plate of the breech-casing, they can easily be drawn out with the finger.

If one gets out of order, it can be taken out and the firing proceed without it, there being however one shot less for each turn, and one cartridge falls to the ground undischarged.

Cartridge.

The arrangement of direct-acting crank from the rear, and the placing of the hopper exactly on top of the gun, at the same time improving its shape, so that cartridges may fall quickly by gravity without the necessity of forcing, has greatly increased the rapidity of firing, the new gun being capable of firing up to twelve hundred shots per minute, whereas the army reports claim only about four hundred and fifty shots per minute with the old gun. The new type of gun is very light, weighing only ninety-seven pounds, and it can easily be carried on mules or horses over rough country and operated at short notice.

Guns are made of 0.42, 0.43, 0.45, 0.50, and 0.55-inch calibre, and the larger calibres have an effective range of over two miles. The gun is reported by a Board of the War Department as "capable of maintaining uninterruptedly for hours a most destructive fire at all distances, from fifty yards up, being beyond all question well adapted to the purposes of flank defence at both long and short ranges."

Cartridge.

The MILTIMORE CAR AXLE COMPANY OF NEW YORK, exhibits a patent COMPOUND CAR AXLE, the invention of MR. GEORGE W. MILTIMORE, which it is claimed fully meets the difficulties experienced from the sliding of wheels on the rails, whether caused by curves, irregularities in the track or differences in

Gatling Gun, New Style Frame.

Limber Carriage: Gatling Gun Company, Hartford, Connecticut.

Cartridge.

the circumference of wheels, and inseparable from the use of the ordinary rigid axle. The improvement commences with a radical change from the ordinary arrangement, in that the axle is kept stationary while the wheels revolve, thus eradicating at once all tendency to torsional stress. The axle, which may be either of steel or cold-rolled shafting, is of the same size throughout, and passes at each end into a cast-iron pedestal-block, in which it is firmly secured and rendered immovable by a horizontal steel bolt passing through both axle and block. The axle is encased in a loose revolving sleeve of wrought-iron pipe, having cast-iron ends, on which seats are formed for the wheels, which are loosely mounted, each wheel being held to gauge on the inside by a shoulder in the casting, and on the outside by a cast-iron nut screwed to the end of the sleeve and fitting against the hub. Oscillating cylindrical boxes of brass fit in between the sleeves and the axle, forming the only points of contact, the bearing surface being on the under 'side. These boxes are made with a curved bearing on the outside to allow them to adjust themselves freely to the spring of the axle, and thus insure a perfect bearing on the interior for the whole length of the box, and avoid wearing at the ends. A box-ring fits closely to the outer half of the curve, and the sleeve-casting is turned to fit the inner half, sufficient room being left at the ends for oscillation.

The action of the device is as follows: When drawn forward, the wheels in moving, although loose on the sleeve-bearings, carry the sleeve round with them, the friction being much greater than on the axle-bearings; and on a straight track with wheels of the same diameter there is no motion whatever on these outer bearings. When, however, owing to the slightest curve, or an irregularity of track or other cause, one wheel is required to move faster than the other, instead of sliding one wheel, as is the case with the ordinary arrangement, either wheel is perfectly free to accelerate or retard its motion independent of the other, according to the space over which it has to move. No tensional strain can be thrown on the sleeve, for if a wheel should be forced slightly out of the perpendicular, as when the flange strikes the outer rail of a curve and thereby cramps the hub on the wheel-seat, it at once turns the sleeve with itself and gains the necessary increase in motion at the opposite hub where there is no cramp.

It is claimed that the following advantages are gained by the use of this axle: A reduction in power required to haul the train, consequently a saving of fuel; increased durability to wheels and axles; saving of wear on road-bed; increased comfort and safety to passengers; great economy in lubrication; freedom from hot boxes; less expense for repairs; and ability to use wheels of larger diameter. The results of practical experiments which have been made on the Vermont Central and other railroads for considerable lengths of time would seem to justify these claims, and there are at present seven cars equipped with these axles in daily service on the West End Passenger Railway in the Exhibition grounds, operating with great success. Trucks with these wheels have been running on the Vermont Central and on the Chicago, Dubuque and Minnesota Railroad for a considerable time, and the results give a durability of at the very least double that of the rigid wheels. The Miltimore wheels, after a service of sixty thousand miles on roads of heavy curvature, show exceedingly light flange wear, evincing an equivalent saving of wear on the rail. It is stated that axles now in service, running fifteen months at a rate of one hundred and fifteen miles per day, have consumed but one pint of oil per month, and when grease is used the saving is still greater. In addition to this the use of cotton-waste is entirely dispensed with. A great advantage exists in the facility with which a wheel may be changed and a new one substituted should the breaking of a flange or any other cause require it. Two men with a jack can easily remove a wheel and replace it in a short time without disturbing the car. The removal of the torsional strain from the axle affords greatly increased safety to the train and also allows the use of larger wheels, resulting in a smooth, even motion to the car and saving in power to draw the train. A fast passenger-train of five cars with forty-inch wheels has been running on the Vermont Central Railroad from one hundred and fifty to two hundred miles per day for eighteen months with great success. The wheels being triply cushioned, the hammering so destructive in the case of the ordinary axle is very much reduced. Even if an axle should break, the sleeve acts as a protection, and it would be almost impossible for the wheels to get out of place. If all that is claimed for this axle continues to bear the test of practical use, it is destined to effect an entire revolution in railway equipment.

MESSRS. RICHARDS, LONDON & KELLEY, of PHILADELPHIA, and LONDON, ENGLAND, make a fine exhibit of machinery for working in wood, from which we select one of their BAND SAWING MACHINES, the front and side elevations of which are shown by the engravings on pages 106 and 108. The machine is very substantially constructed, the frame being of cast iron in one piece, with a rectangular cored section; The wheels are sixty inches in diameter, made of wrought iron covered on the circumference with wood faced with leather or gum, and are warranted to stand the tension of blades up to three inches in width, and resist safely any centrifugal strain. A vertical adjustment of sixteen inches is provided to the top wheel, which is carried on a steel shaft two and a half inches in diameter, with bearings on both sides of the wheel, and saws may be used up to thirty-two feet in length and three inches in width. The supports of the shaft rest on springs, which equalize the tension on the blades, allowing them to expand and contract freely.

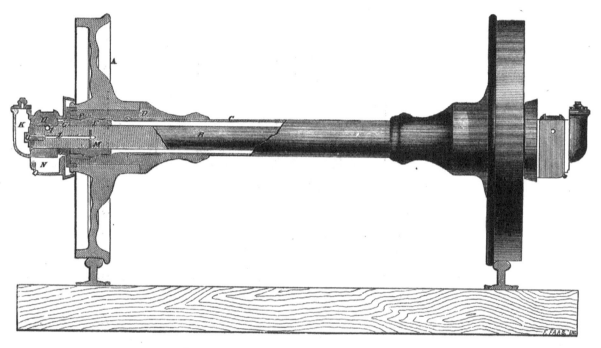

Patent Compound Car Axle: George W. Miltimore.

The machine has feed-rolls adapted to take timber of twenty-four inches in width and ten inches in thickness, or to cut from one side of a plank five inches thick. The method of imparting motion to the feed-rolls is novel and very superior, being accomplished as follows: A revolving plate with its axis at right angles to the feed-shaft comes into rolling contact with the circumference of a wheel on the feed-shaft, which slides on a spline of the shaft, and may be moved to and fro each way from the centre of the revolving-plate. The action of the revolving-plate causes this wheel to turn with greater or less rapidity, depending upon its distance from the centre, and its movement operates the feed, the speed of which is regulated accordingly. The feed will be either forward or backward, depending on which side of the centre of the revolving-plate the wheel is placed, and the direction can be changed at a moment's notice. The power being frictional makes it a safeguard against breakage, and at the same time it is sufficiently tractive for all practical purposes. Attempts have been made previously to use feeding appliances of this kind in moulding

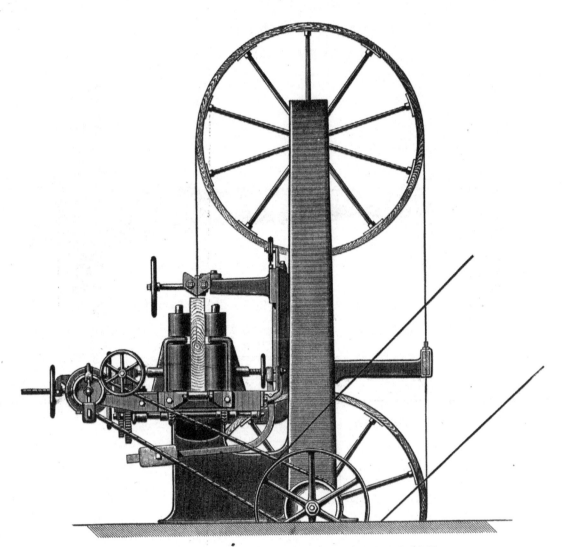
Band Sawing Machine: Richards, London & Kelley, Philadelphia.

and other machines, but the conditions were for some strange reason always reversed and the result was a failure. The arrangement here adopted seems to accomplish all that is wanted, and the rate of feed may be increased from zero to forty feet per minute or the reverse, the feed being started or stopped at pleasure, and made either forward or backward.

The saws used are those of M. M. Perin & Co., of Paris, France. Band saws were invented nearly seventy years ago, William Newberry, of the city of London, England, having in 1808 constructed and patented a band sawing machine, which, judging from the illustrations preserved of it, appears to have been a very good machine, possessing nearly all the capabilities of those of the present time. The pivotal table, the parallel gauge, the feeding rolls, and radius link were all provided, the great material difference being the inconvenient manner of removing and replacing the blades. Circular saws were hardly in use at that time, and the opportunity would seem to have been exceedingly good for competition against the reciprocating saws of the day. Little or no use was made of the invention, however, and it lay dormant until within the last twenty years, when the subject again came forward, and saws of this kind were first exhibited as a novelty at the Paris Exhibition of 1855. The cause of this is believed to have been due to the difficulty experienced in the manufacture and joining of the blades, which could not be made to stand the flexion and strain to which they were submitted in working, and it was not until M. Perin, of Paris, undertook the manufacture of blades some twenty-five years ago, and by perseverance triumphed over every difficulty, that the success of the band saw was achieved.

The blade is the principal part of the machine and the only part from which difficulties arise in its operation. France has had the monopoly of the manufacture of saw-blades, and will probably keep it for a long time, unless some of our American firms come forward and spend the money and time on experiment, and bestow that care and attention on the work which have produced their results in France, trusting not to any present remuneration, but rather to what may come in future years. The impetus given to the manufacture by the efforts of M. Perin, the special knowledge requisite, much of it kept secret; the tedious hammering process required, necessitating skilled labor, which may be obtained at so much less cost in France than elsewhere, and many other reasons have all combined to give her the supremacy.

There are various causes for the breaking of saws, such as crystallization, extreme or irregular tension, heat generated by friction on the guides, or careless use. It is well known to all those interested in such matters that a certain temper is requisite in deflecting steel springs, and that if this temper is obtained they will last for years or for a life-time. When one remembers, then, how difficult it is to obtain this temper, even with short springs like those in gun-locks, it is easy to conceive the almost insuperable difficulties in the way of obtaining this temper with bands of steel twenty to thirty feet in length; and the least variation in this temper for even an inch in the length of the saw

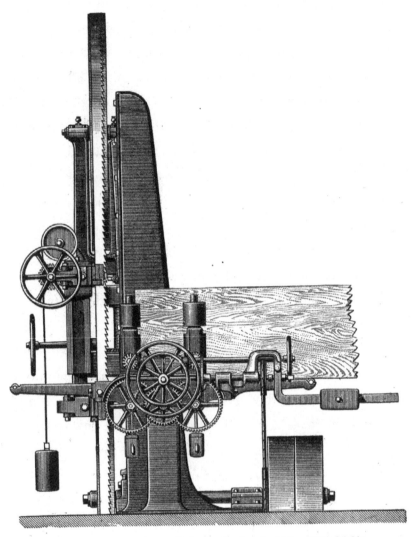

Band Sawing Machine: Richards, London & Kelley, Philadelphia.

destroys the value of the blade. In addition to this, as if to increase still further the difficulties already quite sufficient, there appears to be no reliable method for ascertaining in a finished blade if the quality and uniformity of the temper are correct. The buyer must depend on the good faith of the manufacturer, the value of the saw depending not on its appearance, but on the care with which it has been made and the perfection of the processes used;

and the blade should be completely finished ready for use by one firm, so that what may leave the hands of one party in good condition may not be spoiled by the bad work of another.

Thousands of Band Saw Machines are now in use, and occupying the high position that they do in reference to economy of both labor and material, they may well be classed among the prominent machines of the day.

In connection with the subject of band saws we would draw attention to an exceedingly effective "BAND SAW SETTING MACHINE" on exhibition and manufactured by the same firm under the patent of Mr L. O. Orton, the inventor. This machine is intended to accomplish two objects; to furnish a method of rapidly and accurately setting saw teeth and to do so by impact or blows just as would be done by a hand hammer, thus giving a permanent set to the teeth without liability to change as when set by springing or bending. The illustration accompanying shows the machine and its method of working, the saw being held in a filing frame, such as usually employed, to which is attached the setting device which is to all intents and purposes really a hammer in the hands of the operator. The frame is formed of two rails or bars connected by cross rails on which are wheels, which receive and stretch the saw blade in position. The setting mechanism consists of a pivoted swinging frame carrying two dies or hammers so arranged that when

Band Saw Setting Machine and Filing Frame: Richards, London & Kelley.

the operator by means of a handle on top swings the frame back and forth, they will strike right and left, giving alternate blows against two die-blocks placed on opposite sides of the saw teeth, the saw blade passing through a groove, and the alternate teeth coming under the hammer. By a simple mechanism a hook or pawl engages with the saw teeth and at each movement of the swinging frame draws the saw forward the distance of two teeth so that

Filing Vise.

the teeth are brought automatically into the proper positions to be struck, one pair after another. An adjustment is provided to regulate this movement of the saw in a moment to any pitch of teeth. Where large saws are under operation or where the teeth are far apart two pawls are used, one for each single swing, but with ordinary saws one pawl is sufficient and is preferred.

Band Saw File.

The whole of the setting mechanism attached to the frame may slide to any part, or be secured if desired and used independently. The degree of force of the blows and the time in which they are given are in direct control of the operator, the action on the teeth being the same in effect, but more perfect than can be attained by a hammer in the ordinary way.

Soldering Tongs. Scarfing Frame.

A filing vise is also attached to the frame, although it may be used independently, and is arranged with an improved clamping device consisting of two volute faces, one formed solid with the vise and the other with a handle, there being in this case with a long vise, three of these with the handles connected by links, and all actuated by one movement. By turning the handle right or left the jaws of the vise are instantly closed or released. A band saw file with round corners is recommended and used, giving a circular form to the bottom of the spaces between the teeth and preventing fracture. A scarfing frame and tongs for soldering the two ends of a saw-blade together are also exhibited. The ends of the saw are first scarfed or tapered for a length of one to two teeth, depending on the pitch, care being taken to make the scarfing

Reciprocating Mortising Machine: Richards, London & Kelley.

true and level. The silver solder of the jewelers is generally used, rolled into thin strips so that a piece of the size of the lap can be cut off and laid between. The joints are cleansed with acid, the solder placed between and the whole then clasped with the tongs which must be at a full red heat. The tongs are removed as soon as the solder runs, and a wet sponge applied to restore the temper, the joint being afterwards filed up into proper shape.

Messrs Richards London & Kelley, also exhibit a strong heavy "RECIPROCATING MORTISING MACHINE," arranged for railway car and other similar work which is deserving of notice. Motion in machinery may be divided into two classes, rotary and reciprocating, a few exceptional cases combining both motions. There are various difficulties arising in the employment of reciprocating motion that render its use objectionable wherever it can be avoided. These difficulties obtain especially in wood-working machinery on account of the speed at which it is necessary to work. In consequence rotary motion is

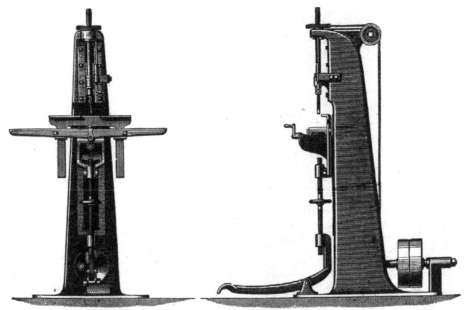

Elevations of Reciprocating Mortising Machine: Richards, London & Kelley.

every day coming more into use, and new applications being made of it. In England and France mortising is done almost entirely by rotary machines or by hand, but in this country reciprocating machines have been extensively used. The great variety of designs from different makers give evidence of the imperfections encountered and the efforts constantly made to overcome them. The machine which we illustrate belongs to that class in which the reciprocating parts are all brought down towards the timber operated on, the chisel having a continuous motion with a uniform range and a positive eccentric. Chisels of any width are received, and there are two boring spindles, one fixed and the other to traverse twelve inches. The feed movement is actuated by a treadle and may be locked to prevent jarring the foot of the operator. All joints are compensating and operated without noise. The distinctive feature of the machine consists in its being direct acting, and having the crank shaft not on top but near the bottom in the base of the column, the machine standing upon a foundation without top-bracing. The crank shaft, chisel bar and boring spindles are of steel.

We also give an illustration of another mortising machine exhibited by this same firm belonging to that modification in which the wood is moved up

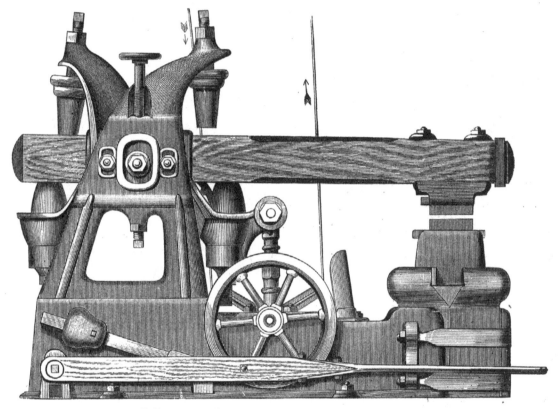

Rubber-Cushioned Helve Hammer: Bradley Manufacturing Co.

or fed to the chisel, the operating parts consisting of a crank shaft, a plain chisel bar and connection. It is well adapted for joiner and cabinet work, carriage work and general purposes, and is capable of being driven at a high rate of speed—four hundred to five hundred revolutions per minute, like the previous machine requiring no top bracing, the crank shaft being placed in the base near to the foundation, avoiding vibration and jar. The table is raised by a foot treadle to feed the lumber to the chisel which has a uniform stroke of five inches. The chisel is provided with the automatic reversing device of

H. B. Smith, allowing it to be reversed by power with a friction band and at the same time holding the chisel bar firmly while in motion and preventing any possible deviation from its proper place owing to wear or loose joints. The escapement is performed by hand so that the chisel can be reversed at will, independent of the treadle. The table is made either as here shown or arranged to clamp the piece of timber to be mortised, and the whole moved by rack and pinion. The firm deserves credit for the manner in which it has endeavored to overcome as far as possible the inherent difficulties in this class of machines.

The trip- or helve-hammer approaches nearer to the hand-hammer in its action than any other mechanical agent of its class, and for this reason is better adapted to certain peculiar kinds of work. There have been various causes, however, operating against its use, one being the difficulty of making proper connection with the driving power. The sudden shocks which it produces on shafting in starting, the irregular motion and the varying power required, all prevent the use of rigid connections, necessitate the adaptation of slipping belts, and require strict application of the principles of elasticity in the entire construction of the machine. Even with all this, the wear and tear in the ordinary hammers, as usually built, is far beyond what occurs with other machines, tending to counteract any inherent advantages that this special form may possess. THE BRADLEY MANUFACTURING COMPANY, OF SYRACUSE, NEW YORK, has placed on exhibition one of its RUBBER-CUSHIONED HELVE-HAMMERS, represented by the engraving, on page 113, which it claims possesses great advantages and improvements over any other hammer of its kind in use. With the exception of the helve, which is of wood, the entire hammer is constructed of iron and steel, so proportioned as to dispose of the material to the best possible advantage. The helve is hung upon two hardened adjustable steel centres and almost perfectly balanced, motion being given to it by a broad eccentric with an iron hub, a bronze shell and a cast-steel strap, all so perfectly fitted as to reduce friction to a minimum and to allow complete adjustment. Rubber cushions are provided and so arranged as to absorb the concussion of the blow of the hammer and materially decrease the strain and jar which ordinarily obtains. Set screws in the upper and lower sockets of the oscillator allow of adjustment to these cushions. The bearings throughout are of the best quality anti-friction metal, except those of the main shaft, which are of bronze; an adjustable eccentric is used, easily regulating the length of stroke required, and a universal joint connection prevents the possibility of binding or heating. By the method adopted of raising and lowering the husk, dies varying an inch in thickness may be used without shimming up either end, thus preserving the key-ways and hammer bolts. In securing the hammer-head to

Milling Machine: Brainard Machine Co.

the helve, rubber cushions are used beneath the nuts and collars of the bolts, absorbing all concussion, preventing loosening or breakage and increasing the elasticity and flexibility of the blow. A foot treadle around the bed of the hammer allows the operator to stand in front or on either side and by a gentle pressure bring the tightener in connection with the belt on the drive-pulley, varying the stroke as desired. On removing the pressure, the brake acts at once on the balance-wheel and the hammer is brought to a stop instantly, with the helve always up. The action is the nearest approach to that of the human arm that it seems possible to obtain, being accurate and powerful, perfectly adjustable in length of stroke, rapidity of motion and weight or force of blow, and entirely under the control of the operator. Water, steam, or any other power may be applied.

It is claimed that not more than half the power is required to do a given amount of work that is necessary in the direct steam hammer. There is no liability to corrosion of steam chest, sticking of valves or freezing and bursting of pipes in winter from non use. For intermittent use it is exceedingly well adapted, always responding to the touch of the treadle, be it once per day or once per week. Its *drawing* capacity and accuracy of stroke give it great advantages over the vertical or dead-stroke hammer. The play required in the guides or ways of the latter for expansion of the ram under heating, allowing it to run loose enough to shuckle, is fatal to nice die swedging. In this hammer, the centres being away from all heat, it strikes equally well whether the head is expanded or not. It is claimed that as a drawing hammer it has no superior, and it is under such perfect control that a block of iron three inches square may be reduced to one-eighth inch square under the one hundred pound hammer without adjustment. A simple device allows adjustment from one power to another, as from a sixty pound to a forty pound hammer, in a few minutes. It is claimed that no hammer in use possesses the elasticity of stroke which this does. Objection is sometimes made to helve hammers because they do not strike perfectly square on different thicknesses of work, but this is obviated in the construction of the dies, and in swedging it gives no trouble.

A certain class of machines technically known as Milling Machines have long been used in a somewhat crude state for a few special kinds of work, such as in the manufacture of fire-arms and sewing-machines, where the cheap and rapid duplication of interchangeable parts was impossible with any other form of apparatus. The name arose from the kind of cutting tool employed, specifically known as a *mill* and consisting of a revolving wheel on the periphery of which the cutters are arranged like cogs to a mill-wheel, the action of the machine being directly the reverse of that in the lathe, the tool revolving on the work instead of the work on the tool.

When attempts were made to apply these machines to general work, defects were revealed so marked as to preclude their employment except in few cases. This led to important improvements and developments on the old type, until of late years their capabilities have been largely extended and much more generally understood, and their use has grown rapidly in favor especially in the United States, and has become a necessity in almost every metal-working establishment.

THE BRAINARD MILLING MACHINE COMPANY and its General Superintendent, Mr. Amos H. Brainard, with whom the subject has been a special study for many years, claim considerable credit for the improvements that have been effected and for the introduction of machines for general use, combining the requisites of capacity, convenience and power, together with beauty of design and perfection of workmanship. This Company manufactures Milling Machines exclusively, of various classes, and makes quite an extensive display in the Machinery Hall. What is known as its STANDARD UNIVERSAL MILLING MACHINE, holds the first position in importance among all the varieties produced, having all the movements and power of the plainer machines, with far greater range and capacity, and being applicable to an almost endless variety of work quite impossible with ordinary machines. Four different sizes of this class are made, all upon the same general plan, and we select for illustration the third, which is perhaps the most desirable for ordinary use, its weight, capacity and power being sufficient for general and quite heavy work without its being too large for quick handling and rapid running.

The engraving on page 115 is taken from a photograph of the machine as it stands at the exhibition, set for cutting a long, conical blank, spirally and automatically, an operation considered one of the most difficult and complicated ever required of a Milling Machine, and necessitating the use of a special mechanism. Its framing consists of a large square or four-sided column, fixed on an ample base, upon the front of which is mounted a knee, which may be elevated or depressed by a screw worked by bevel-gearing and a crank; a dial and finger attached, allowing of adjustment to the one-thousandth part of an inch. The knee supports a carriage which traverses upon it, and on the carriage is mounted a work-table, moving independently and having T shaped slots on its upper face, carefully milled lengthwise and crosswise, exactly in line with and at right angles to the feed, for the purpose of securing work. The table also has an oil channel entirely around it. Upon the top of the column is the driving cone, full geared, giving six speeds. The main arbor or spindle is of solid forged steel, and upon its front end a screw is cut so that a chuck or

Equatorial: Fauth & Co., Washington, D. C.

face plate may be attached. At the extreme top is a projecting arm which carries an outside centre support for the outer end of a mill spindle, allowing the use of cutters to a distance of fourteen inches from the front of the machine. This arm, notwithstanding its solid connection, can be easily removed when not required, or if desired to make other attachments. Automatic feed gearing is provided which is hung upon the back end of the spindle, connecting with a worm and worm gear which drives the feed-screw. The feed work is independent of any movement of knee, carriage or table, and careful provision is made, especially in the feed work, for wear of running parts and for taking up all slack motion. The feed-screw runs in bronze bushings, and bronze collars are interposed between running bearings to obviate wear and diminish friction. The spiral cutter, as shown, cuts a right-hand spiral, but a simple change of gearing causes it to cut a reverse or left-hand spiral, and both were cut upon the same piece of metal in the presence of visitors. Upon loosening three nuts the spiral cutting attachment may be removed, leaving the work-table flush and unobstructed. Various attachments are provided, such as a universal head, by which spur and bevel gears can be cut, and work milled at any angle or position; a rotary vise and many other devices, allowing an almost endless variety of work to be performed; fluting taps and seamers, finishing nuts and bolt heads, key-seating shafting, making all the cutters required for the machine, &c. Even without any of the special attachments the machine is admirably adapted for plain milling, and is in every respect far in advance of the common style of machine. As it appears in the engraving it weighs about 1800 pounds, has a perpendicular range of 18 inches; the carriage will traverse 5 inches, and the work-table has a movement of 18 inches upon the carriage. The feed may be operated either by hand from either end, or automatically, as desired. A door is provided in one side of the main standard which being hollow, furnishes an ample tool closet.

Messrs. Fauth & Co., of Washington, D. C., exhibit in the Main Building some excellent Astronomical and Geodetic Apparatus, among which we would mention particularly a fine Equatorial Telescope, which, although smaller than many others in use, is of a size best adapted for working under all circumstances, and belongs to that class of instruments by means of which with patient labor many of the best results in astronomical research have been achieved. The engraving on page 118 gives a very fair idea of the instrument. It has a clear aperture of nearly seven inches, a focal length of eight feet, and the lens was manufactured by Alvin Clarke & Sons, the celebrated opticians of Boston, Massachusetts. It is mounted on a pedestal which accompanies it, and very little expense is requisite to place it in working position,—a matter of considerable importance to those of limited means. Azimuth and latitude adjustment have been provided, allowing it to be regulated for almost any quarter of the globe, thus permitting great range of locality in its use. The great care that has been taken in designing the instrument, and the close attention that has been paid to the comforts and conveniences of the observer—giving him a perfect control over the whole machinery without compelling him to move from the eye-piece, thereby dispensing with the aid of an assistant—is one of the marked features of the apparatus. It can be turned to any quarter of the heavens with the greatest ease, and the operator without leaving his post, may readily move it in declination and right ascension to find the object he is seeking. Motion is given by clock-work, with which it may be connected or disconnected at will, and the clock may be adjusted to follow stars, planets or the moon with the utmost precision. The hour circle reads to single seconds of time and the declination circle to five seconds of space, two opposite verniers being used to each circle with lenses attached for reading. The position micrometer is a wonderfully accurate piece of workmanship, combining in itself four distinct motions, and is especially adapted to measure minute differences of declination and positions of double stars. One division of the micrometer screw is equal to one ten-thousandth part of an inch. The attached circle permits angles of position to be read off to single minutes. The field of the instrument can be illuminated at pleasure with different colored light, as some stars show best in this way, or it may be left dark for very faint objects and only the spider lines illuminated. Messrs. Fauth & Co. have made every endeavor to bring the construction of this instrument to perfection, as regards symmetry of form, kind of material employed, and style of workmanship, and inspection shows that their efforts have been crowned with success.

THE BUCKEYE ENGINE CO. OF SALEM, OHIO, exhibits one of THOMPSON'S "AUTOMATIC GOVERNOR CUT-OFF ENGINES," as manufactured at its establishment, which deserves close attention from all those interested in the economic application of steam. The consideration of this subject involves an important principle in the use of steam expansively, the energy being much more effectively given out for the same amount of force, exerted on the piston of an engine, if the steam be used expansively at a high pressure, than if worked full stroke on an average pressure, and less steam required for the same work in the former case than in the latter. Of course, taking the same pressure of steam in both cases, more work can be obtained with full stroke pressure than with expansion, but it is at a waste of steam, and the question under consideration is to do the amount of work required with the greatest economy. Where the cost of fuel is no object, certain reasons may make the use of the ordinary engine working at full stroke, preferable, but these need not be considered here. An important advantage in expansive working also obtains in wear and tear, the shocks and jars to the engine being much less than if worked with full stroke pressure.

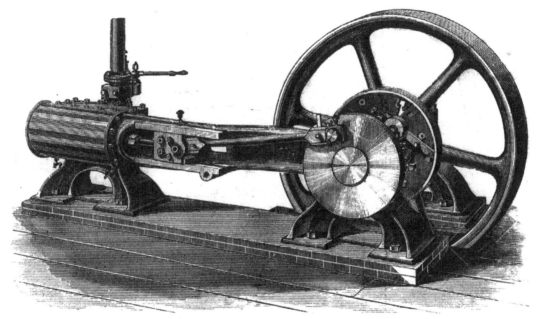

Automatic Governor Cut-off Engine, Fig. 1: Buckeye Engine Co.

In the case of the ordinary slide-valve, working with a continuous motion, there appears to be great difficulty in securing all the advantages of a cut-off, it being impossible to give a full flow of steam and a sudden cut-off. The maximum of economy requires a full boiler pressure to be carried into the cylinder at the commencement of the stroke, and mantained up to the point of cut-off, and the cut-off to be sharp, without causing a gradual reduction of the steam-pressure by what is called wire-drawing. Various methods have accordingly been adopted by means of valve gear, of holding the valve wide open, and suddenly closing when required, by a spring. The Corliss Engine is an example of a very successful method of accomplishing this object. In the case of the engine under consideration, the slide-valve is operated full pressure at full stroke, and a secondary valve is called into action at the proper time, to cut off the steam quickly from a full pressure to zero. When this valve is so controlled by the governor, as to cut off the steam earlier or later in the stroke as required, and maintain a certain desired uniform speed, under variations of load and steam pressure, it becomes an Automatic Cut-off, and as such it is represented in this engine, the action being very different from the wire-drawing or throttling-engine, where the governor performs its duty by throttling the steam more or less, on its passage to its work in the main steam-pipe.

Our illustrations, Figs. 1 and 2, engraved on pages 121 and 122, show

Automatic Governor Cut-off Engine, Fig. 2: Buckeye Engine Co.

front and rear views of the engine on exhibition, it having a horizontal action, with a cylinder of sixteen inches bore, by thirty-two inches stroke. The principle involved in the automatic cut-off appears to be the only true one for the highest economy in the use of steam; it remains that the practical application of it shall be properly carried out. It is claimed by the makers of this engine, that it satisfies all the conditions necessary for this economy, and at the same time is so simple in construction, as to be very little more expensive in cost than an equally well designed throttling-engine, while it may safely be placed in charge of any fairly intelligent and careful engineer. The peculiar points of the engine are the slide-valve and the governor. Fig. 3, which we engrave on page 123, gives a section of the former. This slide-valve is in reality a small moving steam-chest, into the interior of which, the entire supply of live steam is admitted, and passes from thence to the cylinder by ports near its end which are made to coincide alternately with the cylinder ports. The exhaust steam from the cylinder passes out into the steam-chest at the end of the slide-valve, and follows on by ample passages to the exhaust-pipe, going downward freely and directly out of the way, and avoiding, even at the highest speed, that back pressure so often caused by the tortuous passages so common in many other valves. The indicator cards which have been taken of the working of the engine, show with what perfection the valves perform all their functions. The arrangement gives great advantages in allowing the face of the valves to be placed as close to the bore of the cylinder as a proper consideration of thickness of metal, for strength will permit, and by this means reducing the clearance or waste room to a minimum. The openings in the back of the valve admitting the live steam are fitted with self-packing rings, so as to insure a steam-tight connection, and the area of these openings is made as small as possible consistent with the proper holding of the valve to its seat, making it as nearly balanced as practicable or desirable. By removing the top

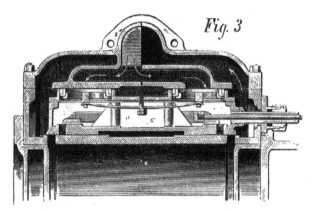

Automatic Governor Cut-off Engine, Fig. 3: Buckeye Engine Co.

of the valve-chest which contains only exhaust steam, the working of the valve may be seen, and any leakage detected and remedied. The main valve is operated by a fixed eccentric, and the cut-off valve works inside of the main valve, the stem passing through a hollow main valve-stem, and is operated by an adjustable eccentric, through the medium of a compound rock-arm device and connections as seen in Fig. 2. A small rock-shaft works in a bearing in the main rock-arm and moves with it, making the movement of the cut-off valve relative to its seat in the main valve, just the same in reference to both time and extent, as would occur if the valve worked in a stationary seat and was attached directly to its eccentric. The main valve eccentric rod works horizontally and the cut-off eccentric rod inclines downward, so that its attachment to its rocker-arm is on a level, or nearly so, with the centre line of main rock-shaft. The cut-off eccentric is automatically adjusted by means of two weighted levers connected with springs, and contained in a circular case fastened on the engine-shaft. This regulator or governor is illustrated by Figs. 4 and 5, which we engrave on page 124, the former showing the position of parts when it is at rest, except spring D, which is not adjusted, and the latter showing their position when the engine is at its maximum speed. It must be stated however, that the two figures are for different governors and adapted to run the engine in opposite directions, one being in arrangement the reverse of the other. It will be noticed that extra holes c, c, are provided in the case so that the same governor may be changed in arrangement in a short time, to allow the engine to be run either way as desired. When the

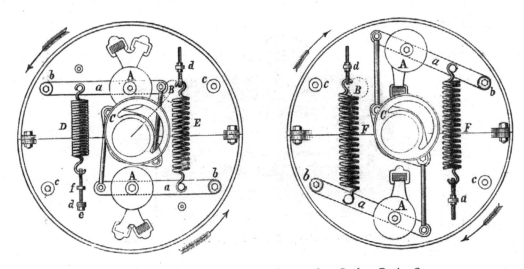

Automatic Governor Cut-off Engine, Figs. 4 and 5: Buckeye Engine Co.

speed of the engine becomes too rapid, the centrifugal force overcomes the resistance of the springs and throws the levers outward, advancing the eccentric forward on the engine-shaft and making an earlier cut-off. When the speed is reduced the springs draw the levers in, and make a later cut-off. Fig. 4 shows the position of parts for the latest cut-off and Fig. 5 that for the earliest cut-off. Set screws are provided to allow adjustment of the tension on the springs, which may be varied to suit the character of the work for which the engine is required. The makers claim great simplicity of parts, a reduction of noise in working to a minimum, very little clearance or waste room in the ports, close governing power, and great economy of steam, also full opened indication ports for all points of cut-off, and a free and unobstructed exhaust.

The Buckeye Engine Company also exhibits one of J. R. Hall's self-acting SHINGLE AND HEADING MACHINES of its own manufacture, which appears to possess considerable merit, and to fully meet all requirements of the trade. Our illustrations shown on this and the succeeding page, present front and rear views of the machine, and show its manner of construction. The cutting is done by a circular saw driven at a rate of thirteen to fourteen hundred revolutions per minute, an automatic device feeding and returning the block of timber under operation, and throwing it back and forth so as to cut alternate butts and points, while at the same time a simple and easily operated mechanism

Shingle and Heading Machine, Front View: Buckeye Engine Co.

permits two or more butts or points to be cut in succession, if desired, and allows the rift of the timber to be kept vertical and in line with the saw, the sawyer having absolute control of the work. The machine may be adjusted in a few minutes to saw shingles of different thicknesses or different lengths without changing the uniformity of the taper or the evenness of the butts and points. It may also be arranged to cut parallel headings without interfering in any way with its excellence as a shingle machine.

Two sizes of machines are built, one varying from fourteen to twenty inches in length of shingle, with widths up to fourteen inches, and the other giving lengths of sixteen to twenty-six inches, the limit of width being the same as in the first. Either the ordinary knife-jointer is furnished, or an excellent form of saw-jointer, which it is claimed increases the yield for a given quantity of timber about ten per cent., requiring however an extra man for each machine. The makers claim a capacity of ten to twelve thousand eighteen-inch shingles per day of ten hours, or twenty-five to thirty-five thousand if the shingles be jointed for the sawyer, these figures being for soft wood blocks or one-third less for hard wood.

In ordinary mechanical operations where a steady motive power of medium or high pressure is required, nothing has yet been found as economical, efficient and ready of application as the steam engine. Where, however, a low power

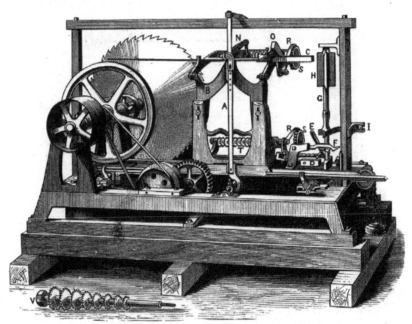

Shingle and Heading Machine, Rear View: Buckeye Engine Co.

is wanted, especially if it is intermittent, the steam engine becomes too expensive, requiring a costly form of apparatus, a skilled attendant, and a certain quantity of steam constantly on hand to bring it into action quickly, to say nothing of the danger almost always attendant more or less on its use. In the progress of the present age, a motor is every day becoming more a necessity which shall combine safety, economy and convenience; shall be always ready without waste while not in use, and shall be simple in details of construction and easy of management without skilled attendance; or, in other words, an engine is required that shall be eminently a domestic motor, and that may be put into our private residences, or used on our great agricultural farms to be managed by the ordinary servant without risk of fire, explosion or increase of insurance fees. The various hot-air, electric and gas engines are all tending to this end, with more or less success. One variety of engine we have already described, and we would now like to draw attention to another, shown at the Exhibition, which appears to merit particular notice. We refer to the BRAYTON READY MOTOR or HYDRO-CARBON ENGINE, the invention of MR. GEORGE B. BRAYTON, as manufactured by the PENNSYLVANIA READY MOTOR COMPANY, and shown by the accompanying illustration. This engine is really a hot-air engine in which the cylinder and furnace are one. It was originally designed for the combustion of gas, but now uses the ordinary crude petroleum, which is mixed with a certain proportion of atmospheric air and burned in the cylinder without explosion, the power so produced being expended on the piston, which works silently with a steady pressure just as with steam in the steam engine. The apparatus consists essentially of a working cylinder in which operates a piston connecting with a fly-wheel; an air-pump for compressing air; two cylindrical reservoirs for compressed air, one serving as a working reservoir and the other as a reserve to start the engine after any length of time it may be stopped; and an oil-pump for forcing oil into the combustion chamber, a few drops at a time, where it is mixed with a supply of compressed air in such quantity as the requirements of the work demand. Several forms of engines have been constructed, the original ones being all single-acting. Recently, however, double-acting engines have been made, our engraving illustrating a ten horse-power vertical engine of this class as shown at the Exhibition, and the latest improvements have been made in a horizontal engine, also double-acting, in which even the slide-rests for the piston are dispensed with, the motion being taken up by a rocker on the upright which supports it. The pistons of both cylinder and air-pump have the same construction with metallic packing-rings as in the steam engine, and the working-cylinder is surrounded by a jacket through which water circulates either by a very small running stream when available, or by continuous circulation from a reservoir, so as to secure a low temperature in its walls. In the single-acting engine the piston becomes quite hot after long working, and it is necessary to keep it regularly supplied with lubricant in order that it may work properly. In the last double-acting engines constructed, however, a great improvement has been effected in making the piston with two rods, both of which with the piston are hollow.

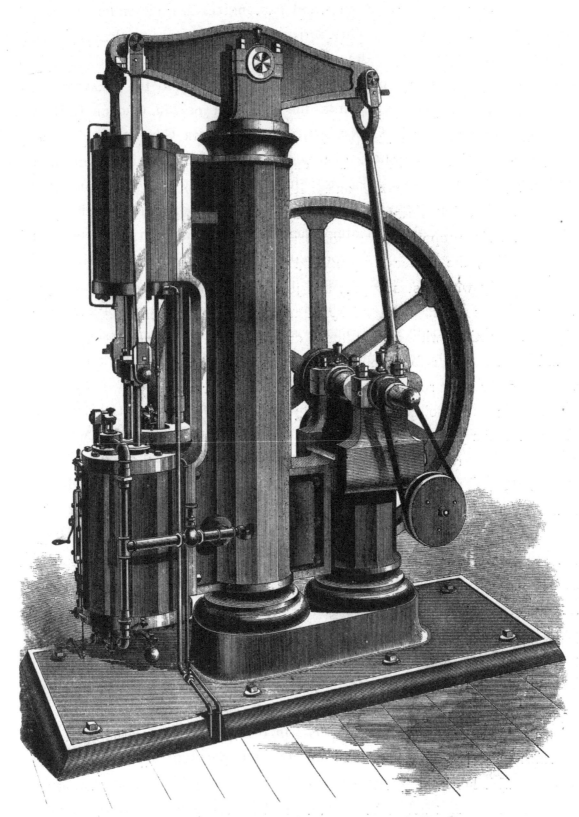

Brayton Ready Motor or Hydro-Carbon Engine: Pennsylvania Ready Motor Company.

These are connected by flexible tubing with the water circulation, which passes first through the hollow piston and rods, then down through the water-jacket of the air-pump, and finally through the cylinder-jacket, keeping everything cool. The tarry products of combustion, it is found, then condense on the piston and cylinder, and provide a quite sufficient lubricant without necessitating any outside supply, the engine running smoothly indefinitely. In these last engines a six-by-nine-inch cylinder has furnished five horse-power net, showing the great capability of the power at command.

The oil-pump is provided with a hand-crank by which a few drops of oil are

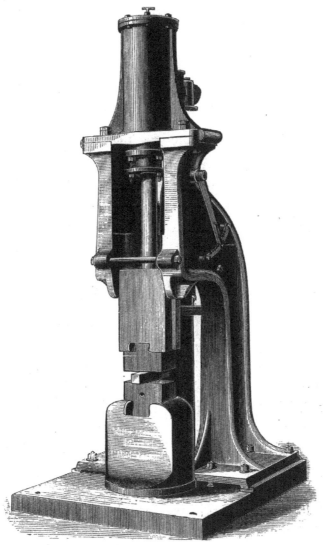

Double-acting Steam Hammer, Fig. 1: B. & S. Massey, Manchester, Eng.

pumped in at the beginning to start the engine. It then works automatically, an adjustment being provided to regulate the supply as required. The oil is forced through a ring of felting and a series of small holes in a state of vapor into contact with the compressed air, both mixing and passing into the cylinder through the meshes of three separate sheets of wire gauze where it is inflamed, a plug being removed from the side of the cylinder for the purpose and re-inserted. The engine starts off at once, and attains its required speed in a few seconds, saving all the time and fuel required in the use of a steam engine to accumulate pressure. The necessary valve is provided for the escape of the products of combustion by an exhaust-pipe to a chimney. A governor regulates the supply of air to the cylinder, and a safety-valve is provided, adjustable to any required pressure by which surplus air escapes from the reservoir.

The efficiency of the engine consists in the expansion by heat of the air introduced and the products of the combustion, carbonic acid and steam, the mixture of air and oil being such as to burn without smoke, thus securing a maximum economy of fuel. The great difficulty of rapidly imparting heat to air has hitherto prevented the most economical application of hot air as a motor, but the question appears to be practically solved by Mr. Brayton's method of intimately mixing the air and fuel before combustion. Explosion is impossible, and as combustion is confined to the interior of the cylinder and is perfect, no sparks being thrown from the exhaust, there is no danger of fire. The machine is of simple construction, not liable to get out of order, readily repaired if it does, and easily managed by any ordinarily intelligent person, the fire being extinguished and the engine stopped by merely closing the throttle-valve. Continuous use for fifteen months appears to establish the fact that no destructive action takes place in the cylinder. The governor acts as a variable cut-off, regulating the speed so readily that but little variation is noticed when the work is thrown on or off, and speed can easily be changed to answer any requirements. The amount of oil consumed it is claimed amounts to an average of one gallon of crude petroleum per horse-power for ten hours' service.

MESSRS. B. & S. MASSEY, of MANCHESTER, ENGLAND, make a fine exhibit of STEAM HAMMERS, which present some peculiarities of design different from the usual steam hammer, and appear to operate with great efficiency. They are double-acting and work without jar or shock, giving blows dead or elastic, and of any degree of intensity, rapidity of action or length of stroke desired, the larger hammers being controlled generally by hand, and the smaller ones arranged so as to work both self-acting and by hand. The action is therefore completely under control, and can be varied according to the kind of work to be done. Generally with self-acting hammers there is great difficulty in obtaining the heavy "dead" blow so often required; but in these, by means of a hand-lever connected directly with the valve, the hammer may be changed instantly from self-acting to hand-working, and perfectly "dead" blows delivered at any time without the least delay. Their small hammers are particularly intended for smiths' work, being applicable to the lightest kinds of forgings, such as usually done by hand, and their use is rapidly replacing that of handwork, resulting in great economy of labor, fuel and material even in the smallest smith-shops. The hammer shown by Fig. 1 is of a class comprising several sizes, and exceedingly convenient and easy to operate with, allowing ready access on three sides, and, owing to the double standards on the fourth side, with opening between them, permitting long bars to be worked on the anvil in either direction. The arrangement for working the valves in these hammers, as already stated, is a combination of self-acting and hand-worked gearing, and it is different from that ordinarily employed, being without the usual cams or sliding-wedge. As the hammer rises and falls when in action, a hardened roller on the back of the head slides on the face of a curved lever, which rotates about a pin near its upper end, and is held by a spiral spring always in position against the roller. At every movement of the hammer this lever operates a valve-spindle and regulating-valve, the length traveled by the hammer being controlled by another lever attached to the fulcrum-pin of the curved lever, and by which this pin may be raised or lowered by hand, and the points at which the steam is admitted or allowed to escape varied at pleasure. A guard-plate and catch permit this governing lever to be fixed at any point desired. The regulating-valve is hollow through the centre, being really a double piston open at both ends, with a number of ports for the steam to enter and escape, arranged all around on the sides, and holding it in perfect equilibrium. The ports open and close very quickly, and allow great rapidity and force of action to the hammer, as many as two hundred and fifty blows per minute being struck with a pressure of from forty to sixty pounds, with the length of stroke entirely under command from a few inches to nearly two feet, and variable without checking the machine.

Ramsbottom's Steel-packing Rings are used on the hammer-piston, which is forged in one solid piece with the rod, and the head is of hammered scrap-iron. The anvil-block is a heavy casting made separate from the base and turned to fit a bored hole in the base plate so as to assure its being kept to its true position.

Fig. 2 represents a light hammer, only a half hundredweight, intended for forging files, bolts, cutlery, etc., and operating with a foot-treadle, so that the workman may have both hands free for the proper manipulation of his work. The foot-treadle is omitted in some cases. This hammer has been worked up to a speed of four hundred blows per minute. Fig. 3 illustrates one of the large size hammers running up to a ton or more in weight. Fig. 4 represents a steam stamp intended especially for die-forging, and regulated either by the foot or by hand. When steam is turned on, the hammer rises to the top of stroke and keeps that position until directed downwards by the action of the operator. It then descends with a single dead blow, performing its work, and rises again into its original position, which it retains until the workman is ready for another stroke. It is wonderful how many articles formerly so expensive are now made by die-forging, being stamped out from the red-hot iron nearly ready for use, requiring in most cases very little work to fit them up, and resulting in great saving of labor. Bolts, rivets, nuts, screw-keys, wrenches, and other tools, and even such articles as sewing-machine shuttles, are made in this way with the greatest accuracy, economy and despatch.

Steam Hammer, with Treadle, Fig. 2: B. & S. Massey, Manchester, Eng.

In the manufacture of cars and in heavy railroad timber-work generally, the joining of the frames, instead of being accomplished by means of a mortise and tenon, as in ordinary building, is done by letting the ends of the cross-timbers into the longitudinal ones, thus giving much greater strength at the point of junction. The depression or groove in which the end of the cross-timber rests is technically called a gain, and a machine that is used for the purpose of cutting these grooves is called a Gaining Machine. MESSRS. C. B. ROGERS & CO., of NORWICH, CONNECTICUT, exhibit a timber GAINING MACHINE

Large-size Steam Hammer. Fig. 3: B. & S. Massey, Manchester, England.

of their own manufacture, which has attracted considerable attention and does them great credit. The engraving on page 136 gives a fair view of the machine, which, being intended for the heaviest kind of work, has been made strong, heavy and substantial in every part, its weight being five thousand pounds. It is furnished with a table, to which the timber to be operated on is clamped, this table moving on ways and having adjustable stops to indicate the points at which gains are to be made. A revolving cutter-head is attached to a frame joined by sliding-gibs to a standard placed at right angles to the table, and an arrangement of gearing and belting is provided for the necessary rotary motion. The cutter-head is made in two sections, which are adjustable longitudinally on its shaft to any width of gain required, the desired depth being regulated by means of a balanced lever set at the proper elevation by adjustable stops. The head and frame at the will of the operator are made to move transversely across the table, reversing automatically and stopping after return. In operation, the timber being clamped to the table, the cutter-head is set by the lever at the proper elevation indicated by the stops; a lever in front throws the feed in gear, and the head moves across the timber, cutting the gain and returning to its first position, where it stops automatically. The table is then moved on its ways to the point marked for the next gain, and the operation is repeated. The machine is readily controlled by the operator, and after being once adjusted for any particular class of work it becomes almost automatic in its action. The points of excellence claimed for this machine by the makers are, its extreme simplicity, combined with every requisite for accomplishing its desired purpose; the great ease with which every motion is controlled, even when operating on the heaviest classes of work, and the automatic precision with which this work

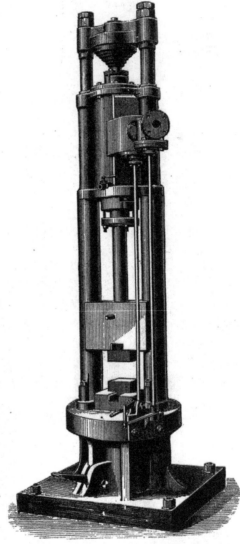

Steam Stamp, Fig. 4: B. & S. Massey, Manchester.

is performed. Its peculiar feature is the reciprocating motion given to the revolving cutter-head.

We have mentioned before the fine exhibit of astronomical instruments made by MESSRS. FAUTH & CO., of WASHINGTON, D. C., and we refer to this exhibit again to call attention to an exceedingly perfect ALTITUDE AND AZIMUTH INSTRUMENT, which has been purchased by the United States Coast Survey Department for triangulation and determination of azimuth. The altitude of a star or other body is its height above the horizon expressed in degrees, the greatest altitude of course being ninety degrees. The azimuth is the angle made by the meridian and the vertical circle in which a heavenly body is situated, and is measured along the horizon from the north or south towards the west, according as the north or south pole is elevated above the horizon, to the point where a circle passing through the zenith and the body cuts the horizon. The instrument of which we speak, and which is represented by the engraving on page 137, from the nature of its construction is employed in the measurement of vertical and horizontal angles, and may be used as a transit for time observation, also for double zenith distances for latitude, and will determine the astronomical position of any station. For geodetic purposes it is used in primary triangulation to measure the angles with the utmost precision. It has two graduated circles and a telescope, the planes of the circles being at right angles to each other, one called the azimuth circle being connected with a solid support, on which it is leveled and kept in a horizontal position, and the other called the altitude circle, being mounted on a horizontal axis, which also carries the telescope like a transit. The design and construction of the instrument are quite novel, combining all the advantages of a repeating instrument without its defects. The horizontal limb, which is thirteen inches in diameter, is graduated to five minutes, and may be read off by means of three microscopes at different points to the nearest single second, these microscopes being illuminated by prisms which derive their light from overhead, and are effective for any position of the circle. This circle may be shifted if desired, so as to bring different parts of the graduation under the microscopes, and thereby eliminate any error or eccentricity in division. The vertical circle is ten inches in diameter, and is graduated the same as the horizontal circle, but is read by only two microscopes, a very sensitive level, reading to single seconds of arc, being affixed to the microscopes to note any deviation from the vertical. This circle may also be shifted for position. Both circles are entirely free from clamps, these being attached to the centre, by this means avoiding the great risk of strain. The clamps and slow motion have differential screws. For time observation a striding level of the utmost perfection is supplied, which is set over the hard pivots of the telescope axis, so as to note any deviation from the meridian, the level being ground to a radius of about two thousand feet, each

Gaining Machine: C. B. Rogers & Co., Norwich, Connecticut.

division of its graduation representing a second of arc. Both this and the level over the microscopes of the vertical circle have air-chambers to correct the bubbles for changes of temperature.

The telescope has a focal length of twenty-four inches and a clear aperture of two and a half inches, its glass being of uncommon excellence, as is proved by the fact of its showing the companion star of Polaris. The micrometer on the eye-piece measures to the one-hundred thousandth part of an inch, and is used for determining differences of zenith distances of stars in computing latitudes. For convenience in observing near the zenith, a rectangular eye-piece is provided. A lamp is placed opposite the microscopes of the vertical circle to throw light through the axis down to the field and render the cross lines visible. This instrument is well entitled to the award which it has received, not only for novelty of design, but for execution and workman-

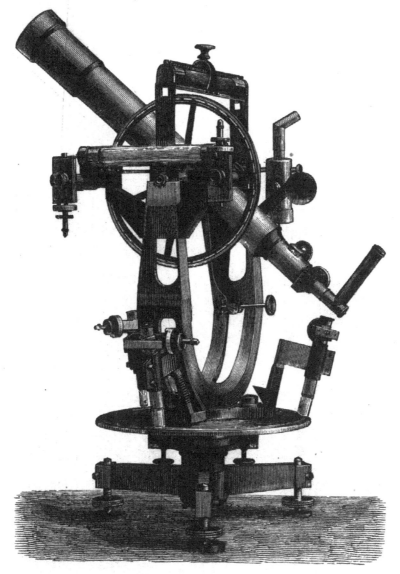

Alt-Azimuth: Fauth & Co., Washington, D.C.

ship, fully confirming the encomiums which have been passed on it by so many astronomers.

One of the most novel inventions that American ingenuity has brought into practical use during the past few years is the GUNPOWDER PILE-DRIVER of

Mr. Thomas Shaw, of Philadelphia, and well worthy of its place in the great Exhibition. The originality consists in the adaptation of a material as a motive

Gunpowder Pile-Driver: Thomas Shaw, Philadelphia.

power, ordinarily so violent and destructive in its action as to be generally considered almost uncontrollable. Yet machines for driving piles by the use of gunpowder have been constructed under Mr. Shaw's patent, and have been in practical use in various parts of the country for several years with the greatest success, demonstrating high economy and efficiency.

The machine is constructed of a strong frame-work of upright timbers, with inclined braces, formed into a ladder in the rear, the whole stiffened by horizontal struts and diagonal ties between. On the inner opposing faces of the uprights, guides are formed, in which a steel or iron gun and a ram move vertically. The gun rests on the top of the pile, being recessed on its under face for this purpose, and it is bored in its upper end to receive a plunger or piston, fitting nearly air-tight, which is fixed to and projects below the ram placed above. The upper end of the ram is also bored to receive a fixed piston projecting below a cross-head at the top of the guides, creating an air-cushion to check the upward movement of the ram should it be subjected to the force of an excessive charge of powder. In the rear of the uprights and placed parallel to them, running their whole length, are powerful double friction-brakes, operated by a compound lever near the foot of the machine, and used to check and hold the ram at any required point of elevation.

For expeditious working it is desirable to have a double-drum hoisting-engine in connection with the machine, and when so used the operation of driving proceeds as follows: A wooden block is placed across the mouth of the gun, and the ram and gun are hoisted simultaneously by one drum, which holds the gun, while the friction-brake is used to hold the ram. The pile to be driven is now raised to a vertical position by the second drum, and lowered in place until its foot rests on the ground. The gun is then lowered upon the top of the pile, keeping it firmly in place; and the block of wood being removed, a cartridge is dropped into the bore. The brake is now released, allowing the plunger to fall. On entering the bore it starts the pile downward, as in the ordinary pile-driver, and an explosion of the cartridge immediately following, the motion of the pile is vastly increased, the ram being at the same time projected upward, to be arrested by the application of the friction-brake. A second cartridge is now introduced, and the ram released as before, resulting in another explosion, which drives the pile still further, projecting the ram again upward. This operation may be repeated with great rapidity as often as necessary to force the pile to the proper depth. The plunger, by its sudden descent into the gun, compresses the confined air into a narrow stratum or cushion, preventing actual contact of the metal and at the same time generating heat, which fires the cartridge, the force of the powder being assisted by the expansive power of the air under the additional heat, and the principles of the hot-air engine called into action. The combination of all the forces developed creates an immense power, which pushes the pile down at the same time that it overcomes the momentum of the ram and projects it back to its original position.

With this machine, by successive explosions of cartridges, each composed of an ounce to an ounce and a half of common blasting-powder, a pile forty feet in length and fourteen inches in diameter may be forced its entire length into firm ground in one minute of time without the slightest injury to the timber, and entirely obviating the necessity of banding the head before driving. There appears to be no blow or concussion, the cushion or stratum of air in the gun acting as an elastic medium, and the pile being, as it were, forced into the ground as if by hydraulic pressure, instead of being pounded down as by the old methods. The sound condition in which the pile is preserved gives it greater sustaining power and lessens liability to decay. A large number of piles have been driven by this machine in the most satisfactory manner at the improvement works of the United States Naval Station, League Island, Philadelphia, both in the water and on shore. In wharf-work the superior alignment of piles driven by this method over the old plan is a great advantage, very much facilitating the work of capping and reducing the cost in labor and material. The machine possesses great simplicity of construction, controllability and readiness of manipulation, rapidity of work, and economy and efficiency of power.

The BROWN & SHARPE MANUFACTURING COMPANY, of PROVIDENCE, RHODE ISLAND, with the same spirit which characterized its exhibit at Vienna, makes an exceedingly interesting and instructive display, in Machinery Hall, of that high class of tools for which it has achieved so great a reputation. These tools have chiefly developed from the requirements of the Company's general manufacturing business, which is conducted on an extensive scale, and they may therefore be said to represent the results of actual experience. In their construction, the uses for which each machine is intended have been well kept in view and every effort made to perfectly satisfy all requirements. As evidence of the character of work which these machines will execute we may mention that the firm has in its regular business manufactured over two hundred thousand Wilcox & Gibbs' Sewing-Machines, which uniformly attest its excellence.

Our first illustration, Fig. 1, represents the Company's UNIVERSAL GRINDING MACHINE, an exceedingly useful tool for performing a great variety of operations in grinding by the use of solid emery- or corundum-wheels, being

Screw Machine, Fig. 2: Brown & Sharpe Manufacturing Co., Providence, R. I.

especially adapted for the grinding of soft or hardened spindles, arbors, cutters—either straight or angular—reamers, and standards; also, for grinding out straight and taper holes, standard rings, hardened boxes, jewelers' rolls, etc. By means of an additional movable table, adjustable by a tangent screw and graduated arc, straight and curved taper-grinding may be performed with the centres of the machine always in line. The work may be revolved upon dead centres or otherwise, and the grinding-wheel may be moved over the work at any angle, producing any taper required. Graduated arcs are provided for grinding of taper holes and angular cutters. Wheels may be used from one-fourth inch to twelve inches in diameter, and the feed-works and slides of the machine are thoroughly protected from the entrance of grit or dust. A special chuck is provided to hold work in which holes are required to be ground. The spindle and boxes of the machine are of cast steel, hardened and ground. Our engraving shows also the overhead works, consisting of a drum, tight and

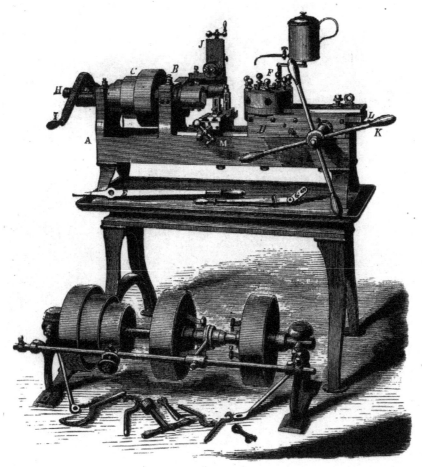

Universal Grinding Machine, Fig. 1: Brown & Sharpe Manufacturing Co., Providence, R. I.

loose pulleys, one iron pulley for driving the work and grinding-wheel, and adjustable hangers with self-oiling boxes. The weight of the whole, including the overhead works, is about two thousand pounds. This machine was purchased by a prominent firm in Alsace.

Fig. 2 shows the No. 1 SCREW MACHINE made by this firm, together with the overhead works, the whole weighing about fourteen hundred and fifty pounds. By this machine may be manufactured all kinds of screws and studs such as usually required in a machine-shop, and nuts may be drilled, tapped and one side faced up; also many parts of sewing-machines, cotton machinery, gas- and steam-fittings may be made at great reduction in time and labor. It is claimed by the makers that as many screws can be made by one man with this machine as by three to five men on as many engine-lathes, and with much more uniformity in size. The size of hole through spindle is one and a quarter inches, and that in revolving head one and one-sixteenth inches, the length that can be milled being six inches. Smaller sizes of the Screw Machines are also made; one size, No. 3, being for the manufacture of screws used in sewing-machines, fire-arms, etc.; and another, No. 4, for still smaller work, such as screws for clockmakers, etc. The last machine has a patented device for opening and closing the jaws of the chuck which holds the wire from which the screws are made, allowing the operation to be performed in an instant without stopping, and effecting great saving of time when making small screws. It is often desirable, in threading screws and in tapping, to cut the thread up to a shoulder or to a given point, or to run the tap in to a shoulder or a given distance, and positively no further. With the ordinary tools this operation is quite difficult, causing great risk of breaking the threading-tool or injuring the shoulder of the screw. By means of a patent die-holder, however, manufactured by the firm, for use in the revolving heads of these machines, the matter can be accomplished without special skill or any risk of damage. Special tools are also furnished, if required, for making screws of any particular form or design differing from those usually made.

Fig. 3 shows the UNIVERSAL MILLING MACHINE of this Company, a tool that was exhibited at Paris in 1867, attracting marked attention and securing a very high award; again exhibited at Vienna in 1873, winning unusual distinction, and now coming forward at our own great Exhibition.

We have before mentioned in another connection the special functions and peculiar features of milling machines, and it is not necessary to repeat them again. This tool, besides having all the movements belonging to plain milling machines, possesses also an automatic movement and feed to the carriage, by which it is moved not only at right angles, but at any angle to the spindle, and stopping at any required point. Centres are arranged on the carriage, in which reamers, drills and mills may be cut, either straight or spiral, the latter right or left as desired, and spur- and bevel-wheels can also be cut. The head holding one centre can be raised to any angle, and conical blocks may be placed on an arbor and cut straight or spirally in either direction. Pulleys fourteen inches in diameter are used on the counter-shaft, the whole width of the three being fifteen inches. The total weight of machine, with overhead work, is about fifteen hundred pounds. The machine possesses an exceedingly

Universal Milling Machine, Fig. 3: Brown & Sharpe Manufacturing Co., Providence, R. I.

large range of work, and performs it with great excellence and accuracy. It is gotten up in very handsome style, and is well worthy of the attention which it has attracted.

In connection with this machine is a gear-cutting attachment, illustrated by Fig. 4, designed for the purpose of cutting gear-wheels with greater rapidity, and also for cutting larger and heavier wheels than can be done with the ordinary apparatus of the machine alone. It has a swing of thirteen inches, and is provided with an index of twenty inches in diameter, containing four thousand two hundred and ninety-four holes, which will divide all numbers to seventy-five, and all even numbers to one hundred and fifty. The screw with set nuts over the spindle is designed as a support to the wheel while being cut, and arbors fitted to the Universal Milling Machine can be used with this attachment.

The Brown & Sharpe Co. also exhibits a very large and heavy Universal Milling Machine, illustrated by Fig. 5 on page 147, which has been designed particularly to meet the wants of steam-engine and locomotive builders or those engaged in the manufacture of heavy machinery. As will be noticed by the weight, which is three thousand eight hundred pounds, it is more than double the size of the smaller machine. It is built with the same essential features and motions, in enlarged parts, the only difference being that it is back-geared, giving six changes of speed, and having also the same number of changes of feed. The same ideas are carried out on a much larger scale than ever attempted before, and as there are more chances of error in a large machine than a smaller one, the errors being magnified as the machine increases in size, it is evident that the difficulties of construction have been augmented accordingly. The spindle-boxes are of hardened cast steel, and together with the spindle-bearings are carefully ground and are provided with means of compensation for wear. A cutter-arbor, projecting fifteen inches, may be carried by the spindles, being supported by an adjustable centre at the outer end. Cutters may be used up to eight inches in diameter. The spiral clamp-bed will move horizontally upon the knee in a line with the spindle of the machine six and one-half inches, and the vertical movement of the spiral bed-centres below the spindle-centres is eight and three-quarter inches. The spiral bed may also be set at angles of

Gear-cutting Attachment: Brown & Sharpe Manufacturing Co., Providence, R. I.

thirty-five degrees each way from the centre line of spindle, and may be fed automatically twenty-two inches, taking also twenty-two inches between the centres, and will swing eleven and a half inches. The hole through the chuck and spiral head is one and a half inches, and the vise-jaws open three and three-eighth inches and are six inches wide and one and five-sixteenth inches deep. The pulleys on the counter-shaft are sixteen inches in diameter, the whole width of the three being twenty-two and a half inches. This machine was purchased by Fried. Krupp, the famous gunmaker of Essen, Prussia.

We must not omit to mention the Patent Milling Cutters which are made by this Company, for milling parts of sewing-machines and other articles of irregular figure, and also for making the teeth of gear-wheels. They may be sharpened by grinding without changing their form, the operation being capable of repetition over and over again until the teeth are entirely worn out, without affecting in the least the standard shape of the work produced by them.

IRA J. FISHER & CO., OF KINCARDINE, ONTARIO, CANADA, represented by ALEXANDER THOMSON, OF FITCHBURG, MASS., as sole agent in the United States, exhibits a BEVEL-EDGE BOILER AND SHIP PLATE CLIPPER, which appears to be an exceedingly efficient and complete working-tool, supplying a want long felt among steam boiler-makers and iron ship-builders. This machine, which is the only one on exhibition, entirely replaces the slow and objectionable process of hand-clipping or the expensive use of planers, doing a much greater variety of work than possible by the old methods and at considerably less cost. Our illustration on page 148 shows very clearly the design of the tool and its mode of working. It will cut the edges of plates to any required bevel, keeping the line of direction of the cut either straight, concave or convex, as desired. This capability of doing circular-work makes it exceedingly serviceable for Iron Ship-building. The plates require no fastening while under operation, and are placed on a truck or table and passed through the shear, the depth of the cut being regulated by a guard, which is adjustable by a screw, and can be readily changed without interference while the machine is in motion and the plate being cut. It is claimed that the capacity of one of these clippers runs from one to two hundred feet per hour, the work being done in a superior manner to that executed by planers, which will only do from two to three hundred feet per day on straight work. A hand-power machine is manufactured by the same party, provided with a concentrated power appliance under Mr. D. L. Kennedy's patent, by which it is said two to three hundred feet of clipping may be done per day on three-eighth inch plate with one man on the lever. The machine appears to combine great simplicity, efficiency and durability, and has received an award of a medal and diploma from the Commission.

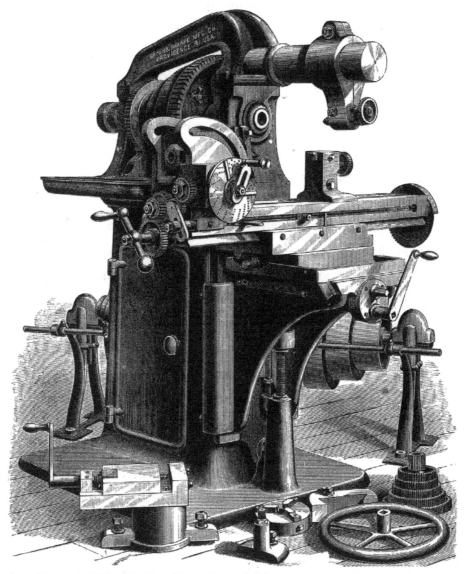

Large Universal Milling Machine. Fig. 5: Brown & Sharpe Manufacturing Co., Providence, R. I.

The STEVENS INSTITUTE OF TECHNOLOGY, now well known both at home and abroad, and attracting considerable attention as the highest representative of a technical school of mechanical engineering at the Exhibition, makes an exceedingly creditable and interesting display in the Main Building, which has been arranged with a view of illustrating—first, the methods of instruction employed, as seen by instruments, models, etc.; second, the results obtained, as shown by work produced by students; and third, the contributions to the progress of science, as exemplified by apparatus used in original investigations, samples of new substances discovered and published papers. There are exhibited a number of instruments of precision such as a linear dividing engine of large size and great accuracy, a spherometer and Dore's polariscope, also an extensive assortment of illustrative apparatus, a large induction coil for producing statical electricity, President Morton's vertical lantern with special attachments, and a number of instruments for delicate researches, Regnault's apparatus for heat, Tyndall's for radiation, and Thomson's for electricity. In the department of engineering are a number of models of elements of machinery and construction, working models of pumps, steam-engines, etc. The collection of engineering relics is exceedingly interesting, and several remarkable historical relics are exhibited. Here is the high-pressure condensing engine, water tubular boiler and screw which in 1804 drove John Stevens's first steamboat eight miles an hour on the Hudson, and the twin screws used with the same engine in 1805. Here also is Fulton's own drawing for the engine of the Clermont and one of his autograph letters. In the Chemical department are found various rare chemically pure substances prepared in the laboratories of the Institute, the new hydro-carbon Thallene and a number of its derivatives first discovered by President Morton, also an extended series of anthracine derivatives, etc., many of these being the only representatives of the substances found in the entire Exhibition. The work of students is very well represented. There are specimens of drawings, designs worked out and in no sense copies, the conditions being given and the work planned, calculated and executed by the student as if in actual practice, being a much higher grade of work than mere copying; also apparatus and machines actually made by the students, as the Odontoscope designed and constructed by Mr. J. M. Wallis, of the class of '76, for the purpose of testing the teeth of spur-wheels as to their being of the correct form, an oil-testing machine built by the class of '77, and many other interesting specimens of work.

Among the instruments exhibited which are used in actual work is Prof. R. H. Thurston's Autographic Testing Machine, an apparatus that has so perfectly fulfilled the conditions for reliable testing of the strength, elasticity, ductility, resilence, and homogeneousness of materials of construction as to attract marked attention both at home and abroad, and win for its inventor a lasting reputation, being in use by the United States Government, by various private firms, where unusual care is exercised in the selection of stock materials, by different scientific institutions, and foreign orders for them being received lately from the Russian Government.

This machine, of which we present an engraving on this page, consists essentially of two strong wrenches, which are supported independently of each other by a substantial framework, and have a space between them adapted to the length of any piece to be tested, this being held in between them, and having its centre and their axes all in the same horizontal line. One of these wrenches is revolved by means of a worm gear, the motion being transmitted through the test-piece to the other, to which is attached a weighted arm or pendulum, and causing this arm to swing outward from the perpendicular, and to react with a torsional strain upon the test-piece by an amount proportional to its weight and the angle it is made to assume with the perpendicular.

A cylinder is attached to and moves with that wrench which bears the worm gear, and around this cylinder is wrapped a sheet of profile-paper, upon

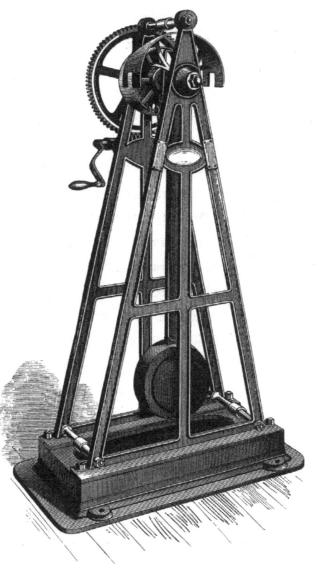

Thurston's Autographic Testing Machine.

which the apparatus makes its record by means of a pencil borne by the other or weighted arm and partaking of its motion. The direction of the motion of this pencil is changed, however, by means of a stationary guide-curve attached to the frame of the machine, the form of the same being such that its ordinates, and therefore the distance that the pencil is thrust forward, are precisely proportional to the forces tending to produce torsion on the test-piece, and developed by the weighted arm while moving up an arc to which the sines of the curve are proportional. The cylinder and the pencil having precisely the relative angular motion of the two ends of the test-piece, the curve described by the pencil upon the profile-paper is such that the ordinate of any point in it measures the force producing distortion at a certain instant, and its abscissa the amount of that distortion at the same instant.

When a piece is subjected to torsion, a range of distortion is obtained so great as to be easily measured, and the stresses are applied in the most favorable way to bring out all the characteristics of the material under test. The record of the test being made by the machine itself, avoids all errors of personal observation. The manner in which the registry is effected makes the automatic record peculiarly reliable and valuable, since, as the degree of distortion of the material and the resistance offered by it record themselves simultaneously and continuously from the initial strain to the point of final rupture, therefore the curve described by the apparatus is a complete record of every condition under which the piece has been tested, and of its exact behavior under these conditions. Such an automatic registry it is believed has never before been made.

Every characteristic of the material is revealed by the diagram, which also affords a comparative measure of the ultimate resistance of the test-piece; its ductility; its homogeneousness as to internal strains, structure and composition, and many other properties of great interest and value that no previous testing apparatus has ever given.

This machine is of special value to the engineer in revealing the laws governing the resistance to stresses variously applied, in suggesting formulæ embodying those laws, and in furnishing the proper constants for these formulæ, as well as in modifying those now in use which are shown not to take into account important elements. The diagram affords unusual facilities for studying the molecular structure of materials; the effect of internal strains and the manner in which they are induced or relieved; the changes in character due to different compositions or to any peculiar treatment in manufacture, and the results of experiments made to determine the ways of improving the qualities of materials.

The inventor, by means of this instrument, has made and published to the world important discoveries in reference to the behavior of certain classes of materials under strains exceeding the elastic limit, and has also been able to throw much light upon the phenomena attending molecular action induced by changes of temperature, by external forces, and by special treatment. The machine is one that in the hands of the careful manipulator may be made to reveal the most valuable results, and is only another evidence of the great good which is accomplished by such a school as the STEVENS INSTITUTE OF TECHNOLOGY in the extensive facilities which it affords for invention and research.

MESSRS. HOOPES & TOWNSEND, OF PHILADELPHIA, well known for many years as extensive manufacturers of bridge-rods, bolts and nuts, railroad splices, etc., make an exceedingly interesting and handsome exhibit in Machinery Hall of the various products of their works, artistically arranged in a neat pavilion, and attracting as many visitors by the beauty of the display as by the excellence in quality of the articles. Here are found car-forgings, truck-irons, washers and chain-links, square and hexagon nuts, boiler- and tank-rivets, bridge-machine-, and car-bolts of every description, all showing a high state of perfection in workmanship and finish.

One of the great specialties of this firm has been the manufacture of cold-punched nuts, and great ingenuity and skill have been brought to bear on this subject, resulting in the punching of nuts of a depth compared to diameter of hole that has never been achieved elsewhere. We have before us now a nut one and three-quarters of an inch in thickness that has been punched cold, with a hole only seven-sixteenths of an inch in diameter, as clean and straight and with as perfect faces as could possibly be desired. It has generally been considered that the limit in thickness of metal for punching was equal to the diameter of the punch, even if the latter was made of the very best steel and hardened in the most thorough manner, and that any greater thickness involved considerable risk of breakage to the punch. Here, however, is a nut four times the diameter of the punch in thickness, punched with apparently the greatest ease, and characterized by superior excellence of finish, allowing it to be used directly on fine work without further attention. This seemingly impossible result has been accomplished by using only the best material in the punch, making it accurately fit the die, and employing machines sufficiently heavy and accurate to insure that the iron being punched shall receive only direct vertical pressure without any lateral or bursting strains, the force applied to the punch being only such as it is capable of sustaining, and sufficient time being given to allow it to penetrate the metal at a rate depending upon the natural fluidity of its particles. One punch is exhibited, made from American steel—Midvale Steel Works—that has punched over two hundred holes, and

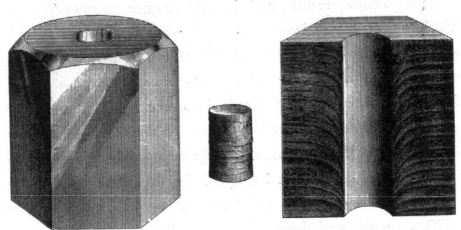

Cold-Punched Nuts: Hoopes & Townsend, Philadelphia.

is still in good condition. It is claimed that the iron after punching retains all the qualities possessed by it when it came from the rolls, and that the fact of its being cold-punched is a guarantee of its high excellence, a proof not obtained in the case of hot-pressed nuts. The metal in a cold-punched nut it is claimed is really strengthened instead of weakened, that portion around the hole being compressed and made more dense and stronger. It has been asserted in objection to cold-punched nuts that there was great roughness of holes from broken fibres of iron, and that the holes were also larger at one end than at the other, and almost always out of perpendicular to the faces of the nut. MESSRS. HOOPES & TOWNSEND claim to have entirely overcome these difficulties, bringing, with their method of working, an experience of over twenty-five years, and they state that they are able to furnish first-class nuts, cold-punched, with holes no larger at one end than the other, perfectly straight, perpendicular to the faces, and almost as smooth as if drilled, without any loosened fibres. Over three tons of cold-punched nuts have been tapped per day by this firm without even one being condemned on account of the threads pulling out, they being equal for all practical purposes to threads chased by lathe-work. In hot-punched nuts a hard and brittle scale is formed, which soon takes off the fine cutting edge of the tap, and the gauge is lost. On page 152 we give an

Steam Riveting Machine: Pusey, Jones & Co., Wilmington, Del.

engraving showing a section of a cold-punched nut, the surface of which has been planed and treated with acid so as to bring out the fibre. It will be noticed that the action of the punch has bent the lamina of the metal all downward, the nut being punched with its top downward, and it is claimed that this gives the iron much greater resisting power, the fibre standing, as it were, at an inclination to the bolt, and taking the stress on end instead of across.

Tests have been made with these nuts, confirming the superior qualities claimed for them. Cold-punching is productive of economy in the fact that there is no expense for heating the metal, and also that no scale is formed as in hot-punched nuts, and the taps are not worn out. It is stated by the firm that it has punched twelve tons of nuts with a single tap without wearing it out or causing any sensible loss of gauge.

We have before mentioned the great importance that power-riveting has attained within the past few years, superseding hand-riveting almost entirely in all large establishments, and accomplishing not only great saving in time and labor, but also turning out work of much increased strength and perfection. These riveting-machines are operated either by hydraulic power or steam, and we desire here to draw attention to a Steam Riveting-Machine as exhibited in Machinery Hall by PUSEY, JONES & CO., OF WILMINGTON, DELAWARE, and shown by the engraving on page 153, that appears to be particularly well adapted for iron ship-building and bridge-work—in fact any work consisting of I beams, channels and angles, or of flat plates in conjunction with them. The parts to be riveted together are first joined and held in position with respect to each other temporarily by bolts at intervals, as is the usual custom, and are manipulated on the machine by crane-power. The rivets are heated in quantities in small furnaces provided for the purpose, and put in place in the rivet-holes by boys, the work being moved along under the riveting-stamp as rapidly as they are inserted. The stamp operates vertically by direct steam pressure on a large piston, the operation being governed by a foot-lever, seen in front on the engraving, the short arm being weighted. The pressure of the foot admits the steam, and when the rivet is driven, the foot is removed, and the action of the weight raises the lever and opens the exhaust.

It is claimed by the manufacturers that the capacity of this machine is equal to that of ten gangs of hand-riveters, the attendance of but one skilled workman, three laborers and a boy being required in working it.

Another exhibit made by this firm, and not mentioned on its list nor entered in the catalogue, is the iron work of the Machinery Hall itself, and all who have examined it, especially that in the roof seen from the galleries at the east and west ends of the building, and decreasing so regularly and beautifully in the distance as the eye glances from truss to truss, will testify to its neat and accurate workmanship, as well as to the care and ability of the contractor, Mr. Philip Quigley, who framed the timber-work and erected it.

Messrs. W. & L. E. Gurley, of Troy, New York, make a large display of the various instruments used by the civil engineer and surveyor, from among which we select two of the most important for illustration.

The Engineers' Transit, as shown by the engraving on page 157, has a telescope of eleven inches fixed in an axis supported by two standards, and turning readily in its bearings, so that it may be rotated in a complete revolution or "transit," and enable sights to be taken in opposite directions. The telescope has connected with it a long spirit-level, a vertical arc of three inches radius, and a clamp with slow-running tangent-screw, these appliances enabling level lines to be fixed and vertical angles measured. The vertical arc is divided upon a silver-plate surface, and by means of a vernier at the lower part may be read to ten seconds of arc. There are two horizontal circular plates revolving freely on each other about concentric vertical axes, the upper carrying the standards for the telescope, and supported by a tapering spindle, which fits into a hollow cone carrying the lower plate. A clamp with tangent-screw connects the two plates, and a compass circle is fixed to the upper plate, properly divided and furnished with magnetic needle. The lower plate, called the limb, is divided around an inside edge into degrees and fractions of a degree, and figured to every ten degrees. The upper plate has two opposite verniers, with divisions to correspond with the lower plate, and will give readings for angles to ten seconds of arc. Spirit-levels are attached to adjust to the horizontal, four leveling-screws being provided for the purpose. The instrument is supported upon the usual three-legged stand or tripod, and so far it does not differ essentially from those in general use for engineering purposes for many years past, being, however, an exceedingly fine and accurate piece of workmanship. The special novelty in the instrument consists in the solar apparatus attached above the telescope, forming an important addition to the ordinary transit. The system of public land surveys adopted in the United States since the beginning of the present century, requires such lands to be laid out in areas, the boundaries of which shall be truly north and south, east and west, or, technically, meridians and parallels. To readily determine these lines the Solar Compass was devised by William A. Burt in 1835, and has ever since been the standard instrument of the Government surveyor. The Solar Apparatus here shown, and which is essentially the same as the instrument just named, consists mainly of three arcs with their appliances: the latitude arc, represented by the vertical arc attached to the axis of the telescope and previously described; the declination arc, a segment in a vertical plane shown in the engraving above the telescope; and the hour circle, a horizontal circle surrounding the little disk above, and fixed to the centre of the telescope axis. The declination arc has a movable arm attached, which has at each end a small rectangular block, containing a little lens with a small silver plate opposite

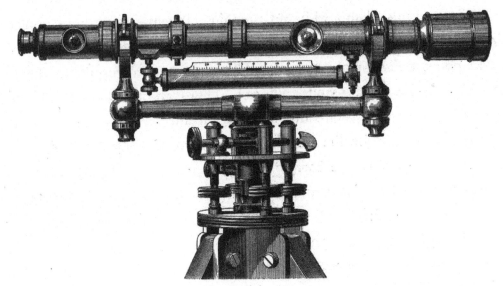

Y Level: W. & L. E. Gurley, Troy, N. Y.

to it, marked with two sets of lines, enclosing a square of precisely the proper size to contain the image of the sun when focused upon it by the other lens. The declination circle is divided and figured from zero to thirty degrees, with vernier attachment. The hour circle is divided into one hundred and twenty equal parts, and figured on each half of the circumference from one to twelve, being read by an index attached to the frame of the declination arc to five minutes of time or closer if desired. In taking an observation for true meridian, the transit is set up and carefully leveled—the latitude of the place, the declination of the sun for the given hour, and the proper time set off upon their respective arcs. The whole instrument is then turned horizontally until the sun's image is brought within the square of the silver plate, and being securely clamped in this position, the telescope indicates the true meridian. Any angle may now be laid off from the meridian, and the bearings of lines determined independent of the needle, which latter shows its variation from true north and south. This solar apparatus may be readily removed and put aside or replaced at will, without interfering in any way with the ordinary use of the transit, thus combining the advantages of both instruments.

The other article selected is a Y Level, shown on page 156, an instrument used in determining levels, fixing grades, etc., and holding a most important position in a list of the essential tools of the civil engineer. The one here exhibited has a powerful telescope twenty inches in length, to which is attached underneath, a long and delicate spirit-level, provided with a scale, so that the bubble may be readily and accurately centered. This telescope, as well as that in the transit already described, has an adjustment of the slide-tube of the object-glass, which insures its moving in a right line, and enables the engineer to take long or short sights in any direction without deviation, and so

Engineers' Transit: W. & L. E. Gurley, Troy, N.Y.

correcting what is termed the travelling of the object-glass. The telescope-tube is supported in a Y at each end, with a clip or band above, swinging on a pin in one arm of the Y, and held in its place in the other by a movable tapering pin. When these pins are loosened, the telescope-tube may be revolved in the Y's, or taken out and reversed in position, end for end, these movements being required in making adjustments of the instrument. The Y's are supported in the ends of a horizontal bar, and provided with two nuts to each for adjustment, so that the longitudinal axis of this bar may be made parallel to the centre line of the telescope. The bar at its centre is connected to a vertical spindle, which revolves accurately in the hollow frustum of a cone in the head of a tripod socket, the latter having two parallel plates with ball and socket connection and four leveling screws. These screws, in both this instrument and the transit, are covered at their upper ends by caps or thimbles, so as to effectually exclude all moisture or dust. A clamp with tangent movement is provided for fixing the instrument in the bearing of the spindle, and when unclamped, a little screw below the upper parallel plate of the tripod head, working in a groove in the spindle, prevents it from coming out unless desired, while still permitting it to revolve freely. The tripod head may be unscrewed from the legs, and both it and the upper part of the instrument may be separately packed away in a box for safe transportation from place to place. The arrangement of the details, as seen by the engraving, allows of a long socket or spindle, while still bringing down the whole instrument as close as possible to the tripod head—both matters of great importance in the construction of any instrument of this kind.

This firm has introduced the use of the new metal, aluminium, so remarkable for its lightness, and now manufactures these instruments of this material, saving nearly one-half in their usual weight, at a cost, however, of about fifty per cent. advance on the ordinary prices.

Quite a number of looms of various kinds are shown in the Exhibition, most of them in practical operation, attracting crowds of visitors, all interested in the curious automatic movements and apparently marvelous results accomplished by these machines. Even those most experienced in such matters find a fascination in watching the passage of the shuttles to and fro, and observing the pattern as it is gradually and so beautifully worked out.

Weaving may be defined to be the making of cloth by the intersecting of threads. Those running lengthwise of the material are called warp or chain, and those across, woof or weft. Taking into consideration a loom machine generally, without reference to any particular type, the essential parts may be described as follows: A substantial framework is used as a support for the whole apparatus, in one part of which is a horizontal roller, called the beam or yard roll, longer than the width of the cloth to be made, on which is wound the warp, all laid out flat, each thread side by side. These threads pass in a plane, horizontal or nearly so, through the heald, a sort of comb dividing them regularly, towards another horizontal roller parallel to the first, and at the other side of the frame, called the cloth-beam or breast-roll, upon which the fabric is wound as it is woven. Between these the warp is kept tightly stretched by the action of weights on the roller, or by some equally efficacious means. An arrangement of tight vertical cords intersects with the warp-threads in a plane at right angles to the plane of the warp, there being one vertical cord to each warp-thread. On the line of intersection, these cords have loops or metallic eyes, and each separate warp-thread passes through the eye of the cord belonging to it. Now it is evident that by raising or lowering the cords, the corresponding warp-threads will be also raised or lowered at the crossing points, out of their plane. In plain weaving, two frames are used, in which these vertical cords are stretched, alternately in one or the other, and the mechanism is so arranged that the frames are alternately raised and lowered by the motive power, whatever it may be, successively elevating and depressing the alternate threads of the warp, and giving a space between them for the passage of the thread of the weft, which being wound upon a shuttle, and perfectly free in its action, is made to fly back and forth for each movement of the warp-threads, the motion being given to it as to a stone from a sling or a ball thrown from the hand. The weft-thread is in this way alternately passed or woven through the threads of the warp, a mechanical arrangement, called a batten, at every throw pressing it up tightly home, and producing a firm finished material. The general principles are the same, either in the simplest or most complicated machines. If it is desired to weave patterns, mechanical movements can be designed by which the warp-threads are not always raised and lowered alternately, but according to some regular law for the particular pattern. In the Jacquard loom this is done by means of perforated cards. Then for the weft-threads a number of threads of different colors can be used, each on its own shuttle, the receptacles holding and throwing these shuttles being so arranged mechanically as at the proper time, depending upon the figure or pattern woven, the particular shuttle desired shall be brought up in position and thrown back and forth until its turn comes to be replaced by another.

Various other modifications are made to suit certain kinds of work, such as the weaving of the different kinds of carpets, etc., and invention has a large field for operation.

We desire more particularly to call attention here to some excellent looms exhibited by MR. THOMAS WOOD, OF FAIRMOUNT MACHINE WORKS, PHILADELPHIA, who has made a specialty of the manufacture of Power Looms; Reeling, Spooling, Winding Machines, etc., and has attained a high reputation for his productions.

Three-Box Sliding Cam Loom: Thomas Wood.

Fig. 1 shows his "Three Box Sliding Cam Loom," which is a roller-loom, with an improved sliding cam twilling motion and a box motion, patented in 1873, and since improved. It has but few parts, is of the simplest construction, very strong, and so easy of management that one overseer can take charge of almost as many three- and four-box as of single-box looms, with but very little more work. The box-motion is supplied with a sliding clutch, so that in case of any interference with the operation of the shuttle-box, the motion will unship, and so remain until the box is relieved, when it passes back into gear, thus preventing breakage of parts and stoppage of the loom consequent on the jamming of the shuttle or the lodging of the picker in the box. The picking motion is very complete, the cams adjustable, and each part is independent in its own adjustment. Wrought-iron crank-shafts are used, with cast sleeves shrunk on them, very much increasing the diameter and gaining the benefit of two cast-iron surfaces wearing together without the sacrifice of the light wrought-iron shafts. These looms can be run at as high a rate of speed as yet attained by any other box-looms, and it is claimed with less expense in

Star Loom: Thomas Wood.

findings, power, breakage, and wear of parts. A four-shuttle loom, with these improvements operating all the boxes, has been working during the Exhibition on heavy cottonades and plaid flannels at a speed of one hundred and thirty-eight revolutions per minute.

Fig. 2. represents his "Star Loom" with "Outside Shed Motion," which weaves the goods face side up and is capable of being worked at a high rate of speed. It is furnished with the same pick and box motions as the sliding cam loom, and is especially adapted to the manufacture of Jeans, Cassimeres,

Alpacas, Delains, Cheviots, Plaid Dress Goods, Sheetings, Shirtings, Osnaburgs and all kinds of light and heavy twilled goods. Two of these looms have been in operation during the whole time of the Exhibition on a great variety of work, a single shuttle one at a speed of one hundred and seventy-six revolutions per minute, and a three shuttle one at one hundred and sixty-five per minute.

Fig. 3. gives one of Mr. Wood's "Patent Bobbin Winding Machines," intended to wind direct from the hank or skein to the shuttle bobbin. It is furnished with double anti-friction cones to form the bobbins, a loose pulley to each spindle, yarn regulators on the traverse bar, and adjustable levers and reels for carrying the hanks. By the use of these cones the friction on the yarn is reduced, thus preventing the burning or shading of even the most delicate colors while at the same time a perfect formation of bobbin is acquired. This machine obviates the labor and waste necessitated by spooling, and the floor space occupied by the spooling machine is gained. Each of these machines gained an award of a medal and diploma.

The Educational Department in the United States in its different branches has been exceedingly well represented at the Centennial Exhibition and has attracted great interest, especially from foreigners, so much having been done by the General Government, by the various States and by munificent gifts of individuals to provide for the education of the masses by furnishing great facilities at very low charges and often entirely free, as to merit special attention. In 1862 Congress granted to each State thirty thousand acres of public lands for every senator or representative to which it was entitled, the income of which was to be appropriated in perpetuity to educational institutions where the leading feature should be the prosecution of such branches of learning as are connected with agriculture and the mechanic arts, including in addition military tactics, and allowing also other general scientific and classical studies; the purpose being to "promote the liberal and practical education of the industrial classes in the several pursuits and professions in life." The grant allowed one-tenth of the appropriation to be used in the purchase of experimental farms, but no portion for buildings. The share of the State of New York was 990,000 acres, and in 1865–67 the Cornell University of Ithaca, having already a gift of five-hundred thousand dollars from Hon. Ezra Cornell as a foundation, was established by charter and the entire income of the land grant secured to it so long as it should use it effectively in aid of the objects intended by Congress. Additional donations of nearly one million dollars have subsequently been given to the institution, its library, collections, etc. have rapidly grown and it stands to-day as one of the representative colleges of the country. Practical and scholarly studies are given equal value, the aim being to furnish the best facilities possible for each, to encourage their combination and to allow a wide liberty of choice, affording the student a mental discipline of varied character, while at the same time bearing directly upon his selected profession for life. The engineering school is very large and the opportunities for practical work in the mechanical department most excellent. The University Machine Shop makes an exceedingly creditable exhibit in Machinery Hall, and we have the pleasure of presenting on page 165 engravings of two specimens of the work of its students made from designs of the Director, Mr. John E. Sweet. Fig. 1. shows two Standard Surface Plates of remarkably perfect construction and so truly surfaced that when one is placed upon the other it will float on the thin film of air held between them, or if this be crowded out, they will adhere so tightly together as to require considerable force to separate them. Fig. 2. shows a Measuring Machine designed for obtaining measurements from zero to twelve inches and reading to the one ten-thousandth part of an inch. It is more especially intended for use in the manufacture of standard gauges and is peculiar in the method of taking the measurements, in its range, and also in the fact that any imperfection in the measuring screw can be corrected. Its construction is sufficiently shown by the engraving, and it is believed to be the first if not the only one of the kind built in this country.

Messrs. Clark & Standfield of 6 Westminster Chambers, London, England, make an interesting exhibit in the British section of the Machinery Hall of working models of their Gridiron Stage Depositing Dock, for which they have been awarded a medal and diploma here, as well as at the Paris Maritime Exhibition of last year.

The manner in which this dock operates may be seen almost at a glance, and in giving a description of it, we think we cannot do better than avail ourselves of a paper read by Mr. Latimer Clark before the Institution of Naval Architects of England, on a large special dock of this form which his firm have recently constructed at their works on the Thames and are now erecting at Nicolaieff on the Black Sea for the accommodation of the circular and other ironclads of the Russian Government. From this paper we make copious extracts.

The Depository Dock differs from any other that is known, either constructed or designed, in that it not only raises vessels out of the water, but, when required, deposits them high and dry on fixed stages of open pile work, where they may be cleaned or repaired at leisure. In its usual form it is adapted to vessels of the ordinary type, but it can be readily altered to receive vessels of any kind, or dock circular ironclads of whatever size.

Fig. 1, on page 167, gives a general view of a dock of this kind in the act of moving a vessel on to the fixed stage on which there are already several vessels under repairs. In its general form the dock consists of a number of pontoons, either of square or circular section, which lie parallel to each other at fixed distances apart and range transversely to the length of the dock, each of these pontoons being permanently connected at one end to a longitudinal structure, which forms the main side of the dock, and projecting outwards from it, in the same way as the fingers of the hand,— the whole structure in plan resembling a comb. When the dock is lowered to receive a vessel, the pontoons are submerged, but the side to which they are

Patent Bobbin Winding Machine: Thomas Wood.

attached is never placed entirely under water, there being left a sufficient height to allow a freeboard of six or seven feet, when the pontoons are sunk beneath the bottom of the vessel. When the dock is raised, the tops of the pontoons are well above the water and the side stands several feet higher than the deck of the vessel which is being supported. Fig. 2, on page 169, gives a plan of the dock, A, A,

Standard Surface Plates: Cornell University

A, being the side and B, B, B, the pontoons, Fig. 3, on page 169, shows an elevation submerged, having an outrigger attached and ready to raise a vessel, and Fig. 4, on page 169, is the same, as raised and floating on the water with the vessel. The form of elevation without the outrigger resembles the letter L. It is evident that such a form as this, viz., a dock with only one side to it, would be entirely unstable when submerged, unless the necessary stability were imparted to it by some such means as is accomplished by the outrigger.

Measuring Machine: Cornell University.

This outrigger consists of a broad flat pontoon, divided into numerous compartments and loaded with concrete ballast until half submerged. Its form gives it immense stability and it carries along its middle line a row of rigid upright columns, which project through the pontoon some distance above and below, and are stiffened by struts; the top and bottom of each column being hinged to a pair of parallel bars or booms C, C, which are also hinged at their opposite ends to the sides of the dock, so that the outrigger remains stationary, while the dock is free to be raised and lowered vertically, the action of the parallel bars always retaining it in a horizontal position.

Each of the fingers or pontoons is usually divided into about six separate compartments, by means of five transverse vertical bulkheads, and the side of the dock to which the pontoons are attached is formed of a series of parallel, vertical columns, as shown, or is made as a long box girder, divided by numerous bulkheads into large water-tight chambers. Its height may vary from twenty to fifty feet or more, its width from ten to fifteen feet, and its length is made about equal to that of the longest vessel intended to be docked. The pontoons are made about twice the length of the beam of the largest vessel, so as to be available for paddle-steamers; their height is from ten to twenty feet, depending on the buoyancy required, and their width from seven to fifteen feet.

The machinery for working the dock is carried in the vertical columns or in the chambers of the side, as the case may be, and consists of a number of powerful pumps worked in the usual manner. When it is desired to submerge the dock, the necessary valves are opened, and water admitted to the compartments of the pontoons, thus gradually lowering it to the required depth, its horizontal position being at all times maintained by its connection with the outrigger. The vessel is then floated over the pontoons, the water pumped out until the keel takes its bearing on the blocks; the bilge-blocks are hauled into place by chains in the usual manner, and the vessel being firmly blocked and shored, the pumping is continued until the whole is raised to the full height required, and the valves are then closed. It will be seen that the dock in this position, with the ship on it, has very great stability, quite independent of the outrigger, which, having performed its duty of controlling the dock when submerged, is no longer of service, and may be entirely removed if desired, as seen in Fig. 5. Without it the dock is much narrower than any other form of dock, and can with great facility be floated through very narrow entrances or channels. The vessel can now be examined, repaired and painted as in any ordinary dock, or may be moved from place to place.

The great feature of this system consists in the fact that the vessel may be easily lowered on to a fixed staging along the shore, and deposited high and dry as seen in Fig. 6, leaving the dock free to raise or lower another

Gridiron Stage Depositing Dock: Clark & Standfield.

vessel, or any number of them, one after the other. This staging consists of a series of piles driven into the ground in rows parallel to each other, and at right angles to the shore, which are capped by horizontal timbers and placed exactly the same distance apart, centre to centre, as the pontoons. (See Figs. 2 and 6.) The height of the vessel above the water is greater than the height of the staging, so that when the dock with the vessel on it is brought alongside the staging, the pontoons can enter freely between the rows of piles, the vessel being carried directly over the staging without touching it. The dock is now slightly lowered by admitting water into the pontoons, until the vessel rests upon the keel-blocks on the fixed staging, when bilge-blocks are placed under the bilges, and the vessel securely shored, the dock being then lowered a little more to clear everything, and finally drawn out from the staging, free for use again to receive other vessels. The vessels may be removed again into the water by reversing the process.

In viewing the question of the stability of this dock during its operations, it is essential to consider it as consisting of two very distinct portions—the side and the pontoons—the latter taking no part whatever in the raising or supporting of the vessel, but simply giving stability during submersion. It must be clearly understood that the lifting power is applied solely and wholly in the pontoons directly beneath the vessel, and not in the side, the pontoons being provided with a certain number of hermetically-sealed chambers, into which the water cannot enter, and which are sufficient at all times to give them a surplus of buoyancy, so that they can only be submerged by forcing them down by the weight of the side, water being pumped into the latter to carry them down. Similarly, in raising the dock, the water is pumped only out of the pontoons, that in the side flowing out by gravity, and the lifting power comes from the pontoons alone, the dock having no tendency to turn one way or the other. Some power is necessary, however, to keep the dock horizontal in all stages of submersion, and this is furnished by the outrigger, which gives a stability of several hundred tons, although, with ordinary management, not more than a ton or two of this should ever be brought into action. The valve-engineer is provided with a spirit-level, and if the dock shows the slightest tendency to depart from the horizontal, he adjusts his valves or shuts off the pumps so as to make the proper correction. There are more than one hundred water compartments connected with the pumps by separate pipes, each pipe with its own valve, and all are brought to the valve-house and divided into four groups, corresponding with the four quarters of the dock, each of which is governed by a principal valve, thus enabling the action to be controlled throughout with the greatest ease.

In docking ordinary vessels it is not absolutely necessary to make use of

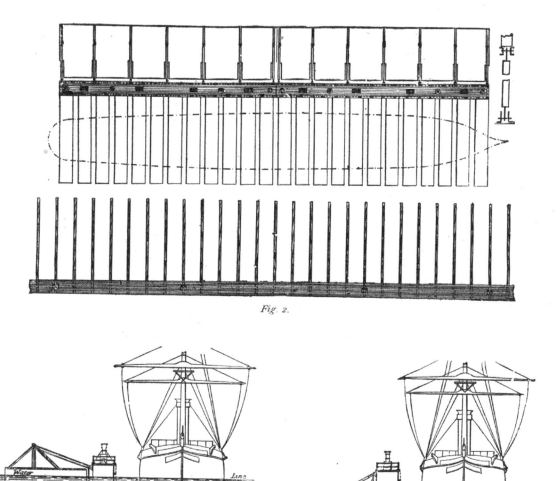

Fig. 2.

Fig. 3. Fig. 4.
Clark & Standfield's Gridiron Stage Depositing Dock.

cradles, although they may be employed with advantage in many cases. These cradles may either have side shoring frames, as in Fig. 5, or may consist only of a platform of longitudinal iron girders, stiffened transversely and provided with the usual keel- and bilge-blocks. In either case, the cradle and vessel are lifted bodily and deposited on the staging together, to be removed again in the same way. There must evidently be sufficient depth of water at the particular spot where the vessel is raised to allow the dock to be lowered to receive it over the blocking; but when docked, only a depth somewhat less than the height of the pontoons is required, and the whole may be floated into shallow water or wherever the staging is fixed, the latter being of indefinite length to suit the requirements of the port.

The Depositing Dock is usually designed in two equal parts, each of which is capable of readily docking the other for purposes of cleaning and repairs, or of being used as a smaller but complete dock. Fig. 2 shows these parts

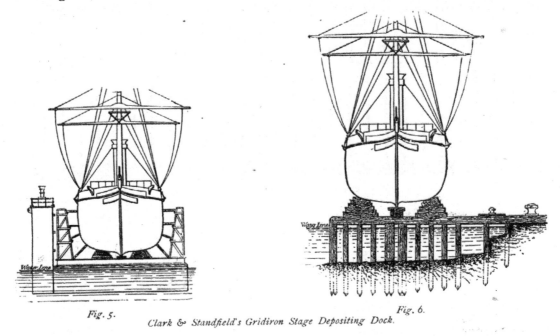

Fig. 5. Fig. 6.
Clark & Standfield's Gridiron Stage Depositing Dock.

connected to receive a vessel of ordinary form, and Fig. 7 represents them as arranged for a circular iron-clad; Fig. 8 showing the same with one side and its outrigger removed preparatory to depositing the circular vessel on the staging. The size of the dock may be increased with great facility whenever the growing trade of the port renders it necessary, by simply adding a third section in the centre, without any alteration of the existing portion, and at an expense not exceeding what would have been incurred if made the full size at first. This constitutes an important advantage over other forms of docks. The dock possesses great economy of first cost and working, requiring no fixed foundations, and being easily removed from place to place. It is very much quicker in its operations than an ordinary graving dock, as in the latter the whole space of the dock requires to be pumped dry, while in this only a weight of water equal to that of the vessel has to be raised, and that through a smaller height. The heaviest vessel can, if necessary, be raised in a little more than an hour. With this dock vessels can also be transported over shallows, and railway trains may be taken across ferries from one side of a river or lake to the other. The complete train is run on to a cradle resting on the

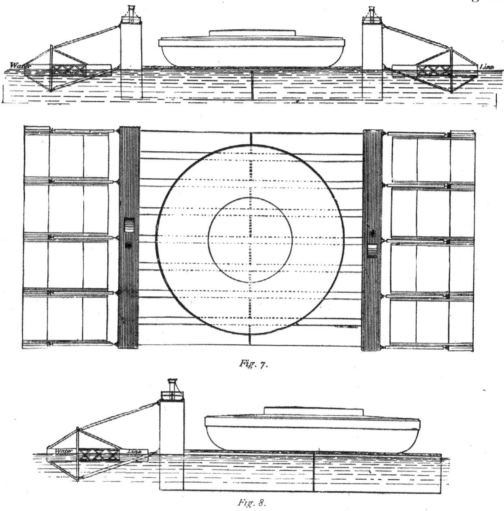

Fig. 7.

Fig. 8.

Clark & Standfield's Gridiron Stage Depositing Dock.

piles, and then lifted away by the dock, which is provided with false bows and fitted with propellers, and carries the whole across to a similar staging on the opposite shore.

The advantages claimed for this dock may be summed up as follows, and are thus stated by Mr. Clark:—

"1. With one dock any number of vessels can be docked and deposited high and dry out of water on wooden platforms, in a convenient position for cleaning and repairs, along the waste sloping shores of a river or wet dock.

"2. The provision for an additional length of staging at a comparatively nominal cost is equivalent to the building of an additional dock.

"3. As the dock is used ordinarily for lifting vessels on and off the stages, it can be kept at all times ready to receive disabled or other vessels, which can be at once deposited on a stage, and the dock left free for further use. In this respect it has a great advantage over all other descriptions of graving docks.

"4. A vessel can be placed upon the staging, cut in two, and lengthened by lifting one-half further along the staging by means of the dock.

"5. Vessels can be conveniently built on these stages on an even keel, and launched without the slightest strain, and without the risk and cost of launching, and without occupying the space required for the formation of ordinary ship-ways.

"6. Vessels, when on the dock or stages, are thoroughly exposed to the action of sun and wind, allowing paint to dry and harden rapidly, and affording great facility for examination and repairs.

"7. The dock, with or without a vessel, may be readily transported from place to place, for the purpose of raising or depositing vessels at different points.

"8. The dock will not, under any circumstances, sink, even if all its valves be intentionally left open.

"9. One-half of the dock can be readily raised level upon the other half for the purpose of cleaning or repairs.

"10. By the use of air, which may be stored in some of the chambers under compression, a vessel may be raised, sighted and lowered again in less than an hour.

"11. These docks, if constructed in the first instance too small for the requirements of trade, can be at any time enlarged to any extent at the same rate per ton as the original cost.

"12. The docks are capable of receiving vessels of any size or length, or of a width too great to pass through ordinary dock-gates—such, for example, as circular iron-clads of one hundred or one hundred and fifty feet in diameter.

"The durability of vessels laid up high and dry on the stages would be immensely increased, and at a few hours' notice a whole fleet might be lowered into the water."

Before the Nicolaieff dock was built, the largest dock-entrances in existence did not exceed one hundred feet, and it is believed that no entrance so wide as one hundred and twenty feet had hitherto been proposed. Circular iron-clads were designed in the Russian navy, however, up to one hundred and sixty feet in diameter, and even larger vessels talked of, and the want of some method of docking such vessels was one of the most serious difficulties to be overcome in connection with the introduction of these circular war-vessels.

Messrs. Clark & Standfield's plan, it is seen, met the point, and the Nicolaieff dock was designed under the supervision of Admiral Popoff, who made numerous modifications to meet the special requirements of his Government, the construction being commenced in January, 1876. It agrees with the general description we have given. Its side is two hundred and eighty feet long, forty-four feet six inches high, and twelve feet broad; the pontoons are seventy-two feet long by eighteen feet deep, and fifteen feet broad, and the clear space between them is five feet. For special reasons it has been made so as to divide into three lengths—two of one hundred feet, and the centre one of eighty feet—each furnished with its outrigger and complete in itself, having separate engine-rooms, engines, pumps, and valves. The pontoons, instead of being permanently fixed to the side of the dock, are made easily removable and mutually interchangeable. They can also be connected end to end, so as to increase the width of the dock to any extent. No divers are required in making these connections. Each pontoon is divided into six compartments by upright bulkheads, and the four central compartments are again divided by horizontal bulkheads into upper and lower chambers, the upper chambers being hermetically closed, so that when all the other compartments are filled with water, the pontoon alone will float perfectly level, with its deck a little above the water surface. The Nicolaieff dock will raise a vessel of four thousand tons in a little more than an hour.

It is hardly necessary for the general reader, to enter further into the details of construction and working of one of these docks, or to describe the excellent arrangements for controlling the buoyancy of the pontoons or fingers to any extent that may be desired. We would refer, for fuller description, to the "Transactions of the Institution of Naval Architects, London, 1876."

Messrs. Clark & Standfield have also in the Exhibition a working-model of a Floating Dock of the ordinary form, but without closed ends, which they call the Tubular Dock. Its great peculiarity consists in its being constructed entirely of cylindrical tubes, making it unsinkable, very strong and capable of standing rough weather, while at the same time very light and easily transported from port to port. It can be furnished with pontoons, on which vessels may be lifted within the dock and then floated away, quite clear of the water, to any convenient place for repairs, etc. Pontoons of different sizes may also be added at any time, each pontoon being of course equal in working efficiency to an additional dock.

The American Bridge Company, of Chicago, Illinois, makes a handsome exhibit in Machinery Hall, the most prominent feature of which is an exceedingly complete and beautiful model of the new Point Bridge, now being erected by this Company across the mouth of the Monongahela River at Pittsburgh, Pa., from designs prepared by Mr. Edward Hemberle, the Company's engineer.

The bridge is constructed on a stiffened chain suspension principle, the design of which is believed to be novel. It very much resembles at first sight that adopted by Mr. T. Claxton Fidler, of Great Britain, for suspension-bridges, but on closer examination will be found to differ from it quite essentially. Various plans have been suggested to stiffen a suspension chain independent of the roadway trusses, and some methods have been put into practical use. These systems all consist of two suspended chains or members with stiffening bracing between them. In one case the chains are hung parallel to each other; in another, the lower chain is hung to a catenary line, the upper chain meeting it at the towers and at the centre of suspension, and the parts between being bulged out on a contrary curve, so as to form two lozenge-shaped figures. The latter is Captain James B. Eads' St. Louis plan reversed. In yet another case, Mr. Fidler's plan, the upper and lower cables meet at the towers and at the centre of suspension, these points being in the catenary line, while from the centre of suspension either way to the towers the upper cable is straight, and the lower cable is hung so as to fall below the catenary line, its maximum distance below being the same as the upper cable is above the catenary. In

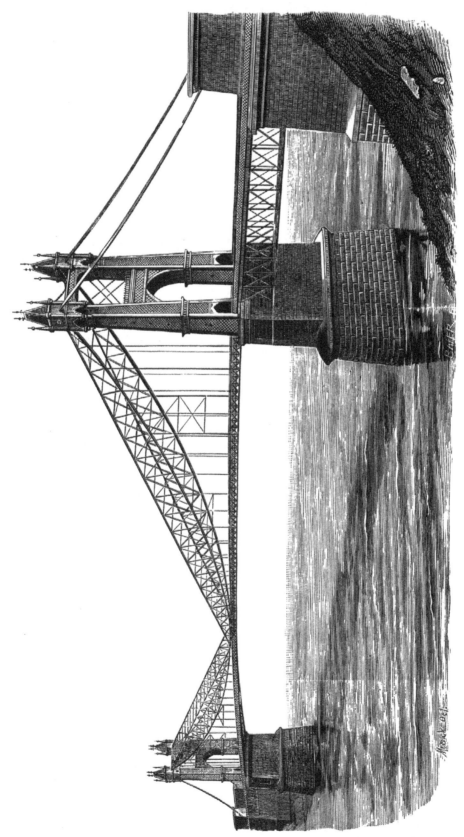

Fig. 1.—Point Bridge, Pittsburgh: American Bridge Co.

Mr. Hemberle's plan, as adopted at the Point Bridge, the lower cable is hung in the catenary curve, the same as in Captain Eads' plan, while the upper cable forms two straight lines as in Mr. Fidler's arrangement.

The system of two parallel chains possesses many disadvantages, and theoretical investigations would suggest Mr. Fidler's plan as the most advantageous and economical, the catenary line passing through the centre of the truss, and the top and bottom cables or chords remaining always in tension, under all conditions of loading, and assisting each other in carrying the full load. In Captain Eads' and Mr. Hemberle's systems, the lower cable forming the catenary line carries entirely the equally distributed load, the upper members being only of service in resisting the tendency to change of form in the curve of the cable from the action of the moving load, and being subjected to strains varying from tension to compression, depending on the location of this load. From investigations made by the AMERICAN BRIDGE COMPANY in reference to these three systems as applied to the Point Bridge, taking a permanent weight of about three thousand pounds and a moving load of sixteen hundred pounds per lineal foot, and an equal factor of safety for both loads, it is claimed that while Mr. Fidler's plan saves material in the chord members, both being subjected to tension only, and assisting each other in carrying the uniform load, it requires long bracing members; and as neither upper or lower cables are self-supporting in their proper curves, the erection becomes difficult under circumstances where false works cannot be readily employed; also that while in Captain Eads' plan the lower member, designed as a catenary chain, can be erected from cables and is self-supporting in its proper position, thus giving facilities for building the balance of the structure, still the top members do not add to the strength for equally distributed load, and are subjected to compressive as well as tensive strains, having also unpracticable shape and dimensions, and necessitating long bracing members. In Mr. Hemberle's plan, however, although the upper members do not add to the strength for equally distributed loads, and have to resist both compression and tension under various conditions of load, still it possesses the same advantages for erection as in Captain Eads' plan, and at the same time affords a more practicable shape and sectional area in the upper members for good cheap workmanship and for the purpose of insuring stiffness, giving bracing between them and the lower member of only about one-half the length required in the other systems. These are the reasons given for the adoption of this plan for the Point Bridge by this Company. It should be remembered, in this connection, that the adopted plan gives only one-half the effective depth for stiffness that is obtained in the other systems. It may be found, however, to work out better practically even with this disadvantage.

The structure consists of three spans, the main span being eight hundred feet from centre to centre of piers, with a clear elevation of eighty feet above low water at the mid channel of the river, and the two side spans one hundred and forty-five feet each, placed on a grade of three and four-tenths feet per hundred, and constructed as iron trusses entirely independent of any connection with the back chains from the cables supporting the channel span. This arrangement gave the advantage of allowing the back chains to be shortened, and thereby reducing the total length of bridge, and the side spans proved of great service as false works for the erection of the back chains, allowing free waterway below for floating ice and barges during high water. The height of the towers above low water is one hundred and eighty feet, and the deflection of the main cable eighty-eight feet. The roadway is twenty feet wide, with double tramways and one track for a narrow-gauge railway, having, in addition, two outside footwalks of six feet each. The piers for the towers and the abutments and anchorages are founded upon timber platforms sunk to a gravel bed. The towers are constructed entirely of wrought iron, except the bases of the columns, and are built up of eight columns each, four columns braced together for supporting each chain, each column being two feet square, and having a sectional area of sixty-four square inches at the base, and fifty-four square inches at the top. The caps are formed of two box-girders five feet in depth, each resting on two columns, and having on top, across from one girder to the other, five box girders seventeen inches deep, on which steel plates are placed to form a bed for eighteen steel rollers, four and three-eighths inches in diameter and fifty-one inches long. On top of the rollers rest other steel plates, and on these are placed the saddles, consisting of twelve wrought-iron plates twenty-six inches wide and two inches thick, the connection with chain-links being made with pins six inches in diameter.

The chain-links are twenty feet six inches long, centre to centre of pins, the latter being six inches in diameter, and the links are two by eight inches in section, except where they bring only a single shear on the pins, when they are made one by eight inches. The width of heads is sixteen inches. The back chains have each twelve links of two by eight inches section, or eleven of two by eight inches, and two of one by eight inches, the sets of links thus alternating from twelve to thirteen in a set. Each back chain has therefore a sectional area of one hundred and ninety-two square inches. The anchor-plates are of cast iron, eight by ten feet area, and the connection is made with the back chain by a six-inch pin. The main chains are composed of fourteen and eleven links in a set alternately, the sectional area decreasing towards the centre. The top chords of the stiffening-truss are composed of channels and plates, forming a rectangular section twenty-two inches wide by thirteen inches deep, with full tension splices. The top and centre joint connections are made by forged bars twelve inches wide, with a head on one end, and the other end riveted to the chord. The last section of the top chord

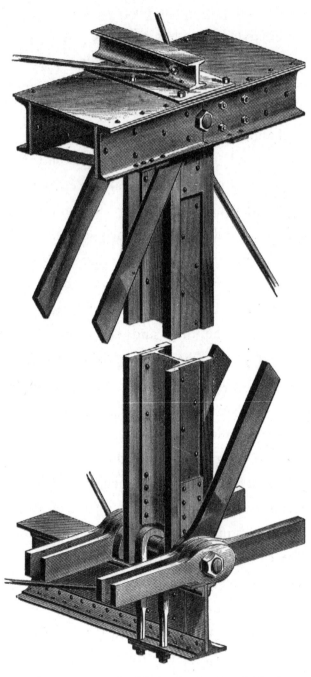

Fig. 2.—*Point Bridge, Pittsburgh: American Bridge Co.*

is not put in until the whole bridge is finished, so as to insure correct length, and the condition that the stiffening-trusses have no strains from dead weight. The posts of the trusses are made of **I** beams and plates, and all tie-rods are adjustable. Our engraving, Fig. 1, shows the general character of the structure and the system of trussing adopted between the upper and lower cables, the details of which are given in Fig. 2. Lateral and diagonal bracing is provided between the top chords and also between the chains, proportioned so as to resist the effects of wind-pressure.

The roadway girders are eight feet high, and form a hand-railing. They

Fig. 1.—Pony Planer: Goodell & Waters.

have expansion-joints every hundred feet, and are suspended from the chains by flat bars at distances of twenty feet apart, struts being used at the expansion-joints to form rigid connections at these places. Cross-girders, three feet in depth, connect the stiffening-girders at intervals of twenty feet, and support two lines of iron stringers, wooden joists being placed on this iron system to carry the flooring. A double system of tie-rods secures lateral stiffness to the floor, and the wind-pressure is taken up by horizontal steel-wire cables underneath and secured to the floor.

The bridge is designed for a moving load of fifty pounds per square foot of flooring for the trusses, towers and chains, and for seventy pounds per square foot for the floor system and suspenders. The maximum stresses allowed are six tons per square inch in the main chains, four and one-half tons in the towers, five tons in the suspenders, and six tons for tension and five tons for compression in the upper chords. In erecting the structure, the main chains were raised from false cables, the links being brought into position by a traveling platform, and one chain was raised at a time. The cables were then shifted to the other side, and the second chain, weighing about three hundred and twenty tons, was completed in thirty working-hours. The balance of the bridge is being raised from the chains, the roadway is completed, and it is expected to have it ready for traffic by March 1, 1877. The total cost will amount to five hundred thousand dollars.

Wood planing machines may be divided into three classes—carriage-planing, parallel-planing and surface-planing machines. In the first class the lumber to be planed is fixed to carriages, which move on guides or rails entirely independent of the lumber itself, either as to its surface or its shape, and the cutting operation is usually performed by means of traversing heads, which revolve in an axis perpendicular to the surface being dressed, although this is not necessarily the method, as cross cylinders, running on an axis parallel to the surface planed, are also used. In the second class the lumber passes through feeding-rollers, which move it continuously between stationary guides and cutters arranged on a cross cylinder revolving in an axis parallel to the surface planed. It is evident that there may be one or more of these cylinders, making one or more cuts of the lumber at one operation if desired, they being set to different gauges. The operation of the machine reduces the lumber to a uniform thickness, or to thickness and width at the same time if required. In the third class a constant gauged amount of material is cut away from the surface of the lumber every time it is passed through the machine, without reference to a reduction to uniform thickness, the principle of action being the same as in the ordinary carpenters' plane, either revolving or stationary cutters being used in a fixed bed, and gauged above the surface of the bed by the amount of the cut.

Machines of the first class are the only ones that can be depended upon to produce material out of wind, and to always plane perfectly true. They require, however, a large amount of space, equal to double the length of the material to be worked, and are very slow in their action, the latter defect however being doubtless capable of much improvement by the use of revolving cross cylinders and improved methods of action. The second class is by far the most important of the three, and comprises at least ninety per cent. of all the machines used for planing in this country, including within its province the extensive list of moulding-machines. For planing and matching of boards, plank and all material that can be bent or sprung into a straight surface, the machines of this class have no equal either in speed or efficiency. Lumber may be dressed by them on one, two, three or four sides all at one operation. Machines of the third class

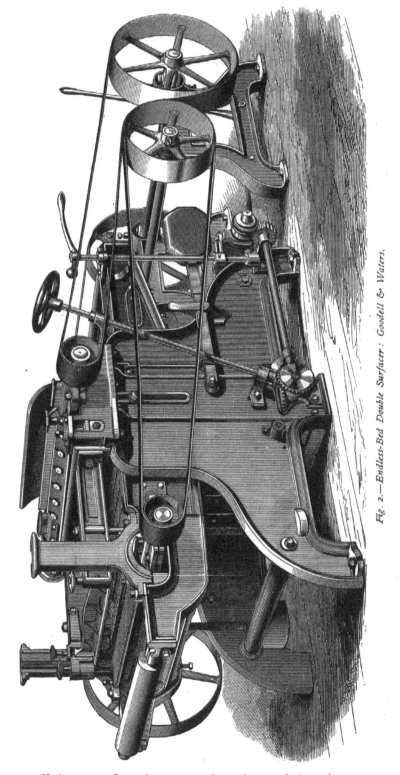

Fig. 2.—Endless-Bed Double Surfacer: Goodell & Waters.

might more properly be designated as smoothing-machines, and are of very limited application, although very useful in the special work for which they are adapted. It will be noticed that when machines of the second class plane on four sides at one time, the operation of the machine for two of these sides must be after the system of those of the third class.

MESSRS. GOODELL & WATERS, OF PHILADELPHIA, make an extensive exhibit of wood-working machinery, and we desire to call attention to some of their planing machines belonging to the class of parallel-planers. Fig. 1 illustrates their IMPROVED PONY PLANER, a machine designed especially for the use of carpenters, box-makers and those doing a general jobbing work, although it is a machine of great value for small work even in a regular planing-mill. It is readily adjustable to various thicknesses of lumber, is light running, starts up and stops quickly, and does the work for which it is intended better than is possible with a large and fast-feeding machine, and generally in about the same length of time. It is a strong, well-built machine, taking up but little room, has a capacity for material up to twenty-four inches in width, and from one-eighth to four inches in thickness, and may be run at a high rate of speed, feeding from thirty to thirty-five running-feet per minute. The cylinder is of steel, five inches in diameter, and carries two knives. It is moved at a velocity of from four thousand to four thousand five hundred revolutions per minute. Special machines with some slight modifications are made for cigar-box work and also for carriage-makers' use. These last machines will plane smoothly down to one-sixteenth inch in thickness without tearing up or clipping the ends of the lumber, and will work equally well on hard or soft wood. The counter-shaft used with these machines has tight and loose pulleys ten inches in diameter and six inches face. The driving-pulley for cylinder is twenty inches in diameter, with four inches face, and the counter-shaft should make from one thousand to eleven hundred and twenty-five revolutions per minute. These machines weigh from thirteen to fourteen hundred pounds.

Fig. 2 shows this firm's heavy ENDLESS-BED DOUBLE SURFACING MACHINE, which has recently been constructed from entirely new designs for the express purpose of correcting, if possible, every known defect in this class of machines. Endless-bed planers, when properly constructed, have been admitted for a long time to be the strongest and fastest feeding-machines known, and, taking the first cost into consideration, also the most economical for the user. The cylinders of this machine are both made entirely of steel, and have long heavy bearings running in the best Babbit metal boxes. The upper cylinder pressure-bar and pressure feed-rolls are mounted in one heavy casting firmly bolted to the sides of the machine, and the boxes of the lower cylinder are also tied together by means of a heavy yoke, so that the cylinder may be raised or lowered without the possibility of twisting or binding the boxes. The raising or lowering of the bed is accomplished either by hand or power, and the power arrangement is very simple and easily operated, consisting of a worm and gear, the worm, which is right and left, being fixed on the end of the feed-shaft and arranged to work endwise

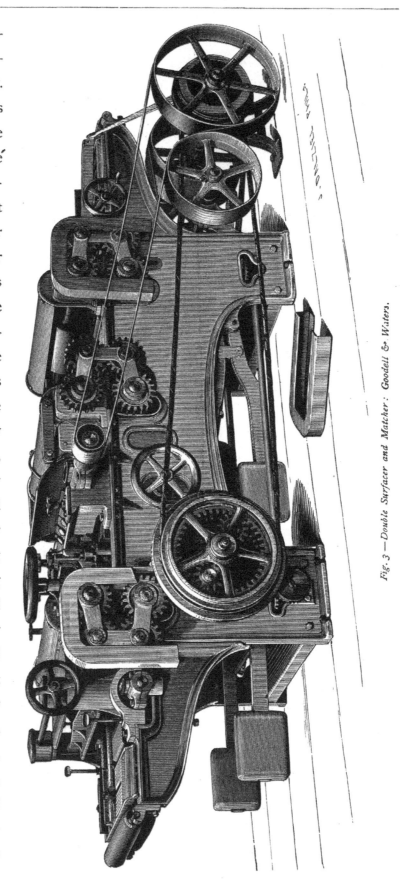

Fig. 3—Double Surfacer and Matcher: Goodell & Waters.

on a spline, so that the operator may throw the right- or left-hand worm into gear at will, moving the bed up or down as desired. The lever for power-movement and the hand-wheel for hand-movement are both conveniently arranged above the bed of the machine and near the scale for indicating the thickness to which the lumber works. The plats of the traveling-bed are very heavy, and are connected by means of a butt-hinge joint, the centre of which is under the centre or opening between the plats. This carries the plats away from the lumber as the bed turns down over the carrying-wheel at the end, and does not allow them to raise or even press against the under side of the lumber, or give any interference with the work of the lower cylinder. When this style of connection is used, only one-half the number of joints occur in the whole bed as would obtain with the use of wrought-iron links. The bed travels on four self-oiling ways—two at the extreme ends of the plats, and two in the middle, the combined width of these amounting to twelve inches. There is no liability of getting out of order or of wearing. The lower cylinder can be reached, if the cutters require sharpening or adjustment, by simply lifting out the front roll, one piece of the adjustable bed, and turning back the pressure-bar, the whole taking less than half a minute to accomplish, and rendering the cylinder as convenient of access as it would be on a bench. The weight of this machine is five thousand pounds.

Fig. 3 represents MESSRS. GOODELL & WATERS' DOUBLE SURFACER AND MATCHER, which is constructed by them in the usual variety of forms, to work two, three or four sides, either as a regular flooring-machine operating to a width of fourteen inches, or as a combined Surfacer and Matcher, with a surfacing capacity of twenty-six inches in width and four inches thick, according to the requirements of the customer.

The machine as illustrated is of an entirely new design, combining some important improvements over the old type. It is adjusted to different thicknesses of material by raising or lowering the bed by means of four heavy steel screws connected to and operated by the horizontal wheel shown near the last set of feed-rolls. It is claimed that several advantages are gained by this method of adjustment:—

1st. The time required to change the machine from one thickness to another is so much shorter than by any other arrangement as hardly to admit of comparison.

2d. The boxes of the upper cylinder can never get out of line, as they are both made in one heavy casting, firmly bolted to the sides of the machine, and forming a part of the permanent frame.

3d. The manner in which the feed-rolls can be weighted is very simple and effectual, requiring but a single rod of iron to make the connection between the weight-lever and the feed-roll boxes.

4th. The details of the machine are very much simplified; a less number of pieces are required in its construction, and there are no nuts or bolts to loosen in raising or lowering the bed, resulting in much less liability of derangement of parts; the height of the machine is less, making it much more solid and its members more easy of access; and finally, there is nothing on top of the machine which can become clogged with shavings or dust.

The manner of driving the feed-works merits special attention, being simple and very strong, enabling three grades of feed to be obtained in a space of only three and a half inches in width, and allowing a feed-belt of three-quarters of an inch in diameter, round, twisted or square, this size running very slack on three-cone pulleys, and giving ample power to feed the heaviest lumber. This great power is obtained by the use of planetary gearing shown inside of the feed-pulley, which increases the force four times, and gives to one small belt an effectiveness more than equal to a five-inch belt on the ordinary machine. This feed is stopped and started by means of an improved friction-clutch, which operates quickly without noise or jar, works freely, and does not readily get out of order. There are six seven-inch feed-rolls, the lower ones driven by a train of twenty-inch gearing, two and a half inches wide on the face, with a heavy and improved-shaped tooth, and the upper ones connected to the lower by heavy expansion gearing, making a very powerful feed. The cylinder- and matcher-heads are all made of steel, and run in self-oiling boxes. The upper cylinder has three knives, and is six inches in diameter, while the lower has two knives, and is five inches in diameter, both being fitted with beading-cutters which can be used or not as desired. The matcher-heads are six inches in diameter, with three knives, and the matcher-spindles are arranged to swing down below the surface of the bed, running in heavy hangers or castings, and supported by a three-inch screw. They will swing down out of the way by simply removing one bolt, and will be found very convenient on the combined machine. The machine is fitted with a good arrangement for holding the lumber to the guides, and is provided with all the pressure-bars, chip-breakers and other accessories necessary to make it complete in every respect. The counter-shaft has tight and loose pulleys twelve inches in diameter, eight inches in face, and should make eight hundred and fifty revolutions per minute. The weight of the machine is from six thousand to six thousand five hundred pounds.

In the various operations of making and printing cotton and worsted goods, it is very necessary to dry the material rapidly and evenly, and at the same time stretch or smooth it as in a process of ironing. For this purpose drying-machines are made in which steam is employed to heat rollers, around which the material to be dried is passed, and at the same time kept in a state of uniform tension.

Messrs. H. W. Butterworth & Sons, of Philadelphia, have attained a high reputation for the manufacture of Drying-Machines, and make an extensive exhibit in Machinery Hall of these and kindred appliances. The figure on page 187 represents one of their Drying-Machines such as are used for print-works, bleacheries and for drying cotton-warps and finishing tickings, osnaburgs, etc. This machine is arranged with twenty-four cylinders, supported by a framing, eighteen of them being on a horizontal and six on a vertical frame. The grouping of these cylinders in a horizontal, vertical or other direction may be modified to suit special requirements; and where the floor-space is contracted, the vertical arrangement is preferred. The frames of the machine are made of cast iron, being quite heavy in their construction, with broad planed surfaces, and hollow passages are cast in them for the transmission of the steam used in heating the cylinders and the return of the condensation, thus dispensing with outside pipes and connections. The steam passes into each cylinder and leaves it again by means of branch passages cast on to the frames and connecting with journals in which the axes of the cylinders run. The stuffing-boxes for the journals are packed from the front by an arrangement introduced by Messrs. Butterworth & Sons in 1867, this packing, however, forming no part of the bearing. The advantages derived from this method consist—first, in the easy access given to the packing, which also lasts longer than in the ordinary arrangement; second, in an allowance of greater freedom for expansion of the cylinders than can be attained in any other way; and third, in furnishing an abundant length of bearing for the axles. This firm formerly packed the stuffing-boxes on the inner side, but this rendered them much more

Drying Machine: H. W. Butterworth & Sons, Philadelphia.

difficult of access, and at the same time there was a greater tendency for them to blow out with the steam pressure. The length of bearing also obtainable for the axles was much less. In drying-machines as usually constructed, the practice has been to introduce the steam to the cylinder by means of a steam-pipe connecting from the exterior through the end of the journal by a countersunk joint. This arrangement did not allow of free expansion and contraction of the cylinder, and caused the end of the journal to press against the end of the steam-pipe with more or less force, depending on the temperature to which it was raised, producing consequently more or less friction.

The cylinders of these machines are made either of copper or of planished tinned iron, more generally of the latter material, as it is the least expensive. Properly speaking, the material used for these cylinders is not "tinned," but is iron coated with a composition invented by Messrs. Butterworth & Sons, and especially adapted to resist the action of acids such as are used in print-works. Messrs. Butterworth & Sons are the only manufacturers of this material in this country. Motion is communicated from one cylinder to another by cast-iron gearing, seen very distinctly in the engraving. The cylinders are carefully made, but no special balancing is required such as is necessary in drying-machines for paper-making, the material to be dried in the present case being of much stronger texture.

In machines with wide cylinders, where more than one width of material is dried at the same time, the steam is so applied that each width is dried uniformly. A uniformity of temperature is maintained throughout the machine by allowing the steam to enter the top cylinder at one end, and the corresponding bottom cylinder at the other. The working pressure of the steam is usually from five to ten pounds per square inch, and it is controlled by a regulator manufactured by Messrs. Lock Brothers, of Salem, Massachusetts. The water of condensation is removed from the opposite end of the cylinder to that at which the steam enters, by means of Collins's Patent Trough, a device very extensively used in England, and quite effective in its operation, causing the water to pass out through the journal in a similar way to that by which the steam enters at the other end. The material to be dried, before entering around the cylinders, passes first through a "stretcher," made of brass, which prevents the edges from turning down, and smoothes out all wrinkles, delivering it perfectly even and regular. The tension of the fabric is controlled by passing it between three rectangular bars, alternating above and below them, one after the other, and around a roller; or in another way by means of a strap and weight attached to the roller, from which it moves on to the dryer.

Messrs. Butterworth & Sons have made nearly five hundred of these machines, which command almost an exclusive use in the print-works, bleacheries, and cotton and worsted manufactories in this country, and have also been sent to South America and to Italy. Their display at the Exhibition also includes a Dyeing-Machine for cotton-warps with three compartments, a Friction Calendar with three rolls, and a Drying-Machine with three cylinders, each sixty inches in diameter by thirty-eight inches long, which is designed to run in connection with a calico printing machine. The exhibit as a whole is exceedingly creditable not only to the firm, but to the manufacturing industries of Philadelphia.

Mr. L. P. Juvet, of Glen's Falls, New York, makes an exhibit of a Time Globe, which has attracted a large amount of attention from its novelty, not only as a curious piece of mechanism, but also as possessing great simplicity of construction and usefulness. The instrument, represented by the accompanying engraving, consists of a terrestrial globe mounted in a vertical circle and provided with adjustment, so that its axis may be placed at the proper inclination and in the proper direction to give it exactly the same position as that of our earth. By means of chronometer mechanism within its interior, the globe is made to revolve once in twenty-four hours, and a fixed hour-circle, or zone-dial, surrounding it at the equator, gives the time of the various meridians or localities on its surface, while the mean time of the place where the instrument is in use is given in the usual way by a dial, with minute- and hour-hand placed, as shown in the engraving, at the north pole. To set the apparatus in operation for any particular locality, the hands of the clock-dial are moved until they accord with the time indicated by the longitude of that place on the zone-dial at the equator, care being taken that the globe occupies its proper sidereal position by the compass, and also gives by the zone-dial the actual time of day or night, as the case may be, for this locality. By means of a sliding vernier on the vertical circle, graduated to degrees, the latitude of any locality may be ascertained, and the proper inclination of axis may be given to the globe for the earth or for any other heavenly body. If the globe be set to the required inclination of axis for the earth, and is then placed in the light of the sun, with its axis parallel to the earth's axis—that is, the poles pointing north and south—then the light falling upon it will give not only the amount of light or darkness in each country, but will give it at the identical time in which it occurs. The instrument therefore notes the time, longitude and latitude of any place in the world, as well as the difference of the same between two or more places, and may well be called a universal time-keeper. The globe may be placed in any position without injury to the works, which are of the same construction as in an ordinary watch, and run eight days, winding up by a stem-winder connecting with a thumb-piece at the south pole of the axis. The globe is mounted on a handsome stand, which contains an aneroid barometer and a thermometer, making the instrument complete for all observations connected with time or weather, revealing at a glance the actual movements of nature at any time an observation is made. This instrument appears destined to be an exceedingly useful piece of school apparatus, bringing down to the comprehension of a child the most complicated movements of the earth, the alternations of day and night, and the connection of these with the divisions of time, and providing a new

Time Globe: L. P. Juvet, Glen's Falls, N. Y.

and valuable aid to those who teach by the object system. Mr. Juvet has received a medal and a diploma of the highest merit for his exhibit.

MESSRS. DAVEY, PAXMAN & CO., OF COLCHESTER, ESSEX, ENGLAND, make an exhibit in Machinery Hall of a PORTABLE ENGINE, which deserves particular notice from its construction, including several special features, and from the good results that it has given in trials abroad. The engine, as represented by the engraving on page 182, has a stroke of one foot, with a cylinder eight and five-eighths inches in diameter, the interior dimensions of the fire-box being one foot ten and one-half inches in length, two feet six and three-quarter inches in width, and two feet six and one-half inches in height above the fire-bars. The boiler is tubular, having thirty-nine tubes six feet five inches long, each, and two and one-quarter inches outside diameter. The total heating-surface amounts to one hundred and eighty-two and five-tenths square feet, of which one hundred and forty-six square feet come from the tubes, and the balance from the fire-box, the heating-surface of the latter being very materially increased by the introduction of ten special tubes, called Davey-Paxman tubes, three of these springing from each side-plate, two from the tube-plate, and two, shorter than the others, from the back-plate. These tubes curve upward immediately after leaving the plate to which they are joined, the whole tube, except the curved portion, being vertical, and they enter the water-space of the boiler again through the top plate of the fire-box. They are provided at the upper end with deflectors to prevent the water which rushes up through them from being projected into the steam space, and thus causing priming. These tubes furnish an exceedingly effective heating-surface, and the deflectors at their upper ends act as intended with great efficiency, as has been proved by working one of the boilers with the man-hole cover off and with the deflectors removed from some of the tubes, thoroughly convincing those observing the experiment of the important duty which the deflectors perform. The application of these special tubes to a locomotive-boiler, as in the present case, not only acts by giving additional heating-surface, but also increases the efficiency of the surface previously existing by improving the circulation in the water-spaces around the fire-box. Salt water has been used in one of these boilers without the least difficulty being experienced, the tubes remaining perfectly clean of deposit after some months of use.

With the exception of these special tubes, the boiler does not differ from the usual forms, but in the engine proper a peculiar arrangement of cut-off is provided, the suppression of the steam being effected by means of an independent expansion-valve actuated by cam gear. The cylinder with its covers is thoroughly steam-jacketed, and an ordinary slide-valve works in a very small valve-chest, another chamber being formed on the cover of this chest, in which the expansion-valve operates. This expansion-valve acts by rotating about an axis, having four ports formed through it, and according to its position it covers or uncovers five ports in the division between it and the main valve-chest, each three inches long by three-sixteenths of an inch wide, the movement of the

Portable Engine: Davey, Paxman & Co., Colchester, England

expansion-valve required for this purpose being only about one-quarter of an inch. The expansion-valve spindle at the end next the crank-shaft is encircled by a light cylindrical casing supported in a fixed position by a bracket attached to the boiler. A helical spring within this casing bears against an enlargement of the valve-spindle, and tends to force the spindle towards the crank-shaft. A sleeve, fitted with two cam-leaves of case-hardened wrought iron, slides freely up and down on that portion of the crank-shaft opposite the expansion-valve spindle, the leaves of the sleeve being broader at one end than the other.

These leaves, acting on a small friction-wheel or bowl on the end of the expansion-valve spindle, which is pressed against them by the helical spring, open that valve for a longer or shorter period at each half revolution, according as their broad or narrow ends are brought into use. The sliding-sleeve on the crank-shaft is connected with the governor, which controls its position, and

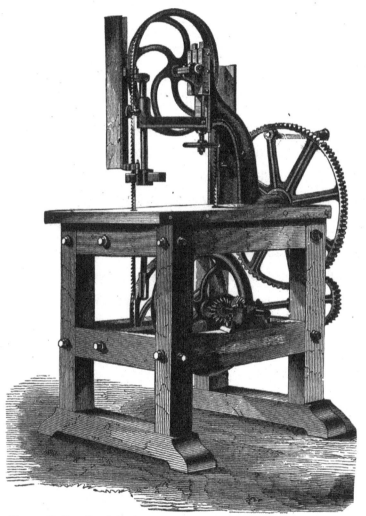

Fig. 1.—Endless Band Sawing Machine for Hand-power: F. Arbey, Paris.

therefore also controls the period of admission given to the steam by the expansion-valve. As the balls of the governor rise, the sleeve is shifted so as to bring the narrower end of the cam-leaves into action, and if the balls lower, the broader end comes into play. An India-rubber pad or ring is placed between a fixed collar on the expansion-valve spindle and a disk abutting on the spring casing containing the helical spring, so as to take the recoil of the valve-spindle when released by the cam-leaves.

The feed-water is heated as follows: It is first raised to a temperature of about ninety or one hundred degrees by leading into the feed-tube a portion of the exhaust steam, and when thus partially heated it is taken by the pump and forced through a heater placed on the top of the boiler, which is traversed by the exhaust-pipe, and from thence it is led through an annular heater situated in the upper part of the smoke-box under the base of the chimney, and finally into the boiler through a check-valve. During a trial at Cardiff, in England, some time ago, the water was by this arrangement heated to over two hundred degrees, the boiler evaporating from this temperature ten pounds of water per pound of coal, at the rate of two and eighteen-one hundredths pounds of water per square foot of total heating-surface per hour.

In that section of Machinery Hall devoted to France an extensive display of wood-working machinery is made by FD. ARBEY, OF PARIS (represented in this country by MR. EUGÈNE L. TOURET, OF NEW YORK), who has attained a high reputation, particularly in Europe, in this department of manufacture. His works, situated at 41 Cours de Vincennes, near la Place du Trône, are very extensive, and here are constructed all varieties of machines required in the fabrication of timber-work, such as sawing-machines, planers, mortising- and boring-machines, veneer-cutting and moulding-machines, wood-turning lathes, etc., etc. The different kinds of Fixed and Portable Log-Sawing Machines made by M. ARBEY are on the most approved models, operating with reciprocal-vertical, band- and circular-saws. They are used for squaring off logs and for cutting them up into scantling or plank, admitting logs up to thirty-nine inches in diameter. Those operating with a band-saw are best adapted for all kinds of curved work, and although applicable to straight work, are not so preferable as the reciprocal-vertical saws, requiring more skilled attendance. The latter machines are fitted with a large number of saws if required, all working at the same time, and cutting a whole log into plank or boards at one operation, as usual at many of the mills in this country. Special bogies are furnished for cutting on a curve, a very useful arrangement for many purposes—for instance, in the case of arches for wooden bridges, which are now very often cut directly from the log to the required curve. The machines are fitted with either wooden or iron frames, and the motive-power ordinarily required averages from four to six horse-power for the usual varieties of timber, the saws being well sharpened and the machinery well oiled. The timber moves on trucks, or on beds running over stationary wheels, being fed either by chain or rack-feed, adjustable to a rapid or slow motion as desired. A finer class of machines is made for sawing squared timber into scantling, floor-boards, veneers, etc. Most of these work with continuous roller-feed, an excellent device, drawing long or short pieces through with great facility, one piece pushing the other, but not powerful enough, however, for very heavy work, when chain or rack-feed must be used. Some machines are made for both heavy and light work, and have a combined roller and chain or rack-feed. In establishments where large quantities of planking are cut, a special adaptation of this system is made, the improvement allowing two pieces of timber to be cut at the same time, as many as sixteen saws being used, and both roller and rack-feed employed. The reciprocating veneer-sawing machine is an exceedingly exact piece of mechanism, the wood to be operated on being glued to a slide which feeds vertically against the saw, and is arranged with lateral adjustment for different thicknesses of cut. The power required is not over three horse-power, the saw making two hundred and forty cuts per minute. The blades are very thin, with finely set teeth, and are tightly strung.

M. ARBEY makes quite a variety of circular-saws, which are separated generally into two classes—those with fixed and those with rising spindles, the former being made for straight- or cross-cutting, and the latter being adapted for grooving, tonguing, rebating, tenoning, and all kinds of joiner's and cabinet-maker's work. The raising or lowering of the spindle is effected by means of a handle placed in a convenient position above the table, and just enough of the saw is left exposed to give the proper depth of cut.

In the matter of ENDLESS BAND-SAWS, all sizes and styles are made by M. ARBEY, the larger kinds operating with self-acting feed, and capable of cutting heavy pieces of timber. Fig. 1 illustrates one of his ENDLESS BAND-SAWS FOR HAND-POWER, and Fig. 2 one operating with FOOT-TREADLE, and provided with a canting iron table for cutting on a bevel. The frames of these machines require to be made very strong to allow them to be driven at the requisite speed without vibration, and to give the saw the required tension so as not to twist while cutting, and to prevent its slipping. M. ARBEY states that after trying various kinds of covering for the wheels, he finds India-rubber to be the best. A knife-saw is used in one of these machines with great advan-

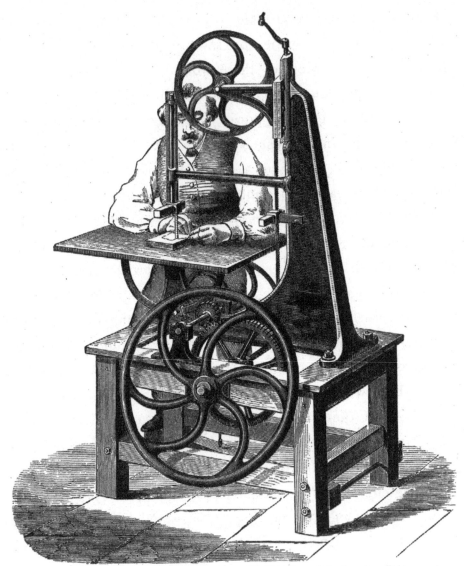

Fig. 2.—Endless Band-Sawing Machine, with Treadle and Canting Iron Table.

tage for cutting leather, cloth, etc.; and with a toothed saw of a certain kind, soft stone, zinc, etc., may be cut. These band-saw machines fill a most important place in wood-working machinery, accomplishing wonderful results at great saving of time and labor. A number of fret-saws are also made by M. ARBEY of excellent design.

M. ARBEY's great success appears to be in the construction of planing

Fig 3.—Planing Machine, with Fixed Table and Continuous Feed, for planing four sides at once.

machines and turning-lathes. We have mentioned, in a previous article on planing-machines, the different types in use. M. Arbey makes planers with movable and fixed tables and with revolving or reciprocating cutters. His great specialty, however, is with a particular form of revolving cutter having helicoidal blades (Mareschal and Godeau's patent), the axis of revolution being in a plane parallel to the surface cut. These machines are of the following kinds: with movable tables for trying up; with fixed tables and continuous feed for planing; with fixed table and continuous feed and side cutters for parquet-planing; and with fixed table and continuous feed for planing four sides at once. Our engraving, Fig. 3, illustrates one of the latter machines. The patent helicoidal or spiral cutters with which these machines are fitted are far superior in every way over the old-fashioned straight knives. The form and arrangement of the blades and their pitch is such that at the moment one edge ceases cutting, the next commences, and the knives are each cutting during the whole of the revolution, the machine making about two thousand per minute, and the shocks ordinarily given to the bearings are entirely done away with, thus eliminating the principal cause of their heating. The equal cutting of the knives does away with vibration entirely, so much so, in fact, that the machines may be erected on a common floor, lessening the expense of massive foundations, and the wear and tear is also correspondingly diminished. The knives always present the same angle to the wood, an angle which allows cutting across the grain, even if knotty, and of planing parquet-flooring already put together. The smoothness of the draw-cut is such that the use of a finishing-tool is not required. Another great advantage with this cutter is that it throws the shavings all to one side of the machine without clogging up the working parts, and entirely out of the way of the operator.

The knives used are very thin, only from four- to eight-hundredths of an inch in thickness, and are quite flat, being pressed closely to the spiral cutter-block by the back iron and nut, projecting only about eight- to twelve-hundredths of an inch. This makes them very economical, not only on account of their cheapness, but also owing to the short time required to sharpen them. The method of sharpening is an American idea, and it is accomplished very simply, without removing the knives from their places, by the use of a traversing emery-wheel, which does its work in a most accurate manner and with the greatest facility. As sharp knives are a necessity for good work, this improvement cannot be too highly valued.

These machines work wood either broad or narrow, thick or thin, along the grain or across it, without vibration, and with very small power, effecting an immense saving in sharpening and in wear and tear over the usual and older forms of machine. Those with fixed tables are used for parquet-flooring, mouldings and thin boards that do not require trying up. It is evident that any variety of cylinders may be used, with blades of corresponding form, to suit various kinds of work. Special machines are made on the same principle for recessing railway sleepers, for cask-making and for cutting long tenons.

M. ARBEY makes a large variety of mortising-, tenoning- and boring-machines in which slot-boring is used, working very much more rapidly than the reciprocating mortising-machines so much used in this country. The tool makes the bottom of the mortise flat, throwing out the shavings, and it is only necessary at the termination of the operation to square out the ends by a double chisel. In some cases, as in carriage- and cart-wheel making, etc., the round ends to the mortise are preferred. The spindle in these machines should run at the rate of at least two thousand revolutions per minute; and although a variety of borers have been tried, a simple hollow bit that can be made anywhere and is easily sharpened has been found to be the best. The frames are made either entirely of iron or of wood. In some machines the tenons are cut with spiral knives, while in others circular-saws are used. The rising spindle saw-bench previously mentioned may also be employed, but does not do the work as promptly or accurately as the special machines. Fig. 4 illustrates a MACHINE FOR SLOT-BORING AND CUTTING LONG TENONS BY MEANS OF CIRCULAR-SAWS. This is a very compact and convenient tool, and different wood-working operations may be accomplished by it, such as tenoning, mortising, edge-planing, and sawing, the tenons and mortises being cut of any suitable size and inclination, and the machine readily adjusted and controlled during its operations. The bed-frame consists of two main parts, one for guiding a horizontally sliding table, and the other for supporting a vertical guide-frame, and, on adjustable carriages, horizontal shafts of two vertical circular-saws. To the side of the bed-frame supporting the sliding-table is secured a vertical guide-frame having a vertically sliding carriage which supports in suitable bearings a vertical saw-

Fig. 4.—Machine for Slot-Boring and Cutting Long Single Tenons with Circular-Saws.

shaft provided at its upper end with a double set of horizontal circular-saws. The space between these saws may be readily adjusted by a suitable changing device and intermediate sleeve or collar to the thickness of the tenon to be cut. The piece of timber is first fed to the horizontally cutting saws, and then to the vertical saws, the latter cutting through the parts at both sides of the tenon, and completing the operation. The extent of the horizontal motion is controlled by stop devices, and the horizontally sliding part may be firmly secured in any position desired by a clamping-clamp-screws at the requisite degree of inclination. screw or otherwise, so that, by detaching the circular-saw and attaching a mortising tool, the revolving shaft may also be used for mortising and grooving, the remaining saws being placed so as to be out of the way. This machine may also be used as an ordinary circular-saw. Tenons and mortises of different angles are produced by placing the saw-bearings on swinging plates connected with the carriages, and securing the plates by Double tenons may also be produced with great facility by sawing them first and then mortising out

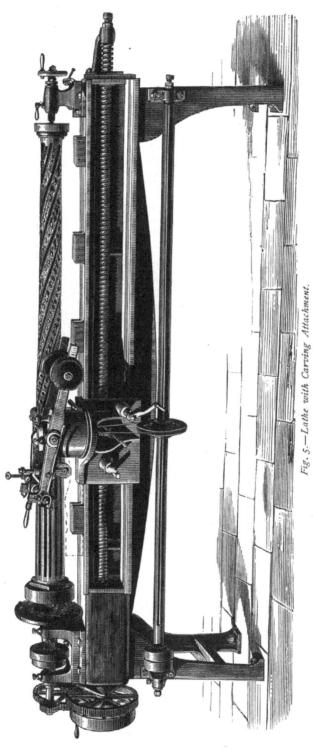

Fig. 5.—Lathe with Carving Attachment.

the intermediate grooves. The saws are revolved by belt and pulley connection with the driving-shaft; and the machine, owing to its compactness and ready adjustability, may be used very quickly and efficiently for the different purposes for which it is intended.

Very convenient machines for cutting double or triple tenons are made

Fig. 6.—Lathe for Turning Irregular Bodies.

with rotary cutters, which work very rapidly, producing from fifty to eighty tenons per hour, depending on their size. The cutters may be shifted to make straight or taper tenons, or to alter the position of the shoulders, as may be required.

M. ARBEY makes a number of special machines for carpenters, cabinet-makers and joiners to execute twisted and straight grooving and beading, for moulding in hard or soft wood, and for cutting veneers. Veneers are cut in two ways—by a sawing-machine, previously mentioned under the head of saws, and also by what is termed a slicing-machine, fitted with a sliding knife block and a thin knife and back iron. The thin knife with back iron does away with the special sharpening-machine required. All that is needed is to unscrew the knife-rest from the saddle, taking care to always keep the three pieces, the knife-rest, knife and back iron, together. These machines cut from ten to fifteen veneers per minute, and from one hundred to one hundred and fifty to the inch of thickness, whereas the sawing-machine will not cut more than twenty to twenty-five to the inch. For very perfect veneers, sawing is better, but for work of the usual quality, these machines answer admirably, and the cut is so clean that for common purposes polishing is quite unnecessary. M. ARBEY also manufactures an excellent copying-machine for making spokes, lasts, gun-stocks, etc.; a number of machines for cask-making, and one for making wooden shoes—the latter of no use in this country, but of great service in France and Belgium, where many wooden shoes are worn by the lower classes.

Quite a variety of turning-lathes are made by M. ARBEY, and we desire to call special attention to a carving attachment, illustrated by Fig. 5, which is intended to be affixed to common lathes for the purpose of grooving, channeling and ornamenting columns, balusters, table-legs, and similar articles of irregular shape, producing perfect work with ease and rapidity. The carving attachment is placed on a traveling carriage and supported on an adjustable cylindrical standard, to which the balanced arms of the cutter-shaft are pivoted, the latter being revolved by a pulley and belt connection with a traveling pulley of the cutter-actuating shaft. The cutter-shaft is movable on its bearings by a lever-handle, while the pulley is retained by a clutch connection with a fixed brace of the weighted arms, and it is raised or lowered by means of a curved arm and guide-roller passing along the pattern of the form. When a table-leg or other object is held in position of rest in the lathe, the cutting tool passes longitudinally along the same, and works out in it a groove or channel. The dividing-disk being turned for the distance of one sub-division after each channel is completed, the next channel is then produced by the return motion of the carriage. By turning the object slowly in the lathe, simultaneously with the revolving and traversing motion of the cutter, helicoidal channels or grooves are formed. For grooving conical parts, the cutter-shaft is guided along an inclined guide-pattern, or its axis is placed at an angle to the longitudinal axis of the lathe. The cutter adjusts itself to the shape of the object, and carves, by its uniform forward motion, an ornamental groove of equal depth throughout the entire length. For the purpose of pearling or doing other ornamental carving, the cutting tool is guided to the work by a handle, while the object is turned in the regular manner by the dividing-disk, so that the pearls may be formed at uniform distances. The adjustability of the cylindrical standard, in connection with the balanced cutter-shaft and handles, admits of the convenient and accurate handling of the carving attachment, so that a large variety of ornamental work may be accomplished on this machine in a quick, economical and superior manner.

M. Arbey makes an excellent lathe for turning irregular bodies, which we illustrate by Fig. 6. By means of this lathe, sword-handles and other bodies of unsymmetrical form may be turned in a perfectly reliable and automatic manner. The supporting frame of the machine carries, in suitable standards, two rotating mandrels and fixed centres. One of the mandrels carries a set of cone-pulleys, to which the power is applied from the driving-shaft, and it is then transmitted by intermeshing gearing to the second mandrel, and by end-gearing to a screw-shaft below the table of the frame. A carriage is moved along the side guides of the table of the frame by a connection with the revolving screw-shaft, so as to travel automatically in both forward and backward direction. A suitable lever mechanism is also provided for moving the carriage back and forth to set the tools to the work. The front mandrel carries the piece of wood to be operated on, and the back mandrel carries a pattern, an exact copy of which is to be reproduced. There are two tools—a cutter which operates on the front piece of wood, and a guide tool, which moves on the finished pattern. A joint motion is given to the cutter tool longitudinally and laterally—in a longitudinal direction by the traveling carriage, and in a lateral direction by the power of a spring, seen in the engraving, that is controlled by and presses the guide tool against the pattern in the second mandrel, while it keeps the cutting tool at the same time against the wood operated on in the first mandrel. In this way the exact copying or reproduction of the shape of the pattern or finished body is accomplished by cutting out the simultaneously revolving piece of wood into a corresponding shape until finished.

Thus one piece after another may be made, all faithful copies of the original, the work being turned out accurately and rapidly.

The crank of an engine in motion acts with the greatest leverage when at right angles to the centre line of the stroke, its power gradually diminishing to zero as it approaches the dead-points. There being two dead-points and two points of greatest efficiency in each revolution, it follows that the power of the engine is communicated to the shaft and fly-wheel in a succession of impulses more or less sudden, depending on the rapidity of the stroke. The fly-wheel from its very nature cannot economically accumulate power applied to it in the usual manner, unless it attains a considerable momentum, a portion of which, more or less, is always absorbed every time the crank-pin moves over the dead-point, and restored when at the points of greatest efficiency. The result is an irregular or vibratory motion very detrimental to the machinery driven, as well as to the engine, and a loss of power much greater than would be at first supposed. The defect is partially remedied by making the fly-wheel large and heavy, and, as it were, cramping the engine into comparatively uniform motion. The advantage gained, however, is only partial, and at a further expense of loss of power. Another expedient, that of driving from the periphery of the fly-wheel, is still more objectionable, giving the machinery a leverage over the fly-wheel, and very effectively preventing it from absorbing and transmitting the full force of a stroke to the centres. The "give and take" in belts will not overcome the difficulty, as the uniform motion to the driven pulley depends upon a constant tension of the driving side of the belt, which will not obtain with the usual arrangement, the principal advantage of the belt being in its doing away with the noise of gearing. A mill-stone, for example, requiring a continuous exertion of ten horse-power, is driven by varying impulses of twenty horse-power.

Mr. John A. Hafner, of Pittsburgh, Pa., exhibits a coil spring which is intended to provide an effective, economical and easily adapted remedy for the defects of the present system, by furnishing an elastic reservoir of power between the rigid and unyielding action of the fly-wheel and the machinery to be driven, fulfilling the same duty as an air-chamber on the ascending main of a force-pump. It connects the crank shaft and fly-wheel with the driver, and allows the engine and fly-wheel to increase in speed, in the ratio of the increased leverage of the crank, as it approaches its most effective point, without increasing the speed of the machinery driven, this increased velocity acting upon and counterbalancing the increased compression of the spring, which expands in a forward direction at the time the crank is about to reach and pass its dead-points, and secures an elastic, smooth and uniform pressure. The result of this is economy, not only in working, but also in first cost, owing to the increased capability of the engine and boiler. It is stated that an engine may be employed fifteen to twenty per cent. smaller with this arrangement than without it. The whole machinery is also relieved of irregularity and vibration, and is therefore correspondingly more effective and durable, while the weight of gearing may be reduced fully one-third.

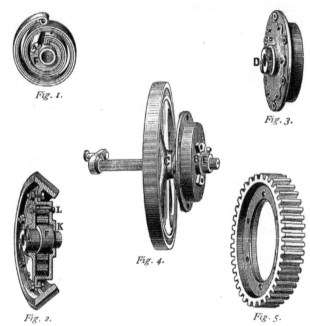

Hafner's Patent Coil Spring and Attachments.

which is represented in Fig. 1, is a spiral coil, consisting of several plates of the best cast spring steel, of varying thicknesses, and its manner of construction is its special feature, rendering its employment possible

The spring, in cases where an ordinary spring would fail. It is well known in the construction of flat coil springs that, the plates being of the same thickness, the smaller the coil, the greater the stiffness, and that therefore when a spring is made of two or more plates of the same thickness, the strain upon the various plates is unequal. This difficulty is obviated by making each plate so much thicker than the one immediately within it, that all will act with the same stiffness. The plates are riveted together only at the inner or hooked end, remaining unconnected at any other point. Owing to the mode of construction, the strain upon the outer plate, which is thickest, is tensile, while that on the inner plates is compressive; and the spring being self-supporting, the grain of the steel is not affected by the strain, while the vibration being obtained merely from the natural tremor or quiver of the steel, the spring will stand and vibrate under a pressure far greater than would be the case with an ordinary single spring.

Fig. 1 represents a front view of the spring and its hub, F, while a sectional view is shown in Fig. 2, giving also the combination with a casing and a bevel-wheel. The inner end of the spring hooks into the hub, F, fitted to play freely in the casing, and the outer end is attached to the casing by a bolt, L. The connection between the hub and casing, and therefore that between the crank or fly-wheel and the driving-wheel, is consequently elastic. The rear view in Fig. 3 shows the extensions of the hub, F, and pawls, D, which fit into the clutch-box (Fig. 11), the latter being keyed fast to the crank-shaft, while the loose hub is

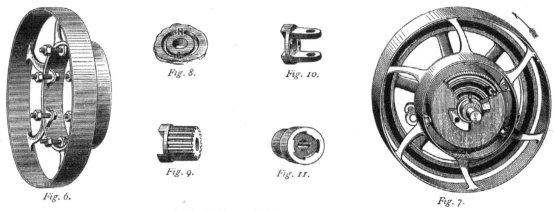

Hafner's Patent Coil Spring and Attachments.

driven by the pawls in a forward direction, its position being maintained by a collar, E (Fig. 4). In case of any accident necessitating instantaneous stopping of the engine, or the slipping off of a belt, the hub rotates freely on the shaft until the acquired momentum is expended, preventing the possibility of injury to either spring or machinery. Another important feature is the facility afforded for starting an engine when on its centre, as the engine may be backed without moving the machinery. As an additional protection to the spring when backing, a groove, G, is formed in the casing (Fig. 8), and receives the lug, A (Fig. 1) of the hub, F, which rests against the stop, H (Fig. 8), effectually guarding the spring from injury.

An exterior flange on the spring casing (Figs. 3 and 4), which is turned and faced to a standard gauge different for each size of spring, admits of the adaptation, in a very short time, of either a pulley, spur or bevel-wheel (see Figs. 2, 5, 6, and 7), as may be required.

This coil spring is not confined in its application to the crank-shaft of a steam-engine, but may be used with advantage on any shaft, to control any portion of machinery where smooth regular work is required. It has been in use for a number of years with great success in flour-mills, both steam- and water-power, for regulating the speed of the burrs; also in paper- and woolen-mills. In cotton- and woolen-mills where excessively high and uniform speeds are required, its use is attended with the most satisfactory results as to economy and increased durability of machinery. Its application is of great importance and utility in the gearing of threshing-machines driven by horse-power. Any one who has seen these machines working with eight or ten horses will bear evidence as to the irregularity of the power, and the absurdity of trying to use this power economically, without the intervention of some highly elastic reservoir or accumulator such as this spring will furnish. One of these springs was tried in a threshing-machine at the Centennial Field Trial, at Schenck's Station, in July, with the most satisfactory results, and a diploma and medal were awarded the inventor. It is claimed that the use of one of these springs will save twenty per cent. in the working of a ten horse-power machine, and fifty per cent. in wear and tear.

IRA J. FISHER & CO., OF KINCARDINE, ONTARIO, CANADA, represented by ALEXANDER THOMSON, OF FITCHBURG, MASS., as sole agent in the United States, exhibit a Bevel Edge Boiler and Ship Plate Clipper, which appears to be an exceedingly efficient and complete working tool, supplying a want long felt among steam boiler-makers and iron ship-builders. This machine, which is the only one on exhibition, entirely replaces the slow and objectionable process of hand-clipping, or the expensive use of planers, doing a much greater variety of work than is possible by the old methods, and at considerably less cost. Our illustration accompanying shows very clearly the design of the tool and its mode of working. It will cut the edges of plates to any required bevel, keeping the line of direction of the cut either straight, concave or convex, as desired. This capability of doing circular-work makes it exceedingly serviceable for Iron Ship-building. The plates require no fastening while under operation, and are placed on a truck or table and passed through the shear, the depth of the cut being regulated by a guard, which is adjustable by a screw, and can be readily changed without interference while the machine is in motion and the plate being cut.

It is claimed that the capacity of one of these clippers runs from one to two hundred feet per hour, the work being done in a superior manner to that executed by planers, which will only do from two to three hundred feet per day on straight work. A hand-power machine is manufactured by the same party, provided with a concentrated power appliance under Mr. D. L. Kennedy's patent, by which it is said two to three hundred feet of clipping may be done per day on three-eighth inch plate with one man on the lever. The machine

Bevel Edge Boiler and Ship Plate Clipper: Ira J. Fisher & Co., Ontario.

appears to combine great simplicity, efficiency and durability, and received an award of a medal and diploma from the Commission.

MESSRS. GREENWOOD & BATLEY, OF ALBION WORKS, LEEDS, ENGLAND, exhibit in Machinery Hall an exceedingly interesting TYING-IN MACHINE of their manufacture, the invention of Messrs. J. P. Binns and J. Shackleton, designed for tying-in the new warp to the old in weaving operations, or connecting each end or thread of the new to the ends or threads of the old warp, so as to allow them to be drawn through the mails or eyes in the healds or harness and sley or reed. Until the present time this operation of tying-in has been done altogether by hand, either by taking the two ends of each thread separately and tying them together, or, as in the manufacture of fine goods, having a light and elastic warp, by twisting them together. The object is satisfactorily effected by this machine, working entirely automatically and by power, making a secure knot, and thus performing what was previously a tedious hand-operation with accuracy and dispatch. The machine, well illustrated by the engraving on page 211, Fig. 1, has a suitable frame-work, on one side of which is placed the warp-beam with the new warp upon it, and on the other side the healds or harness and sley or reed, with a portion of the old warp in them, the ends or threads being secured to rails, and extending across or lapping over each other sufficiently to allow for the forming of the tie. A carriage is mounted upon the frame-work, sliding upon rails or rods, and capable of being moved laterally or crosswise of the warp-threads. The frame-work also supports a rotary horizontal driving-shaft, which the carriage slides upon, and which carries and gives rotary motion to a barrel having several cams attached to it for operating the various levers which control the movements of the working parts of the machine. This carriage supports a vertical reciprocating slide-bar carrying a needle or hooker by which the threads are seized and placed in position to be tied together, the bar receiving its motion from one of the cams already referred to and a suitable lever. Another barrel or hollow shaft is supported on this carriage, receiving rotary motion from the driving-shaft, and on one side of it is a boss. A sliding finger is mounted on the outside of the barrel, carrying a cam at one end, and a reciprocating sliding needle within the barrel is operated by another cam and lever. Intermittent motion is given to the carriage across the warp by a screw-shaft which is worked by a ratchet-wheel and catches. Knives are furnished for cutting the ends or threads of the warp, at the time required, to the proper length for tying together. There are four horizontal shafts, each carrying a half flange or finger for freeing and separating the warp-threads. These shafts are geared together at one end, and intermittent rotary and reciprocating motion is imparted to them by means of a cam with lever and rod, working in connection with ratchets and catches, and operating the screw-bushes or bearings of the shafts. In the frame-work of the machine are two guide-pins for the vertical needle to pass and draw the warp-threads between after being cut, keeping them together and holding them while the finger on the rotary band turns them around a curved groove in the boss on the end of the barrel to form the loop. When the loop is formed, the sliding needle within the barrel draws the ends of the

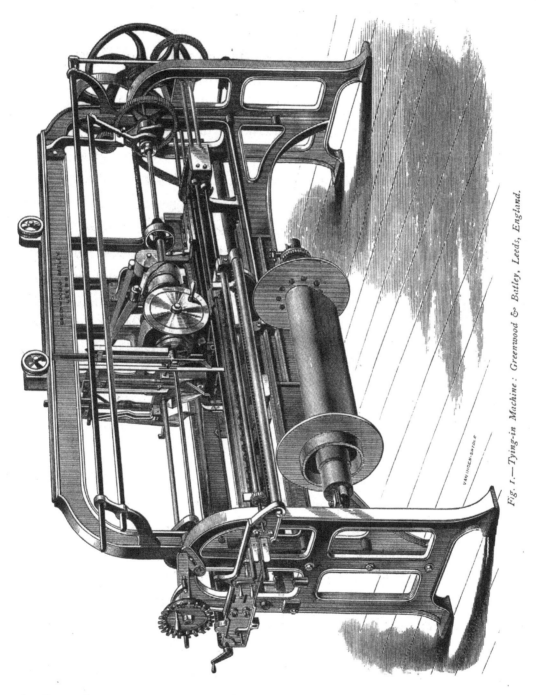

Fig. 1.—Tying-in Machine: Greenwood & Batley, Leeds, England.

threads through the loop, and a lever is brought into operation, pushing the loop off of the boss, drawing the threads tight against the holding of the needle, and forming the knot. The knotted threads are removed by a finger, the next threads are in turn taken up, and the operation is continued until the whole of the warp is tied in. A self-acting stop-motion is attached, which stops the machine should a knot be missed, thus securing good work, and insuring that all the ends of the new warp are attached separately to those of the old.

Another noteworthy exhibit of MESSRS. GREENWOOD & BATLEY is the IRON SHOEMAKER, manufactured by them, and the invention of Messrs. J. Keats, A. Greenwood and A. Keats. This machine, shown by Fig. 2, is of the greatest interest particularly to shoemakers, and also to all harness-, leather-belting-, and portmanteau-makers, and to other branches of the leather trade where a perfectly fast lock-stitch, made with two waxed threads, is required. The work produced is fully equal to that of the very best hand-work in quality, and much superior to it in regularity of stitch and tension. A great many thousand pairs of boots and shoes have been manufactured on this machine in England, including all descriptions, from the lightest ladies' white kids to the heaviest class of best welted walking-boots, also those for laborers and miners, and all varieties having sewn-through soles. A very large number have been made for the British army upon it, giving the most perfect satisfaction in both workmanship and wear. The machine produces a twisted lock-stitch with two waxed threads, making the firmest work known, and it is adapted for sewing the soles on boots and shoes, sewing through the outer and inner soles, sewing the outer sole to a welt, also for sewing harness, siding up heavy boots (closing), belt-sewing, and all leather work generally.

The chief characteristic of the machine consists in the combination of a hook and shuttle instead of a needle and shuttle, the advantages derived by this arrangement being twofold—first, the thread, not having to pass through an eye, can be thoroughly saturated with wax, which is neither squeezed out nor scraped off as is the case when an eye is used; and secondly, as the hook in descending has no thread in it while piercing the leather, the hole made is no larger than the size of the hook, and a thicker thread may be used so as to fill the hole with the thread and wax. The shuttle and hook are both situated above the work, enabling the operator to always see that it is perfectly sewn. A special feature in the machine is the combination of a loop-divider and spreader with the hook. The machine is driven by power, and has a stop-motion actuated by a foot-lever, by means of which action may be suspended instantly in any part of the stitch, at the same time giving the operator both hands at liberty to make any necessary adjustment or perform other duty. The stitch is formed as follows: The hook in descending pierces the leather, and

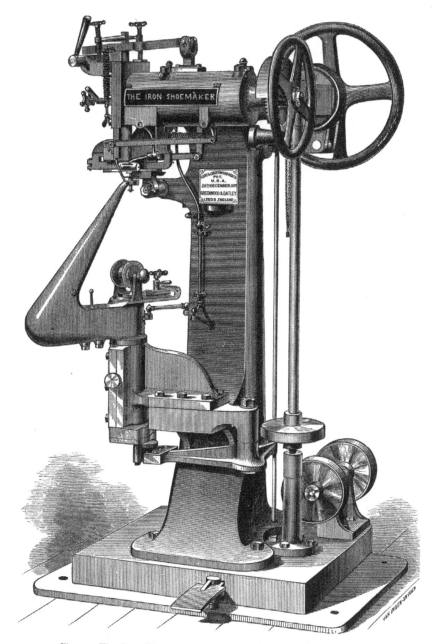

Fig. 2.—The Iron Shoemaker: Greenwood & Batley, Leeds, England.

when at the bottom of its stroke the thread is supplied to it by means of a rotary looper placed inside of the top of the swivel-horn, and actuated by gearing driven from below. The thread passes to the interior of the swivel-horn from a reel conveniently situated below, and is provided with a very simple and perfect tension arrangement. The hook in ascending brings up the thread, and the loop-divider advancing, partially opens it out to form the loop, removing it from the hook, and, in concert with the loop-spreader, preparing it for the entrance and passage of the shuttle. After the feed has taken place, the hook in bringing up the thread for the next stitch, pulls tight the one previously formed, and while this is being done, the action of the presser foot on the material provided for the purpose holds the parts tightly together and insures solid work. The machine is provided with all necessary adjustments for feed, tension and for the thickness of the work to be operated on, etc., etc., conveniently and simply arranged, and easily controlled by the operator. The working parts through which the waxed threads pass are kept heated, so that nothing interferes to prevent a proper tightening of the stitching.

MESSRS. GREENWOOD & BATLEY also exhibit a half-medium, square motion, platen, "SUN" PRINTING MACHINE for fine work, which has been kept continuously in operation, and has proved itself capable of doing excellent work, having great strength of impression and a very large ink-distribution. It has been employed in printing the small colored views of the Centennial Buildings, so many of which have been sold to visitors, and which bear evidence of the quality of the work which the machine will perform.

Among those manufactures which have received special development in America, and whose highest type is to be found in this country, are what may be termed, generally, "instruments for optical projection," including under this title both the ordinary magic-lantern or stereopticon, and those elaborate and refined instruments used for the projection, as images on a screen, of countless phenomena in the subject of spectrum-analysis, polarized-light, electricity, heat, acoustics, and the like.

It has been in this country that these instruments have received their most general use and have attained their highest perfection in efficiency and variety of applications. Their use in illustrating popular lectures on science, and in regular collegiate courses of instruction, has been far more general here than elsewhere, and has led to a remarkable development of their capabilities.

MESSRS. GEORGE WALE & CO., OF HOBOKEN, NEW JERSEY, give a most complete series of specimens of this development, in their collection exhibited in connection with the display of the STEVENS INSTITUTE OF TECHNOLOGY, of whose special exhibits we have previously given some account. In looking over this exhibit, attention is called to what is styled the COLLEGE LANTERN, which, as its name indicates, is specially designed for college use. An excellent illustration of the lantern is given by Fig. 1. The body of the instrument consists of a brass box mounted on brass columns, resting on a mahogany base, and provided with leveling-screws. Within this box is arranged apparatus for producing either the lime- or electric-light, as the special occasion may require. In the

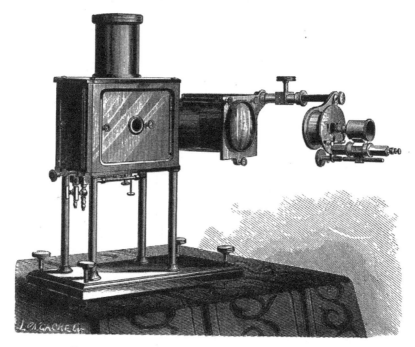

Fig. 1.—*College Lantern: George Wale & Co., Hoboken, N. J.*

front wall of the box are large lenses, whose office is to collect the rays of light emanating from the source employed, and to send them forward in a large parallel beam. Immediately in front of these collecting lenses hangs what may be described as a hinged or swinging plate, carrying another large lens, whose office is to receive these parallel rays coming from the collecting lenses, and to concentrate them so that they will enter an object-glass placed in front, supported by an arm with a rack-work adjustment, which is itself attached to the upper corner of the swinging plate before described. So far this instrument simply represents, in a very complete and convenient form, the ordinary magic-lantern for the projection of transparent pictures. It possesses, however, the additional capacity of allowing all sorts of apparatus, such as tanks to contain liquids in which chemical reactions or other changes are going on, moving slides illustrating the formation and propagation of sound and light waves, and the like, to be placed freely and conveniently in position for proper display. Fig. 2 shows its adaptation for this apparatus. This is however only one of many other arrangements of which it is capable. Thus if the plate we have just described as hinged or swinging is rotated upwards on its pivots, and a triangular box having a glass mirror on its longest side is placed beneath it as shown in Fig. 3, we have the apparatus transformed into what is known as the "vertical lantern" of Professor Morton. In this arrangement the light rays, after being rendered parallel by the two lenses attached to the front of the lantern-box, fall on the mirror of the triangular box, and are thus thrown vertically upward upon the condensing lens in the "swing-plate," which is now horizontal. By this they are collected and made to enter the objective, and, after passing through it, fall on a small adjustable mirror, by which they are directed upon the screen. The purpose of this arrangement is to permit the use of an object such as a tank of water, a delicate galvanometer needle, a magnet and iron filings, a glass Chladni plate with water on it, and similar apparatus which require to be horizontal.

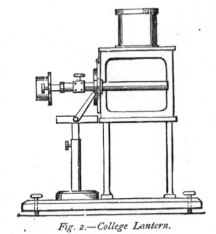

Fig. 2.—College Lantern.

Again, in order to exhibit the magnificent phenomena of polarized light, another arrangement is made which is shown in Fig. 1. By partially unscrewing, with the hand, one of the pivots on which the swing-plate hangs and turns, it can be removed from the front of the lantern-box, carrying with it of course the rack-work arm and objective. An elbow of peculiar form is then attached in its place, having a bundle of twenty-four pieces of thin plate-glass, which acts as a polarizing mirror. To the front of this is again connected the swing-plate with rack-work objective, etc., and the instrument still exists as a magic-lantern, having, however, polarized light for the illumination of its objects instead of the ordinary light. If these objects are large, they are placed, like ordinary pictures, close to the condenser; but if they are small, a microscopic attachment is employed, as shown in Fig. 1, by this means enabling even microscopic objects to be shown with all the splendid colors developed by the polarization of light. If this microscopic attachment is used with the lantern in its first position, it is converted into a gas-microscope of unusual efficiency. Finally, for the projection of all the phenomena of spectrum-analysis, an adapter is applied to the front of the swing-plate in the first arrangement of the lantern, which carries an adjustable slot or diaphragm-plate, and one or more prisms are placed on the adjustable table which accompanies these lanterns, at the proper distance for operation.

In addition to this lantern and some thirty special attachments for the production of various experiments with it, this exhibit of GEORGE WALE & CO. contains several simpler forms of lantern, three varieties of spectroscopes, and a large collection of the apparatus and tools employed in blow-pipe analysis. The exhibit reflects great credit.

Fig. 3.—College Lantern.

The HARRISON BOILER WORKS OF PHILADELPHIA, PA., make an exhibit of the HARRISON SECTIONAL SAFETY BOILER in the Machinery Hall, and also show in Boiler-house No. 4 one of the same boilers of one hundred horse-power, supplying steam for the purposes of the Exhibition. These boilers possess remarkable advantages as respects safety, economy, ease of transportation, and facility of erection and repair, having obtained first-class medals at a number of previous exhibitions, and are well worthy of particular mention among the specialties of our Centennial display. They are the invention of the late MR. JOSEPH HARRISON, JR., a Philadelphian well known for his great mechanical and inventive abilities and for his enterprise and success in business connections some years ago with Russia.

Nearly twenty years ago MR. HARRISON brought to the notice of his friends the subject of this invention, which he had had under consideration for some time, in reference to steam-boilers, its object being to prevent the danger of disastrous explosions. Starting with the fact that the strength of any structure is the strength of its weakest point, he aimed to construct a steam-boiler built up of units of some given strength; and taking a sphere of some metal, as cast iron, Mr. Harrison claimed that a thickness of metal, say three-eighths of an inch, might be first assumed, and then such a diameter be given to the sphere as would establish its bursting pressure at a certain fixed amount, say one thousand pounds per square inch—this ultimate or bursting pressure to be

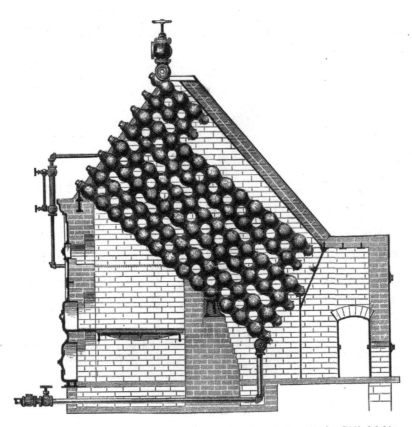

Side Elevation of Fifty Horse-power Boiler: Harrison Boiler Works, Philadelphia.

taken such that the sphere would be safe for what is usually required by those who employ high-pressure steam. It was the development of this idea that led to the design of steam-generator which we are now considering, and which is shown by the accompanying illustrations. The boiler is formed by a combination of cast-iron hollow spheres arranged in groups of twos and fours, each sphere being eight inches in external diameter, connected one with the other by curved necks, having rebate machine joints, and held together by wrought-iron bolts with caps at the ends. The illustration on this page gives a side elevation of a fifty horse-power boiler, showing also a section through the brick-work casing and fire-box. That on this page shows details of a four-ball unit. Fig. 1 gives a transverse section; Fig. 2, a longitudinal section, showing bolts with caps and nuts; Fig. 3, a sectional plan, showing head of connecting-bolt; and Fig. 4, a section of one of the joints, half size. Every boiler is tested, before being used, by hydrostatic pressure up to three hundred pounds to the square inch. Under pressure which might cause rupture in ordinary boilers, every joint in this becomes a safety-valve, and it is claimed that no other steam-generator possesses this property of relief under pressure. In an experiment which has been made, steam raised to eight hundred and fifty pounds per square inch has failed to rupture the boiler. It is easily seen that in special

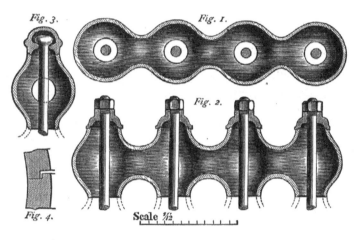

Details of a Four-Ball Unit.

cases, where an extra amount of steam-room is required, it may readily be obtained by adding spheres to the top of the boiler, thus maintaining the system of uniform parts of small dimensions, and avoiding the use of large steam-drums, which, if added to sectional boilers, would destroy their claim to safety. The boiler is easily transported from place to place, and may be taken apart so that no separate piece will weigh more than eighty pounds. Owing to the small size of these pieces, the largest boiler may be put through an opening as small as one foot square, allowing of its introduction into places difficult of access for ordinary boilers. No special skill is required in its management, and the parts being uniform in shape and size, any that may be injured can be replaced with the greatest facility. A boiler may be increased to any extent by simply adding to its width, and as it is the multiplication of a single form, its strength remains the same for all sizes. It may be placed, for the same power, on less than one-half the ground-area required for the ordinary cylinder-boiler. The use of cast iron possesses a signal advantage in that it is not easily affected by corrosion, which so soon impairs the strength of wrought-iron boilers. When the heating-surface becomes entirely worn out and rendered useless, it can be renewed and the boiler made as good as it was originally at half the cost—another great advantage over a wrought-iron boiler, which is scarcely worth the cost of removal when worn out. This boiler makes steam quickly, produces superheated steam without separate apparatus, and under ordinary circumstances it may be kept free from permanent deposit by blowing the water entirely out, under pressure, once a week. The design of this boiler is so novel and so different from anything previously known, its advent marking so distinct an area in boiler construction, and preceding all other forms of sectional safety-boilers, that it is no more than proper to assert that its invention may be justly considered the crowning achievement in MR. HARRISON's most useful life, entitling him, even had he given to the world no other work but this, to have his name enrolled high on the list of the benefactors of mankind.

The subject of rock-drilling by machinery has attracted a large amount of attention within the last ten years, the Hoosac tunnel in this country, that of Mount Cenis in Europe, and various other late important works in rock excavation having stimulated invention in this direction, and brought forth quite a number of machines for this purpose. Among those whose attention was early called to the matter was MR. CHARLES BURLEIGH, OF FITCHBURG, MASSACHUSETTS. The State of Massachusetts had expended a large amount of money on the subject of machine-drills at the Hoosac tunnel without success, and had returned to hand-labor, when MR. BURLEIGH took the question up, and, after much labor and expense, solved the problem, and about nine years ago introduced the first BURLEIGH DRILL into the Hoosac tunnel, marking a new era in the history of rock-drilling throughout the world. These drills, manufactured by the BURLEIGH ROCK-DRILL COMPANY, OF FITCHBURG, MASSACHUSETTS, have now been very extensively employed not only in this country, but also in Europe, and we desire to call attention to the fine exhibit made by this Company in Machinery Hall.

The main elements of the drill are the cage, the cylinder and the piston.

The cage is merely a trough with ways on either side in which the cylinder is moved forwards as the drill cuts away the rock. The piston moves in the cylinder, propelled and operated either by steam or compressed air, like the piston of an ordinary steam-engine, and the drill-point is attached to the end of this piston, which is a solid bar of steel, and rotates as it moves back and forth. An automatic feed regulates the forward motion of the cylinder in the trough as the rock is cut away, the advance being more or less rapid as the variation in the nature of the rock allows the cutting to be fast or slow. The shock of the blow is received by the drill-point and solid steel piston alone, and the piston-rod, arranged with a double annular cam and spiral grooves, performs in its movements three important functions, as follows: first, the movement of the valve admitting steam or compressed air to the cylinder; second, by the operation of the annular cam acting upon a feed device, the movement of the cylinder forward in the cage or slide as the rock is penetrated; and third, an automatic rotation of the piston-bar by the spiral grooves and a spline in a ratchet, a partial revolution taking place at each upward movement, the ratchet remaining

Fig. 1.—*Drill Mounted on Iron Tripod for Rock-Cutting.*

perfectly stationary while the rotating movement occurs, and moving only as the piston again descends. After the cylinder has been fed forward the entire length of the feed-screw, it may be run back and a longer drill-point inserted in the end of the piston to continue the work. An ingenious peculiarity in the form of the cutting-edge of the drill-point insures a perfectly round hole, thus giving it greater area, and allowing a larger percentage of the powder near its bottom. The regular rotation of the drill brings the delivery of each blow at the point of greatest efficiency, each wing of the drill-point striking the rock just sufficiently in advance of the cut of the preceding blow to chip away the rock lying between. The yielding of the chip saves the edge of the drill-point, and the advance in the rock, without sharpening, is ten times greater than possible in hand-drilling, where the hole is formed by the crushing and pulverizing of the rock.

Fig. 2.—*Stoping Drill mounted on Stoping Column.*

The drilling-machine is attached to a clamp by means of a circular plate, with a beveled edge cast upon the bottom of the cage near its centre. This plate fits a corresponding cavity in one side of the clamp, and is held there firmly in any required position by the tightening of screws. The clamp is clasped about a bar of iron, to which it may also be tightly held by screws. The following movements may be made: that upon one plane, by the motion of the plate in its cavity; and that upon another plane at right angles to the first, of the clamp upon the bar, and the sliding endwise of the clamp upon the bar. By these movements any position and direction of the drill is attainable. All that is now required is to attach the bar, of any reasonable length, to a convenient carriage or frame, and the machinery is ready for operation.

These drills are applicable to all kinds of rock-work, whether mining, quarrying, cutting, tunneling, or submarine drilling, combining simplicity, strength, lightness, and compactness. They are easily handled and require but few repairs. Holes may be drilled with them from three-fourths of an inch to five inches diameter, and to a depth not exceeding thirty to thirty-five feet, the rate of penetration being from two to ten inches per minute, according to the nature of the rock. Either steam or compressed air may be used as a motor; and at a pressure of fifty pounds to the square inch, two to three hundred blows can be given per minute according to the size of the machine.

Fig. 1 illustrates one of the varieties of this machine for surface-work, shaft-sinking and channeling. The drill is mounted on a tripod with adjustable telescopic or screw legs, the tripod being constructed entirely of iron, and this form is exceedingly convenient for use upon uneven surfaces and for shaft-sinking. Various sizes are made, suitable to the different purposes for which it may be employed. This drill is very useful in cutting slate, marble or granite, where blasting cannot be done on account of its breaking up the stone. Holes two inches in diameter are drilled in a row, two inches apart, and the connection between them broken down by simply throwing off the rotating action of the machine and placing a flat bar of steel in place of the drill-point. To any one who has not witnessed the operation, the rapidity of its execution is something marvelous. In all open-air work, steam is preferred as a motor, and it is conveyed from any ordinary form of boiler by iron pipes connecting with the drill by a short piece of flexible rubber pipe, so as to permit an easy adjustment of the machine to its work. Of course it is desirable to have the boiler as close to the drill as possible to avoid loss of steam from condensation.

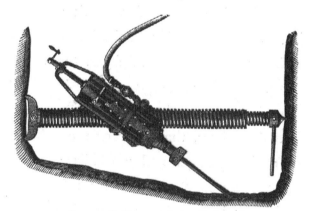

Fig. 3.—Drill mounted for Shaft-work.

In tunneling, the use of the BURLEIGH DRILL is peculiarly advantageous, simplifying and rendering easy of execution, works, the magnitude of which, with only the old appliances of hand-labor, would render them almost impossible of accomplishment. There is not only a saving in actual labor in the use of these machines, but also a great saving in time, a matter so important in these fast days. Fig. 2 represents a stoping-drill mounted on a column, with a claw-foot and jack-screw at the ends to secure it in position. This variety is particularly adapted to small tunnels, adits and stopes from four and a half by six to six by six feet, or even larger drifts, and is the simplest, cheapest and best arrangement for the purpose. A large number of these have been used in the mines of California and Nevada. Fig. 3 shows the same appliance as employed in sinking small shafts.

For larger tunnels the drills are mounted on mining-carriages. One variety of these consists of a vertical column to which the drill is attached, there being a screw-nut for holding, lowering and raising the drill, and the whole mounted upon a platform which slides, by means of a crank and gear, across the carriage, allowing adjustment to any position in the heading. A jack-screw on the top of the column, acting against the roof of the tunnel, secures the machine in position when drilling. Another arrangement, designed specially for railroad or other large tunnel-work, consists of a carriage on which four drills may be mounted upon two horizontal bars, the lower one capable of being raised or lowered by means of chains, pulleys and a windlass. The carriage runs on rails and is constructed with an open space of from four to sixteen feet between the two sides, so that it may be run in as soon after a blast as the track can be cleared by throwing the rock into the centre, allowing the drilling to be resumed with very little loss of time, the removal of the broken rock proceeding simultaneously in the rear of the drill. By means of jack-screws, the machine is raised from the wheels during the drilling and held in place, as with the other drills, by screws running out from the ends of the upper bar to the sides of the tunnel, or from the frame of the machine to the roof. This style of carriage is used in the great Sutro Tunnel of Nevada. In another form also for tunnels, only one horizontal bar, with a vertical swing movement controlled by chains and windlass, is arranged for holding the drills, its position being fixed at any desired point by jack-screws in the ends of the bar.

It has been a point with the manufacturers of these drills to reduce the cost of repairs required in operating them, as much as possible; and as experience has shown that a large proportion of the repairs are due to the automatic feeding devices, they have discarded this entirely in all the smaller machines, arranging for feeding the drill only by hand. This is the case now for all below the tunnel-drill, and it is a great advantage, dispensing with complicated machinery that can only be operated by experts. A machine to go into the hands of unskilled laborers should be as simple in detail and as substantial in form as possible.

We have before spoken of the motors in use for these drills, and that

Fig. 4.—Burleigh Patent Air-Compressor.

steam is preferred for open-air work. In deep shafts and tunnels, however, steam cannot be employed, not only on account of the difficulty of carrying it to any distance, but also because the discharge of the exhaust in the heading would render working impossible. To meet this difficulty, the use of compressed air has been introduced, a medium that possesses all the advantageous properties of steam, can be carried to any distance without material loss of power, and at the same time, when discharging, gives an abundant supply of pure fresh air to the workmen, and accomplishes perfectly the requirements of ventilation. Fig. 4 represents one of the BURLEIGH PATENT AIR-COMPRESSORS, consisting of a steam-engine connecting by means of a crank-shaft with two single-acting air-pumps. It is a compactly-built, close-working machine, the parts nicely balanced, and the whole well-designed. Seven sizes of these compressors are made, capable of running from one to eight of the tunnel-size drills, and they work with great economy and efficiency. The air when compressed is taken into a tank or air-chamber, and thence carried to any desired point in pipes, in the same manner as steam, connection between the permanent pipes and the drills upon the carriages being made by flexible rubber pipe, which is uncoupled when the carriage is run back for a blast. Furnishing compressed air for drilling is only one of the uses to which this machine may be applied, as it may evidently be used in the ventilation of buildings, transmitting of messages and packages by pneumatic tubes, supplying a convenient and safe motor for machinery, etc. The feasibility of compressing air at any convenient point, and conveying it from place to place in the streets of our cities, to be tapped and used wherever required, has already been demonstrated, furnishing a cheap and safe power without dirt or obnoxious gases, and when exhausted into a room, really providing an ample and desirable ventilation. At the Hoosac Tunnel this power was carried over three miles without material loss.

The subject of locks early absorbed the attention of inventors. There have always been valuables in the world, and always those who coveted what belonged to others; hence also the necessity has always existed of a means by which such valuables could be closed up and kept securely in receptacles accessible only to the owner. Researches among antiquities reveal to us that the use of locks was known even in pre-historic times, and a form of lock comes down to us from ancient Egypt that in a primitive shape shows quite a similarity to a noted lock of the present day. It is only within the last century, however, that any great improvement has been made in the manufacture of locks. No doubt all remember the famous Bramah lock which was exhibited at London in 1851, and the reward of two hundred pounds offered to whoever should succeed in picking it, a feat hitherto never performed, but accepted and achieved at that Exhibition by Mr. Hobbs, the American lock-maker, who, by his efforts and exhibits after that, fully established the supremacy of American locks. In this connection our notice has been specially called towards the display of American locks made at our own Exhibition by the YALE LOCK MANUFACTURING COMPANY OF STAMFORD, CONNECTICUT, the products of which firm bear a high reputation for absolute security, and show great ingenuity of design and excellence of material and workmanship.

The Yale locks, as first made by the inventor, the late LINUS YALE, SR., nearly thirty years ago, were a great improvement, but it is to his son, LINUS YALE, JR., that we are indebted for the modern Yale lock, now in so universal use, so moderate in cost and so well adapted to every purpose for which locks are employed, from a lady's jewelry-case or a tradesman's cash-drawer to the

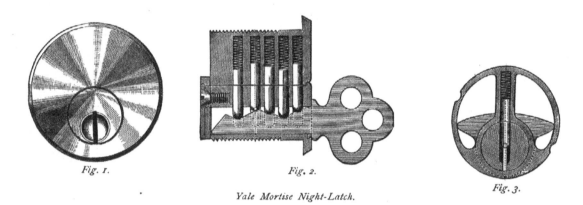

Fig. 1.   *Fig. 2.*   *Fig. 3.*

Yale Mortise Night-Latch.

heaviest house- and store-doors. This lock, which is known as a "pin lock," is similar in some respects to the old Yale lock invented by the father, but possesses a distinctive and most desirable feature in having a key of thin, flat steel, less than an inch and a half in length, and weighing but a small fraction of an ounce. The arrangement in this lock of the escutcheon or tumbler-case containing the mechanism operated by the key, is also a most important characteristic, it being placed close to the outer face of the door, so that the length of the key in all cases remains the same, no matter how thick the door may be. The new lock, therefore, entirely dispenses with the old style of heavy keys formerly in use, to the great satisfaction of those in charge of them. The arrangement of the parts acted upon by the key, and the shape and size of the key itself, form the special peculiarities of the Yale lock, the particular form of lock and the details of construction of course varying with the use for which it is intended. We cannot better explain these prominent features than by a description of the YALE MORTISE NIGHT-LATCH, the escutcheon of which, illustrated on page 227, is applicable to almost any style of mortise or rim-lock; flush-locks for drawers, desks, etc., differing only in that the mechanism here shown in a separate escutcheon is there contained in the body of the lock.

In the illustration, Fig. 1 shows an exterior view of escutcheon with entrance for key; Fig. 2 a longitudinal section with the key entered; and Fig. 3 a cross-section locking towards the rear of the lock. The escutcheon consists, as will be seen from an examination of these cuts, of an exterior shell of cylindrical form, containing in its lower part a small cylinder, from which rises a rib of metal containing the "pin-chambers," and within which is the "plug," attached to the inner end of which is the cam that imparts motion to the bolt. The escutcheon contains five holes or "pin-chambers," each formed partly in the shell and partly in the plug; therefore a pin which fills one of these holes will prevent the rotation of the plug, unless this pin is cut in two pieces, the joint corresponding with that between the plug and its hole, in which case the plug will revolve freely, carrying with it one piece of the pin and leaving the other piece in that part of the pin-chamber contained in the shell. This is precisely the construction of the lock, and forms its great element of security. Each pin is in two parts, the upper termed the "driver," the lower the "pin," and above each driver is a light spring, tending to press both driver and pin downward. In this position the drivers intersect the joint between the shell and the plug, completely preventing the rotation of the latter. To be able to rotate the plug, the joint between each pin and its driver must be put exactly on the line between the surface of the plug and its hole in the shell, and it is very evident that this joint may be cut in any part of the pin's length, requiring therefore for the different pins different elevations, and allowing only that key to open the lock which will give to each pin its proper height. A difference of one-fiftieth of an inch in the elevation of either of the pins will prevent the opening of the lock. This fact gives to the command of the manufacturer an almost incalculable number of variations, and enables him to furnish an immense variety of keys and great range of permutations. Any one not having the proper key and attempting to discover the proper heights to raise the pins and therefore to pick the lock, has such a small chance, according to the theory of probabilities, that his case may be considered hopeless.

Some three hundred different styles and varieties of Yale locks are made, adapting them to almost every conceivable use; but the form of key and construction of the parts acted upon by it are essentially the same throughout the whole series, included in which are many specialties, such as locks for safe-deposit vaults, prisons, hotels, freight-cars, etc.

This Company carries the manufacture of burglar-proof bank- and safe-locks to great perfection, one of the most important of its various kinds being the YALE DOUBLE-DIAL LOCK, the peculiarity of which consists in the double principle of one common bolt being controlled by two entirely independent locks, which may be set on different combinations, thus affording entrance to two different persons, and avoiding danger of being "locked out," a great trouble hitherto with combination-locks. Each dial operates a distinct, four-tumbler lock capable of one hundred million changes.

Yale Time-Lock: Yale Lock Manufacturing Co.

The most notable invention of the Company however in this direction is the YALE TIME-LOCK, an apparatus that locks and unlocks automatically, and in which the hour of locking as well as unlocking can be regulated at will. This invention provides an absolute protection against masked burglars, and at the same time guards a bank from the surreptitious handling of its funds or the falsification of its books by inside parties. No one unauthorized can possibly tamper with it, as its door is fastened with a fine key lock, and need never be opened unless it is desired to change the hours of locking or unlocking. The winding may therefore be entrusted to a subordinate, and the lock be placed in the hands of one person, rendering all accidental or intentional interference with it or its devices for adjustment to the required hours, impossible. The illustration on page 229 shows the mechanism of this lock. It is provided with two separate independent and jeweled chronometer movements, made expressly for this purpose by Messrs. E. Howard & Co., of Boston, Mass. There are two dials, one operated by each chronometer, each revolving once in twenty-four hours, and having near its circumference a series of pins twenty-four in number, marked to correspond with each hour of the day, and spaced at regular intervals around the whole circle. These hour-pins are pushed in for those hours during which the lock is to be locked, and drawn out for those in which it is to be unlocked. The two dials revolve in opposite directions, as shown by the arrows in the engraving, and the co-operation of both the independent chronometer movements is necessary for the locking, although either or both will unlock, thus preventing accidental locking out by the stoppage of either of the movements. The mechanism is cushioned between springs, and no jarring or sledging of the door can affect it. When once set, no further care is required, except winding. This may be done through the eyelets or posts in the glass face without exposing the machinery, and all danger of injury from dust and dirt is avoided. The hours of locking and unlocking can be changed, if desired, in a few seconds, and a special "Sunday Attachment" is furnished, automatically preventing unlocking during Sundays if so required. There are four separate independent devices for unlocking, either of which will open it. A most important point consists in the fact that it can never run down in a locked position. If from any carelessness in setting the lock, or through omission to wind it properly, the dials should be permitted to stop when in the locked position, the running down of either movement will open the lock, thus protecting the user against any possible carelessness or inattention on his part. As now constructed, the YALE TIME-LOCK obviates every objection heretofore made to locks of this class. Once set, so long as it is duly wound, it is *absolutely automatic*, performing, unaided and unattended, every duty assigned to it with a lifelike action almost human.

Weston Differential Pulley-Block.

The YALE LOCK MANUFACTURING COMPANY also makes a specialty of some other manufactures, among which we must mention that great boon to all of those whose business necessitates the lifting of heavy weights, the WESTON DIFFERENTIAL PULLEY-BLOCK. These pulley-blocks, represented by the engraving on page 230, possess great advantages over the ordinary double block. The arrangement consists of an upper fixed pulley, having in it two parallel grooves, one of these being of a less diameter than the other by a small, or so-called differential, amount; and of a lower movable pulley attached to the hook holding the weight; an endless chain passing first over one of the grooves of the upper pulley, then down through the lower pulley, and then up again over the other groove of the fixed pulley. This endless chain is doubled, as it were, one loop connected with the weight by the lower pulley, and the other loop slack. Lugs in the grooves of the upper pulley catch in the links of the chain and prevent its sliding. It is seen that the two ends of the loop holding the weight both pull on the upper pulley in opposite directions, on a very slightly different leverage, it is true, but not enough to overcome the friction, etc., of the different parts of the apparatus, and as this pulley cannot possibly move both ways at one time, the weight is therefore held stationary wherever it is placed. By pulling the slack loop of the chain one way or the other, the weight is gradually raised or lowered, the power required and the rapidity of action depending on the difference of diameters of the two grooves in the upper pulley. These pulley-blocks give absolute safety, and no accident of a weight falling can possibly happen unless something should break. Various sizes are made, plain, geared and with sprocket-wheel, ranging from one-quarter of a ton to ten tons, and they are exceedingly easy of manipulation. One man can operate them without difficulty, although, of course, if great rapidity of action is required with very large weights, more power must be applied as with any pulley. The construction of these blocks prevents any danger from the chain twisting or mounting the sheaves.

We have previously described several exhibits of MESSRS. GREENWOOD & BATLEY, OF ALBION WORKS, LEEDS, ENGLAND, and we now desire to mention one of VINCENT'S PATENT BOLT AND RIVET FORGING MACHINES, made by this firm and illustrated by the accompanying engraving. This machine is specially constructed for the rapid production of bolts, rivets or spikes from bar-iron, any desired shape of head being produced, and the whole formed at one blow or squeeze of the machine. As will be seen by reference to the engraving, the machine is driven by two friction-wheels acting alternately on a third friction-wheel covered with leather, the latter wheel being firmly attached to a vertical screw which raises the tup by means of a heavy brass nut into which it works, coupling-rods connecting the nut and tup together. The blow is upward against the end of this screw, and the shock is contained altogether in the machine, requiring no heavy foundations. The weight of the brass nut, coupling-rods and tup is counterbalanced by weights hung at the back of the machine from a swinging lever, so that the whole may move up and down on the guides with the greatest freedom. The top driving-shaft is kept in position longitudinally by balance-weights, and the machine may be made to work in either direction by movement of this shaft, thus bringing over one or the other of the friction-wheels into action. This movement is effected by a long rod connected with a lever and handle below, and having at its lower end a strong coiled spring. The machine reverses itself at the top of its stroke, after making the bolt, by means of a tappet on the large brass nut striking against lock-nuts on a vertical rod which connects with the lever and handle just mentioned, and it returns to the correct position for placing another piece of iron in the die, stopping where required, by striking a lower pair of lock-nuts. In doing this the weight of the tup, etc., is deposited gently on several India-rubber washers lying on the top of a strong adjusting screw, which goes through the bottom cross-piece of the machine, and carries the knocker out for the bolts. The socket into which this screw fits is only held up by two safety-pins, which will shear if the top should descend violently through carelessness. The fall of the tup is then broken by a second lot of washers placed directly on the cross-piece. The amount of metal in the head of the bolt or rivet can be readily fixed by a screw and dividing-plate attached to the under side of the tup, a small catch engaging in notches in the dividing-plate, and serving to hold the screw in position.

A very ready means is provided for changing the tools, only about one turn of the holding-down screws being needed, when the whole can be lifted out. A sheet-iron trough is placed on the top of the tup surrounding the die, for carrying away the water used to keep the dies cool. The speed

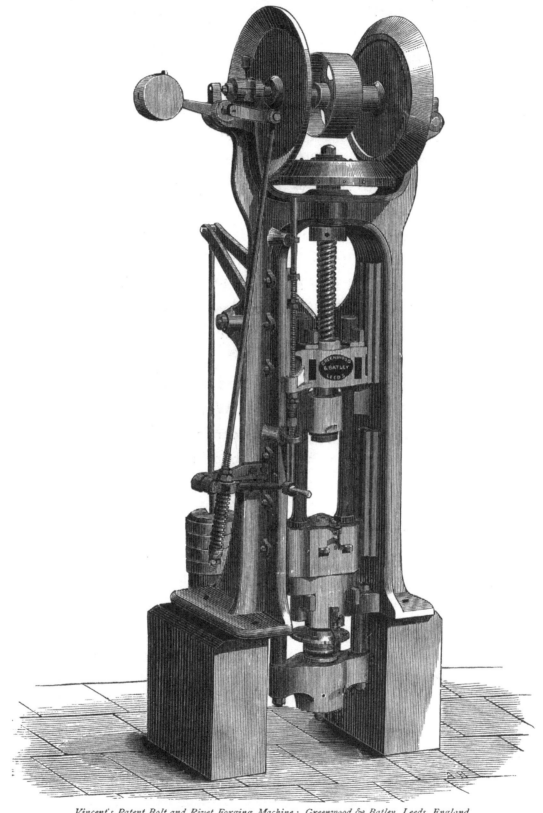

Vincent's Patent Bolt and Rivet Forging Machine; Greenwood & Batley, Leeds, England.

of the top driving-shaft should be from four to five hundred revolutions per minute.

The manufacturers claim as special advantages in this machine, that the die being made in one piece, all the heads are alike and true with the shanks, the sides being nearly perfectly parallel; that the top die can be very readily adjusted to form heads of any required thickness; that the machine possesses great lightness compared with the great power exerted by the screw pulling against itself; and that the speed of production is very great, one man being able to make up to as many as thirty bolts or rivets per minute, according to size. A small furnace of special design is made for use with this machine. The pieces of iron are cut to lengths and placed in holes in the sides, the furnace being square, and all four sides may be used, the body swiveling round as the operator desires. Special retort-furnaces are also used for small bolts and rivets, in which the iron does not come in contact with the flame and does not scale or burn.

The VALLEY MACHINE COMPANY, OF EASTHAMPTON, MASS., exhibits in the Pump annex of Machinery Hall several of WRIGHT'S PATENT BUCKET-PLUNGER STEAM-PUMPS, which appear to possess considerable novelty, furnishing an apparatus for the purpose of raising water at once simple, compact and reliable. Fig. 1 shows a perspective view, and Fig. 2 a sectional elevation of one of these pumps. Referring to Fig. 2, A is the steam-valve, and B the steam-cylinder; C is the upper and smaller portion, and D the lower and larger portion of the peculiarly constructed plunger; E is the water-cylinder, F the suction-valve, and G the discharge-valve; H is a hand-hole for access to the water-valves; I is a passage in the upper end of the water-cylinder, through which water is taken in on the down- and discharged on the up-stroke; J is the vacuum, and K the air-chamber. The crank-shafts, crank and pin are in one forging. The valves are simply circular pieces of metal, rubber or leather, rising on a stem which is fastened to the valve-seats; and should a valve fail, it can be temporarily replaced by a circular piece of leather or rubber-packing until a new valve can be obtained. The packing-rings in the water-plunger are made of gun-metal, the inside ring being made thinnest at the cut, thus giving more elasticity with less friction. If preferred, fibrous packing or a solid bronze metal end can be used instead of the rings. The water-valves can be removed by simply unscrewing two nuts and withdrawing the wedge that rests on the discharge-valve stem, the suction-valves being directly under the discharge-valves in the base of the pump. As is well known, very little of the power of the engine of a steam-pump is expended in drawing the water into the pump-cylinder through the suction-valves, most of it being employed in discharging it. For this reason, in this pump the water-cylinders are made of twice the area in comparison to

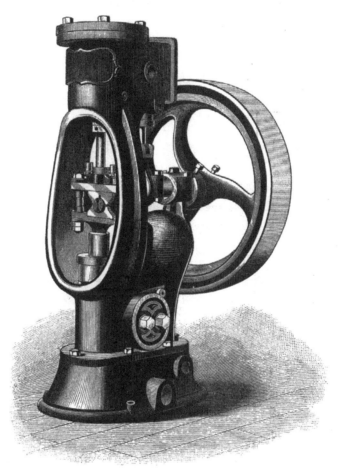

Bucket-Plunger Steam-Pump: Valley Machine Company.

the steam cylinders as in the usual make of double-acting steam-pumps. One-half the number of water-valves is dispensed with, and the quantity discharged on the upward stroke is thrown out through an opening, I, at the top of the pump-cylinder, and does not pass through the valve opening. This it is claimed is the great advantage of the bucket-plunger. With but two valves the same advantages are achieved in regard to a steady delivery as with the ordinary double-acting pump. The water is received only on the upward stroke, the amount being equal to the full capacity of the cylinder. Only one-half, however, is discharged, owing to the small area of the upper side of the piston. On the downward stroke, the water in the cylinder is forced out by the piston, one-half being discharged, the other half flowing into the upper end of the cylinder. It is claimed to be as good a pump as one of the four-valve double-acting pumps, with the advantage of only two water-valves to keep in order.

In describing some of the exhibits previously, we have already given a partial account of the method of weaving. The introduction of the power-loom towards the end of the last century gave an impetus to improvement which has been so great in the various processes of weaving as to almost make it a new art. The fly-shuttle had been invented before that, but otherwise very little had been accomplished to relieve the weaver of the drudgery of hand-labor. We know the art of weaving to be very old. The Egyptians nearly four thousand years ago excelled in it, as the mummy cloths now found in their tombs will testify, but no great improvement was made over the primitive apparatus until comparatively modern times. In 1679, M. Geunes, an officer of the French Navy presented to the French Royal Academy a model loom of his invention, in which the shuttle is shown to be carried through the warp by being inserted in the end of a lever, a corresponding lever meeting it half way, and receiving and delivering it on the other side. M. Geunes termed his invention an "engine for weaving linen cloths without the aid of a workman," and this may be said to be the first known practical idea of a loom driven by motive power. John Kay invented the fly-shuttle in 1733, and this with his previously improved method of making the reed of thin slips of metal instead of slips of reed, may be considered the

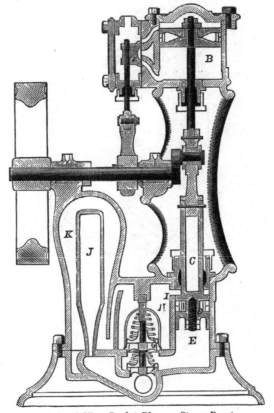

Sectional View Bucket-Plunger Steam-Pump.

most important inventions ever made to the loom. Kay & Stell applied the "tappet shaft" to the "Dutch loom" in 1745, rendering it capable of being worked without the use of treadles; and in 1760, Robert Kay, a son of John Kay, invented the drop-box, by which several shuttles with different colored weft could be used in the fly-loom. Then followed Cartwright's inventions, and although he was unsuccessful in his undertaking, yet his ideas were good, and the subject was too important to be neglected. The matter was taken up by other inventors with more or less success, and every detail which Cartwright had stumbled over was ultimately surmounted, his machine becoming one of the most perfect of the present day. Next came Jacquard with his perforated cards for the weaving of figured fabrics. Whether he was the actual inventor, or whether he merely combined together the ideas of his predecessors, succeeding for the first time in carrying them out practically, seems to be a disputed point. However that may be, the improvements which he introduced were of exceeding importance.

When the power-loom was fairly established, improvements began to follow in rapid succession. The Jacquard machine was first applied to the power-loom about the year 1830, and numerous modifications have been made in it from time to time. The take-up motion and weft stop motion in the loom have been very much improved. Shedding motions for the productions of small patterns have been introduced, many of them displaying great ingenuity. Circular- and drop-boxes have also received great attention, in order to adapt them to the speed of the loom, which has increased to double what it was formerly. Many minor parts of the loom have been greatly improved; but with all this, the same old arrangement of flying-shuttle has been in operation since the time of Kay with little or no change until quite lately.

To substitute a positive, absolute and uniform motion to the shuttle by means of some external appliance not actually connected with it, and exterior to the sheds of the warp, has been a very serious problem, requiring something entirely new and different from a mere modification of existing ideas. The shuttle has been a mere projectile, entirely out of the control of the weaver during its passage across the warp. In the earlier stages of weaving, it was thrown by hand. Then Kay connected pickers or hammers at the end of the shuttle-race, with a cord along the front of the loom, having a handle

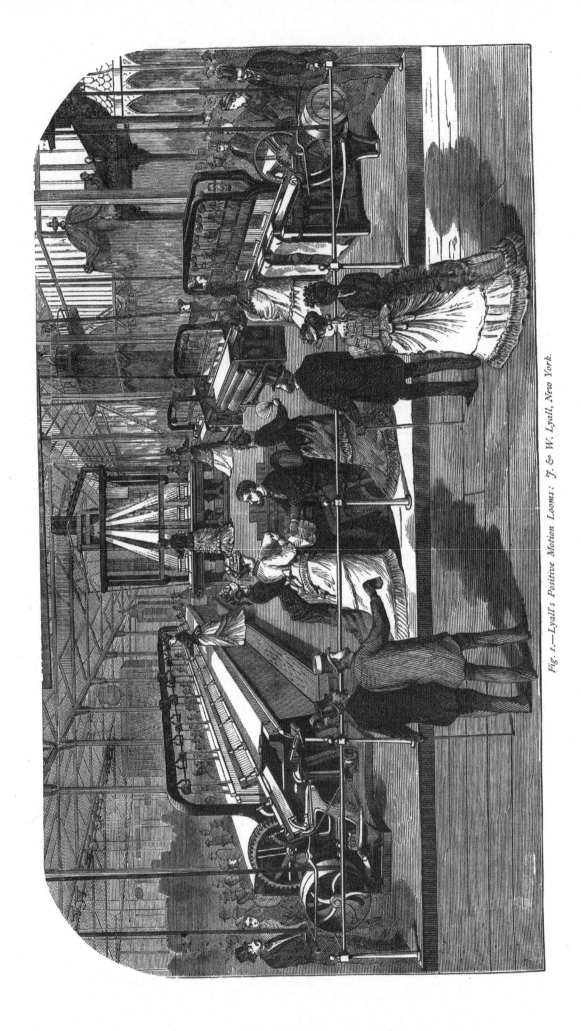

Fig. 1.—Lyall's Positive Motion Looms: J. & W. Lyall, New York.

or picking-peg attached to the middle of it, which the operator could jerk to one side or the other, causing the pickers to strike the shuttle and drive it back and forth. In the power-loom this is done by mechanism, but the principle in all cases is the same. The disadvantages of this arrangement are so many that it seems singular so many years should have elapsed without any essential improvement being made.

The ordinary shuttle, which is of a long cylindrical shape, pointed at the ends, slides over the warp-thread on its slightly convex bottom at a very high velocity, and must therefore produce a large amount of friction, resulting in a frequent breakage of the warp, and in the case of such delicate fabrics as silk, cambric, etc., a constant injury to the filaments. A limit in the fineness of such goods is consequently speedily reached, beyond which the fly-shuttle is

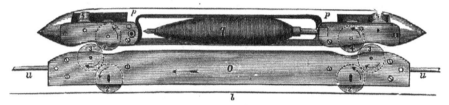

Fig. 2.—The Shuttle and Carriage: J. & W. Lyall.

practically useless. The width of the fabric is also limited by the distance to which the shuttle can be thrown. The farther it has to be thrown, the greater must be the force to propel it, and the more difficult will it be to stop and reverse its motion at the opposite end. The variable action of the propelling force is also a great trouble. If it acts too strongly, the shuttle may rebound, slackening the weft, or perhaps doubling or folding it up. If to obviate this the speed is slightly reduced, the shuttle may in some cases fail to pass entirely through. The irregular action of the force causes some threads to be drawn tight and some to remain loose, making an inferior quality of material, which soon gives way in the tight threads, and is worn out long before its time. If the shuttle should rebound to the threads or fail to pass through, it is struck by the lay, and either the delicate dents of the reed are broken or the threads of the warp destroyed. A delay of perhaps hours ensues, and the work may be repaired only to break again in a few minutes. It is also very difficult to make a perfect selvedge with the ordinary fly-shuttle, and as this is an important qualification for some varieties of goods, it is an essential matter. Another trouble is that the shuttle does not always fly in the direction it is intended to do, and sometimes takes an erratic course, resulting perhaps in great injury to the attendants.

Several devices have been tried to remedy these defects. It has been attempted to drive the shuttle by compressed air without success. Another arrangement consisted in a long needle or arm carrying the shuttle through the warp, and then withdrawing until the lay is beaten, when it again returns to catch the shuttle and draw it back. This has been more successful, but is only suitable for special purposes as in carpet-looms. Two hook-arms have also been used, and no shuttle, each one working half way. A rack- and pinion-movement under the shuttle-carrier has also been employed, working very well for short throws only, and used in weaving ribbons, etc.

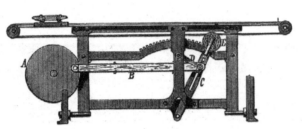

Fig. 3.—Shuttle Motion: J. & W. Lyall.

The question of solving these difficulties was taken up by Mr. James Lyall, and has resulted in his remarkable invention of the Positive Motion Shuttle which is exhibited in the magnificent display made in Machinery Hall by Messrs. J. & W. Lyall, of New York City. In Mr. Lyall's arrangement the shuttle rests on a carriage, o, as shown in Fig. 2, to which motion is given by a short band, u, passing over grooved pulleys fixed to the ends of the lay, and communicating with a single large pulley underneath the loom, this pulley having the proper movement imparted to it by a special mechanism which we will describe presently. The wheels, 2, of the carriage are pivoted to the ends of short horizontal arms, and the wheels, 3, of the same size as wheels 2, are simply journaled in the carriage, at the same time resting on top of wheels 2, from which they receive a counter and exactly equal motion, equal also to the motion of the carriage along the raceway, l. Wheels 4, also of the same size as the others, are attached to the shuttle, and rest on wheels 3. Now supposing the lower sheet of parallel threads of the warp to be stretched above the carriage and below the shuttle, p, the only points where the threads can be in contact with the carriage and shuttle are evidently between the wheels 3 and 4. If we move the carriage so that the wheels 2 revolve to the left, wheels 3 will turn to the right, and 4 to the left. It is evident that as wheels 3 move back at precisely the same speed as wheels 2 go forward, there can be no lateral movement of the threads as they successively rise over wheels 3. The shuttle above cannot affect this in any way, as wheels 4 are rotated by wheels 3 at precisely the same speed, and the successive threads for the unappreciable instant of time that they are between shuttle and carriage sustain no disarrangement otherwise than a very slight elevation. The friction is practically nothing, and as the sheds are constantly alternating and being bodily moved away as the weaving goes on, it is never applied twice at the same points in horizontal succession from thread to thread. In the upper part of the shuttle are two independent wheels, marked 5, which do not engage with the others, but roll along the under surface of a beveled rail, holding the shuttle down to its work. The shuttle is dovetail in section, and cannot be removed after being once in place, unless by drawing it out at the end of the lay.

Fig. 4.—Dwell in the Lay: J. & W. Lyall.

A number of novel movements and mechanical combinations are used in connection with the above-described device to properly carry out the whole work. Two of the most important of these are illustrated by Figs. 3 and 4. It is essential at times to produce a period of rest either in the shuttle or in the lay. The shuttle must stop long enough at the end of its run to allow the lay to be beaten, while the lay must delay its beat during the time that the shuttle is making its passage. In all cases the motion of the shuttle should be fastest midway in its course, and slowest at the ends. In Fig. 3, A is a crank-disk from which motion is imparted by a connecting-rod, B, to a sliding-block in the slotted vibrating arm, C. A link, D, is attached to the sliding-block and pivoted to the frame. The arm, C, has a wheel at its upper extremity, operating the shuttle-band and rotated by a rack and pinion device. When the crank-disk starts from the position as shown in the engraving, the shuttle being at the end of the race, the sliding-block is at the upper end of the slot in the arm, C. A very slow motion therefore takes place, increasing to a maximum as the sliding-block on the connecting-rod descends, until the shuttle is midway on its course, when the motion gradually decreases again. The shuttle never returns until the lay is home, and no matter what the position of the shuttle when the loom stops, it is drawn out of the way of the lay the first thing on starting again.

The dwell in the lay is an essential matter in the manufacture of heavy goods and in all cases where the shuttle has to travel a very long distance. It is obtained by a device shown in Fig. 4. A is a slotted pulley-wheel to which is attached the crank of the shaft, B, which gives motion to the lay. The crank-wrist is eccentric to the pulley and moves radially in the slot as the latter revolves, imparting an extremely slow or no motion to the shaft, B, when nearest the centre, and a quick motion when out towards the circumference.

The exhibit made by this firm comprises five great looms, which are represented in our engraving, Fig. 1. That on the left claims to be the largest loom in the world, and its operation fully establishes the fact that the width of fabric that can be woven by such a loom is unlimited. It has hitherto been a most serious undertaking to weave wide fabrics, such as oil-cloth foundations, with the hand-loom, requiring the services of three men, one at each end to drive the shuttle with heavy hammers, and a third to aid the others in beating the lay. Here before us is a mammoth machine, weaving a fabric eight yards wide and forty yards long in ten hours, or three hundred and twenty square yards per day, and hardly requiring the attention of even a young girl to keep it to its duty. The loom is made in two yard sections, the lay being beaten up in four places at once, and the strength of the material is equal to that which would be obtained from four single looms side by side. The back beams are sections of one yard each, so that they can be made on all ordinary warping-machines, and they are united in the loom by male and female clutches. The action of the apparatus shows at once how easily and beautifully the shuttle operates, no breaking of the yarn or stopping for repairs, everything proceeding smoothly and evenly for hours. The manufacturers state that shoddy almost too weak to stand its own weight may be woven by this machine, and that all the oil-cloth foundation that is used in the United States is now made by these looms to the entire exclusion of the imported material, made by hand-looms in Scotland, which was formerly employed.

Our engraving, Fig. 1, shows on the right, in the foreground, opposite to the great loom, a bag-loom, next to it a ten-quarter cotton-loom, and back of that a heavy jute carpet-loom, all operating with the LYALL improvements. The bag-loom manufactures four seamless bags at one operation, four shuttles, connected by rods, operating in the single race-way, travelling so that each in passing to one side or the other fills the place formerly occupied by that next to it. The bottoms of the bags are closed in the loom, and it is only necessary after they are made to cut them apart. The operator is able to examine both sides of the work while being performed, and holes and defects in the underside that occur with some other looms can here be avoided. The rate of the machine is about one hundred and twenty picks per minute. The ten-quarter cotton-loom, which contains some exceedingly ingenious new mechanical arrangements in the shape of compound let-off motion, variable dwell-crank, etc., has been weaving unbleached sheeting, producing a material unexcelled in point of fineness and level. It operates at the rate of about ninety-four picks per minute, and one attendant can take care of three looms with ease. The carpet-loom runs at the rate of about one hundred and ten picks per minute, manufacturing about one hundred yards of carpet per day.

In the centre of our engraving is a corset-loom, which combines in itself the positive motion and power-loom with the Jacquard apparatus, being a marvel of ingenuity, and will weave four webs of corset at once, perfect in form, with every gore, gusset and welt in place as if done by hand. It will turn out eighty-four corsets per day. The quantity of warp to be kept in action is governed by the Jacquard cards, thereby regulating the shape of the different parts of the corset to adapt it to the shape of the body.

The LYALL loom may be said to comprise in its invention the following advantages: First, the abolition of the picking-sticks; second, a positive motion to the shuttle from any point in its course; third, the unlimited width of the fabric which may be woven; fourth, the unlimited variety of fabrics which may be produced, from the finest silk to the heaviest carpet, from jute oil-cloth foundation to exquisite woven embroideries; fifth, the almost total absence of wear, through the small motion of the reed, which thus wears but little on the warps, through the small opening of the heddles, which thus offer less strain on the same, through the absence of friction of the shuttle on the yarns, and the non-subjection of the weft to sudden pulls on starting; and sixth, the extremely small amount of power required to operate the looms. The eight-yard machine has been operated by a three-and-a-half inch belt, and is easily moved by hand-power exerted on the gearing, it being stated that only one-half horse-power is required to work it.

Messrs. S. C. Forsaith & Co., of Manchester, New Hampshire, exhibit a patent Newspaper Folding-Machine, by which paper of all sizes may be folded

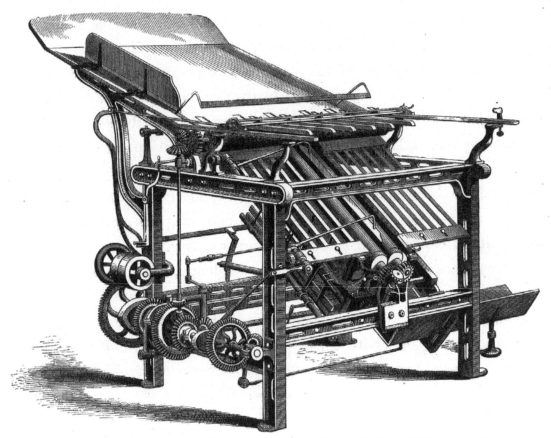

Newspaper Folding-Machine: S. C. Forsaith & Co., Manchester, N. H.

at an exceedingly rapid rate and in the most perfect manner, supplying a great want in the machinery of newspaper establishments, and attracting the admiring attention of visitors owing to its beautiful action. By this machine three-fourths of the expense of hand-labor is saved, the work being accomplished in a highly superior manner, twenty-five to thirty-five hundred sheets folded per hour to any desired size, and the services of only a single operator required. It is evident that in a newspaper establishment of any importance some machine of this kind is a necessity. The accompanying engraving gives a very good view of the machine, showing almost at a glance its manner of working. The papers as they come from the press are placed all together on the table at the top of the machine, and an attendant pushes them forward and downward, sheet by sheet, one after the other, each one until its edge, passing beyond the table, is seized by a pair of reciprocating fingers or grippers. These are operated by a horizontal rod connected at its rear end to the upper extremity of a vibratory lever, seen in the engraving on the left, and worked by suitable devices actuated by the shafting below. The paper is drawn outward by them, upon a support formed of horizontal slats or bars, until the centre of the sheet is brought directly underneath a transverse striker above it, which immediately descends and doubles it up, forcing it downward between two rollers arranged parallel to and under the striker. These rollers draw the sheet in between them and pass it downward, folding it once and delivering it upon an inclined set of slats below. It slides forward on these slats, an adjustable transverse stop preventing it from going too far. Its centre comes directly over another pair of rollers parallel to the slats, and another striker, properly arranged, forces it through the rollers, making a second fold. The paper is then guided to another incline with rollers, receiving here another fold, and the operation is repeated by successive folding-rollers and strikers until the required number of folds are made, when an automatic pusher delivers it into a box or trough. A packing apparatus may be connected with the folding-machine, packing the folded sheets in a box as rapidly as they arrive, ready for shipment.

An important feature in the machine is the readiness with which it may be changed so as to fold sheets of different dimensions, its capacity varying from six to twelve inches each way. It appears to be constructed of substantial and enduring materials, and requires no tapes or belts in its operation, possessing very little liability to get out of order. The sheets of paper in passing through the machine in the process of folding are smoothed and greatly improved in appearance, the results produced being similar to those by the dry press.

The art of Watch-making dates back as far the latter part of the fifteenth century, when watches of some form or other were, it is believed, made in Nuremberg. The business soon spread to England and France, but finally settled chiefly in Switzerland, principally in the cantons of Geneva, Neufchâtel,

Vaud, Berne, and Soleure, where watches have been made extensively ever since. The manufacture in these districts at present employs nearly forty thousand skilled workmen, many of them veritable artists, who turn out annually about half a million watches, of which number about four-sevenths are silver and the balance gold, the entire product being valued at nearly ten millions of dollars. So long as these ingenious, precise and delicate little instruments were made by hand alone, the art of watch-making possessed an interest even for the most casual and inexperienced observer. It seemed really wonderful that the human mind and the human hand could unite to produce such curious mechanism. When, however, machinery came into use, and the art became more or less mechanical, the charm was lost.

There still remain, however, some experts who are devoted to the old art of skilled workmanship in watch-making by hand, and certainly one of the most remarkable of these is M. MATILE, OF LOCLE, SWITZERLAND. The Swiss manufacturers are well represented in the Main Building of the Centennial Exhibition, their productions being classified under the head of "Instruments of Precision." M. MATILE has produced some of the most perfect watches for observation ever made, and his exhibits here shown, and represented by the well-known firm of L. & A. MATHEY, OF NEW YORK, for perfection in workmanship and accuracy of result are quite exceptional in their character. In this collection is displayed a most marvelous watch constructed by him and bearing his name, a piece of workmanship that presents the most extraordinary handiwork in the best known instance of its representation, it being certainly the most intricate watch ever exhibited in this country. It occupied M. MATILE more than two years in its construction, and designed as it is for positive and absolute accuracy in the recording of time, it denotes this to the fifth part of a second, while it is also accompanied by a

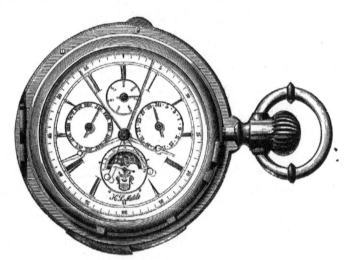

Fig. 1.—Face-Dial of Watch: M. Matile, Locle, Switzerland.

first-class certificate from the Observatory of Neufchâtel, where the test is unusually severe. The watch is a minute repeater, striking upon a most melodious series of bells, the hour, half hour, quarter, and minutes of the unexpired quarter. It is a perpetual calendar, displaying the days of the week and of the month, and the months of the year. It also offers an exceptionally interesting feature by presenting, through the medium of a wheel which makes only a quarter of a turn every year, the recurrence of the twenty-ninth of February once every four years. In addition to this, by means of a double chronograph combination, a double observation can be taken at the same time, as in boat-racing, or in the case of two horses starting at different times. There are two hands on the face of the watch for this purpose, both at the zero point. A movement of the stop starts the first hand, then another movement starts the second hand without stopping the first, and finally a third movement of the stop throws both hands back to the starting-point. There are many cases in which this double arrangement would be a very great desideratum. The watch is also arranged to give the phases of the moon very prettily, by the movement of a miniature planet over a minute horizon in exact agreement with the motions of its planetary prototype.

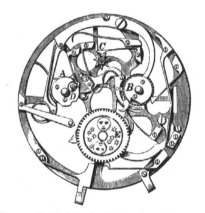

Fig. 2.—Matile Watch—Mechanism of Perpetual Calendar: M. Matile, Locle, Switzerland.

Not the least extraordinary fact in connection with the construction of this truly marvelous instrument is its having been put together with such economy of space and such exact balance of judgment that all of its phenomena proceed within a case not much larger than an ordinary gentleman's watch, while the draft upon the main train for power is but comparatively little. Our illustrations present quite clearly the leading characteristics of the Matile Watch. Fig. 1 represents the face-dial, and Fig. 2 the mechanism of the perpetual calendar. In the latter, the wheel, A, regulates the days of the month, B the month of the year, C the days of the week, D leap-year, and E the phases of the moon. From even this brief and imperfect description it will be seen that this watch may be ranked as one of the most remarkable exhibits in the Exhibition, combining within itself evidences of profound mechanical and mathematical knowledge, exact and experienced skill of hand and eye, and exceedingly great patience, industry and ingenuity.

Messrs. Robert Wetherill & Co., of Chester, Pennsylvania, exhibit a Corliss Horizontal Steam Engine of their own manufacture, illustrated by the engraving on page 249, which possesses some special improvements introduced by them, and has attained an extended reputation among those employing and interested in steam motors. The engine is particularly noticeable, in that all the parts are easily accessible and readily kept in order, the valve-gearing with

Fig. 3.—Matile Watch—Mechanism of the Minute-Repeater: M. Matile, Locle, Switzerland.

Fig. 4.—Matile Watch—Main Train and Mechanism of the Chronograph: M. Matile, Locle, Switzerland.

its connections being outside, and one plain eccentric moving the whole arrangement. There are no cam motions or no complicated details, and the risk of being thrown out of service on account of repairs is reduced to a minimum. The valves for admission of the steam are situated on top of the cylinder, one at each end, and at a distance from the bore not greater than one inch. The exhaust-valves are placed at the bottom of the cylinder opposite to the steam-valves, and can therefore easily free the cylinder of water without lifting it, or the use of special valves or cocks. They are at the same distance from the bore as the steam-valves, giving very little waste space. The exhaust-valves are so arranged that they commence to open their ports when the eccentric is producing its most rapid movement, and by means of the wrist-plate connec-

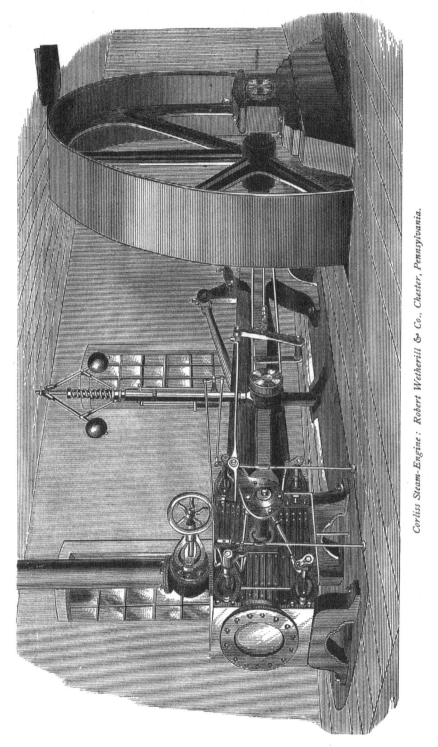

Corliss Steam-Engine: Robert Wetherill & Co., Chester, Pennsylvania.

tion the speed of the opening is kept at a uniform rate, the declining movement of the eccentric towards the end of the throw being compensated for by the increased speed from the wrist-plate. At the same time the steam-valve at the opposite end of the cylinder commences to lap its port by the same motion of the eccentric, doing so by a reverse or slower speed. The same throw of eccentric and the same movement of wrist-plate opens the exhaust-valve one and a quarter inches, and laps the steam-port half an inch in the same time. The exhaust-ports are one-fourth wider than the steam-ports, and the travel of the exhaust-valves being much faster and greater than that of the steam-valves, the engine is cleared of the exhaust steam easily and rapidly. The continual variations of load which usually occur in the operation of an engine are in this case communicated instantly and directly to the steam-valves by the governor, regulating the cut-off by tripping the valve, breaking its main connection, thereby closing it instantly, while at the same time giving the full force of the steam to the last moment, and then working it expansively to nothing and utilizing its full power. The governor in any case has little or no resistance, but is free to move, and only indicates the change of speed, operating by levers moving a small stop which comes directly in contact with the crab-claw or trip-motion which opens the valve. It is therefore very sensitive, and insures a perfect regulation of the speed. The valves are circular sliding, of brass, having stems fitted in them, to which, by means of a lever at the end, the motion is transmitted, leaving the valve itself free to find its own seat with the least possible friction. The valves and chest-covers are so constructed that a valve can be removed in an exceedingly short time by merely taking out four bolts. Each valve has a separate connection, and each one may be adjusted independently and with great accuracy to suit the power required. The indicator-cards which have been made from this engine show great perfection, corresponding very closely with theory. The whole valve-gearing may easily be worked by hand under any pressure of steam, an advantage which is often quite a necessity. The qualification of quickly opening and closing the valves at the proper time, as provided for in this engine, prevents wire-drawing, and is one of the great points of superiority in the Corliss movement, which has attained so widely extended a reputation throughout the world for economy and efficiency in the use of steam.

This engine is provided with WETHERILL'S PATENT PISTON-PACKING, a peculiar system of segmental packing-rings arranged to break joint, and with an elliptical spring attached to each. This packing has been in use for several years in a large number of engines and also on locomotives, giving universal satisfaction in every respect. Engines have been tested with it, operating them as single-acting, with the back cylinder-head off, the packing not leaking in the least, the cylinder being kept polished as smooth as glass, and with no perceptible wear. It is claimed that a piston packed in this way will last for years without the least attention or repairs. The following are a few of the advantages claimed for this piston: It operates with ease and certainty, giving perfect tightness and durability, and is self-acting, without any attention or setting out of springs. It has no follower with bolts or nuts to come loose, is simple and light in construction, combined with proper strength, and there is little or no liability to get out of order. It is not set out by steam, therefore it is not subject to any of the faults of the so-called steam-packing, such as cutting the cylinder by great pressure and wearing it large at the ends, with an unnecessary amount of friction, often causing the rings to break. The piston-head is made in one piece, cast hollow with ribs, affording sufficient strength with one-half the weight of other pistons, and reducing the wear on bottom of cylinder. The springs attached to the segments are of German silver, which, while it does not corrode, retains its elasticity in the temperature of high-pressure steam, which is all that is necessary to keep the piston steam-tight. This piston has been used on some large horizontal engines with a piston-rod running through the back head, with very satisfactory results, the arrangement relieving the weight of the piston on the bottom of the cylinder, saving friction and keeping the cylinder perfectly sound. If it is specially desired, the piston could be made with a follower, with a loose chuck-ring, to be set up as the piston wears down in the cylinder, but this is not considered advisable or indeed at all necessary with this packing.

The subject of barrel-making by machinery is an exceedingly interesting one, and it is almost a matter of astonishment to note the ease and rapidity with which tight and slack barrels, kegs and casks are now turned out in comparison with the slow process of hand-manufacture as in almost universal use comparatively a few years ago. The peculiar requirements necessitate special machinery entirely novel as compared with that employed in ordinary wood-working operations.

Messrs. E. & B. Holmes, of Buffalo, New York, make an excellent display of barrel-making machinery at the Exhibition, for which they claim great durability, rapidity of working, and the satisfactory performance of the skilled labor of this industry, their use being very extensive in this country, and also obtaining even in Europe, notwithstanding the low cost of the competing hand-labor. The machinery of this firm is divided into that for tight barrels such as are intended for liquids, that for slack barrels to hold flour, sugar, cement, etc., and that for making small kegs and casks.

The first operation necessary in barrel manufacture is to prepare the staves. These are either cut or sawed from the solid timber, cutting-machines being made which will turn out from twenty-five to thirty thousand staves per day. The cylinder-sawed staves are coming into very general use, and machines are made by this firm for sawing them of all sizes, from those for the smallest keg up to the largest hogshead, the barrel size producing about five thousand staves per day.

Fig. 1.—Combined Fan and Stave-Jointer: E. & B. Holmes.

With kegs especially, the staves are so short that there is no risk of their being sawed across the grain sufficiently to injure them, and in some cases they are even sawed into the form which they will occupy in the finished article, cylindrical bilged saws being used for the purpose. The staves are next reduced to uniform lengths for each size of barrel by means of two parallel circular-saws at a fixed distance apart, the pieces being presented to them, in one variety of machine, by a swing-carriage pivoted at the bottom of the saw-frame, and in another by means of a continuous feed arranged by two wheels of the same size running on the same axle and having a series of lugs on their circumferences on which the staves are laid and by which they are carried up against the saws. The operation in this latter machine is one of exceeding rapidity. By means of a conveyor operated by the machine itself, the staves are carried off and delivered wherever required. They are now ready to be dressed. Two machines are made by Messrs. Holmes for this purpose. One, for rived or cut, and sawed staves, operates with revolving cutters, dressing staves of all thicknesses, both sides at once, and without cutting the wood across the grain, but leaving the staves winding and crooked as they are rived from the block. This is accomplished by allowing the frame which supports the cutters to oscillate or rock in all directions, so that the cutters can

Fig. 2.—*Trussing Machine: E. & B. Holmes.*

adapt themselves to the crooks and winds in the material, bringing it to a uniform thickness and leaving it convex or concave as desired. This feature overcomes difficulties which are a prolific cause of failures in machines not so arranged. The other machine of this firm is for sawed staves only. It acts with a rotary cutting head and a carrying or revolving bed, having feed-rollers which compel a strong and positive forward motion, obviating any difficulty with irregular feed. The stave is placed upon the bed and carried forward in a direct line under the feed-rollers, coming out with a smooth finish and uniform thickness, the form being capable of change at pleasure. The work is accomplished at a rapid rate; either one or both sides may be dressed, and the thickness may be changed instantly if desired.

The stave now passes on to be jointed and put into the proper form to take its place in the barrel. Fig. 1 illustrates MESSRS. HOLMES'S COMBINED FAN AND STAVE-JOINTER, used for this operation. A great objection to all stave-

Fig. 3.—Machine for Leveling and Trussing Slack Barrels: E. & B. Holmes.

jointers driven by power consists in the dust and shavings which accumulate so rapidly and require such constant care and expense in their removal. The dust especially, fills the air of the entire building to the great inconvenience and discomfort of those occupying it. The machine here illustrated, however, possesses the great advantage of a fan in combination with it, by means of which these nuisances are blown through conductors to the fuel-room of the establishment and entirely avoided. In working the machine the stave-holder is brought to bear upon the stave, which, by the action of the foot of the operator on the treadle, comes into contact with the jointer, the operation clamping it so that it cannot change position or become loosened until the foot is removed, when the clamp recedes from the cutters and releases it, leaving the machine ready

Fig. 4.—Machine for Leveling and then Trussing Slack Barrels: E. & B. Holmes.

for another piece. The desired bevel and bilge may be given without trouble. Messrs. Holmes also manufacture other varieties of jointers without fans, one a wheel-jointer for slack barrel staves, an excellent machine, very rapid in its action and making smooth and perfect joints with great economy of stock.

The staves now being prepared, they are put up into the shape of barrels on a setting-up form, which consists of two heavy circular rings of iron secured together, one within the other, and bolted to the floor, short standards rising up and supporting a hoop above. The form is made adjustable to suit the size and kind of barrels to be manufactured. The staves being set in between the iron rings and carefully fitted together, iron truss-hooks, previously placed in proper position, are lifted up by hand so as to take hold of the lower portions of the staves and keep them in place, and the whole is lifted out of the frame. The upper portion of the barrel is still open and flaring, but by passing a rope around this part, and by means of a hand-windlass, a few turns of the latter brings the staves together, allowing iron truss-hoops to be slipped over. The barrel is now ready for heating to cause the staves to assume the proper curved shape, and for this purpose it is set over a simple iron cylindrical stove and the top closed by a sheet-iron cover, it remaining there until well warmed through. After that it passes on to be leveled up, or put into such shape as to be square to its centre line and stand in a perpendicular position when on

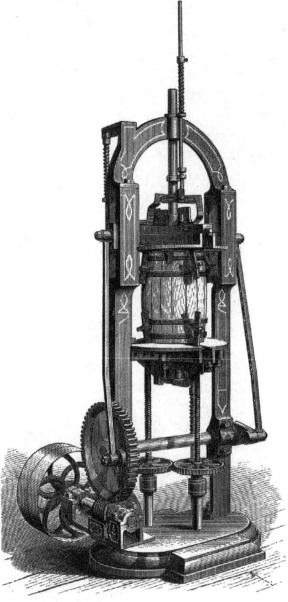

Fig. 5.—Machine for Leveling and Trussing Kegs and Small Casks: E. & B. Holmes.

end. The leveling machine consists of two disks, between which the barrel is placed and held by projections on their inner faces, the disks moving towards each other and powerfully compressing it into shape, the capacity of this machine being three to four hundred casks per hour. The trussing-machine next takes it in hand, driving all the hoops up very tightly. Fig. 2 illustrates one of these

Fig. 6.—*Machine for Chamfering, Howeling and Crozing Tight and Slack Barrels: E. & B. Holmes.*

machines. The barrel is placed on end, and a number of hooked bars pass up through the floor, catching the upper hoops, while notched standards receive the lower ones, and the application of power draws the bars down with great force, wedging the hoops home. This machine will manipulate ten to fifteen thousand barrels per day, and is equally well adapted to tight and slack barrels, the quarter-hoop drives, however, being removed in the former case.

Machines are also provided by this firm to accomplish the duties of both leveling and trussing, one of which we illustrate by Fig. 3, and another by Fig. 4. The machine shown by Fig. 3 will level and drive truss-hoops at the rate of eighteen hundred to twenty-four hundred barrels per day, and that given by Fig. 4 will perform the same work with still greater rapidity, as rapidly in fact as the operator can roll the barrels in, and averaging from four to five

Fig. 7.—Machine for Chamfering, Howeling and Crozing Kegs and Small Casks: E. & B. Holmes.

thousand per day. The latter machine greatly surpasses all others in usefulness in large establishments where barrels are manufactured in great quantities. Fig. 5 represents a very excellent machine for leveling and trussing kegs and small casks, such as lead-, nail- and other kegs, firkins, half barrels, etc. It supplies a want long felt in the business, being very rapid in its operations and having a capacity of from four to five thousand kegs per day with one operator, and is easily and quickly changed from one size of keg to another.

After leveling and trussing, the barrel requires to be chamfered, howeled and crozed. It still has very irregular edges at the ends, the staves varying in length, and it is necessary to cut these ends off perfectly true, then to chamfer

Fig. 8.—Combined Fan, Heading-Jointer and Dowel-Borer: E. & B. Holmes.

or bevel them, and to croze them, or work a groove around the inside for the reception of the head, just below the chamfered edge. In heavy casks, a howel or wide semicircular indentation must also be cut around just below the croze. These operations are all exceedingly difficult and tedious when performed by hand, and Messrs. E. & B. Holmes have done immense service in furnishing a machine to accomplish the same work, not only exceedingly novel and ingenious in its construction, but possessing extraordinary powers, and performing its duties with great simplicity, efficacy and rapidity. This machine, illustrated by Fig. 6, will chamfer, howel, level, and croze a cask of imperfect periphery with

Fig. 9.—Machine for Dressing and Leveling Heads: E. & B. Holmes.

the same exactness as if it were of a perfectly symmetrical shape, finishing both ends at one operation. The three tools required for the various operations are placed in a revolving head, one head for each end of the barrel, and the barrel is passed in between chuck-rings, its ends fitting into the peripheries of cog-wheels which work within these rings. The wheel shown in the engraving as being turned by the workman, governs the backward and forward movements of the near chuck-ring, the other one being stationary. As the barrel passes into place, the chuck-ring is brought to bear against it, and a clutch thrown into gear causes the barrel to rotate rapidly. The revolving heads containing the cutting tools are then brought into rotation at a high rate of speed on the interior edges of the cask, their action being controlled by rests on the outside, compelling a uniform thickness and depth of chime, and at the same time leveling up perfectly. The rests and cutting heads oscillate to conform to the outside irregular form of the cask while revolving, and one revolution of the barrel completes the work. By using chuck-rings of the proper size, any kind of barrel may be operated upon, and the change of sizes is easily made. One man can turn out with this machine from eight to twelve hundred tight, and from two thousand to twenty-five hundred slack barrels per day.

We also show by Fig. 7 another machine very similar to the one just described, but intended for kegs and small casks. It will finish both ends of the keg or cask at the same time, doing its work perfectly and rapidly, and completing from twenty-five hundred to three thousand pieces per day. All sizes may be operated on it, from the smallest lead-keg to half barrels, the change from one size to another being easily and quickly made, and only one operator is required to run it.

Our barrels are now ready for their heads, and Messrs. E. & B. Holmes also manufacture and exhibit very novel and ingenious machinery for making these very necessary parts. The heads, as every one knows, are composed of several pieces jointed and doweled together. These operations are performed by a combined Fan and Heading-Jointer and Dowel-Borer, shown by Fig. 8. In the face of a large revolving metal disk, equidistant from each other, are three cutters, and in front is a stand and rest. The piece of head-board to be operated upon is laid on the rest, and its edges pressed up against the disk, the action of the cutters dressing it down perfectly smooth and straight. Above the disk is a boring apparatus with two horizontal augers and a rest. Transferring the board to the upper rest, the augers are brought forward into action on it by the foot-treadle, and the holes are formed for the dowels. It will be noticed that the periphery of the disk is enclosed and connected with a metal pipe. By this arrangement the disk acts as a fan and blows away all shavings and borings, delivering them to the fuel-room, and saving labor, shop-room and risks of fire, while at the same time adding to the health and comfort of the operator.

The separate pieces forming the head are now joined together with dowel-pins, by hand, and in this shape are taken to the MACHINE FOR DRESSING AND LEVELING, shown by Fig. 9. This machine belongs to that class called surface-

Fig. 10.—Machine for Rounding Heads of all sizes for Kegs and Barrels: E. & B. Holmes.

planers, in which a constant gauged amount of material is cut away from the surface of the material every time it is passed through, there being a revolving cutter in a fixed bed, gauged above the surface of the bed or table by the amount of the cut. Above are four feed-rolls, held firmly against the material being planed, by weighted levers. The action of the feed-rolls carries the work over the revolving cutters, which rapidly smooth off the under side, dressing from fifteen to twenty-five heads per minute.

After this the head-boards are taken to a machine which prepares them finally for the barrel, rounding or cutting them into shape, and making a bevel on each edge, the lower one considerably larger than the upper. This machine is illustrated by Fig. 10, and is so constructed that any size of head, for kegs, small casks and barrels, can be made upon it with the same saw, the necessary change from one size to another being easily and quickly made by the turning of a hand-wheel. It will be noticed that the machine has two standards or uprights. The head is placed between a disk on one standard, and a number of spring-pins attached to a shaft passing through the other standard, and provided with wheel and belt for rotating. The spring-pins are thrown forward by a lever and clamped in that position, holding the work firmly. By the rotation of the shaft the head is carried round for one revolution, at the termination of which the clamp is unlocked by stop mechanism, and the pin-disk springs back and throws off the work. An extra lever provided for the purpose allows the clamp to be unlocked before a complete revolution if desired. The saw is mounted on the other upright, upon a separate carriage, having its own special belt for motion. Two peculiarly arranged knives are fixed so that when the action of the foot on the treadle brings the cutting mechanism up against the edge of the head, both sides are cut at once, while at the same time the work, owing to its rotating, is turned to a circular form. When the treadle is released a counterpoise weight brings the saw-carriage back to its original position ready for another head, that operated upon being thrown off automatically. The saw and cutters pass through the wood upon such lines as to prevent the tearing and slivering of the material, and give a smooth finish to the work. An attachment is provided by which, if desired, the heads may be cut to an oval form to compensate for the shrinkage of material.

The forming the barrel being completed, they are now put together by hand and hooped, and are ready for use. It is wonderful how rapidly, with such machinery as we have described, barrels may be manufactured, some single establishments turning out as many as five to six thousand per day.

Among the various exhibits in Machinery Hall, none perhaps have attracted more attention than that of the STOW FLEXIBLE SHAFT, made by MESSRS. STOW & BURNHAM, OF PHILADELPHIA. This ingenious arrangement consists of a flexible shaft which can easily be bent in any direction, even at quite a sharp angle, without being interrupted in its motion or reduced in effectiveness, and will readily transmit power in any direction. The special application of this invention to dentistry has been in use for several years, and has been already described by us, but it is only lately that the device has been materially modified and enlarged so as to allow it to be made use of for miscellaneous drilling, boring, polishing, etc., operations in which it does its work so beauti-

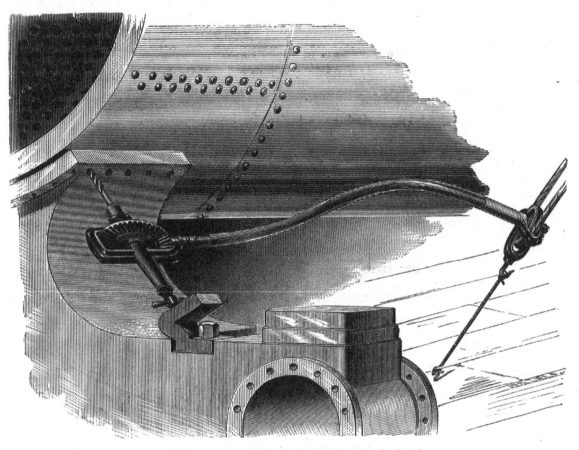

Stow Flexible Shaft—in Locomotive Work: Stow & Burnham.

fully, so rapidly and so readily, even in locations exceedingly difficult of access by other machines for the same purpose, as to commend itself at once to the practical man as a most useful piece of apparatus. The flexible shaft is constructed by winding successive coils of steel wire, one upon the other, each running in the opposite direction to that of the coil below it, the winding being executed at a very slight pitch, and four or five wires generally used simultaneously to make one layer, although the number depends somewhat upon the kind of work which the shaft is expected to perform. When a torsional strain comes on the shaft, its peculiar construction prevents the liability of any coil being twisted smaller, and it is quite flexible, the consecutive wires of each coil opening imperceptibly along the outer radial surface. A distance of about one and a half inches in length at each extremity of the cable is brazed solid, and to these solid ends are fitted sockets, one to receive the tools to be operated, and the other the pulley or other appliance through which the motion is transmitted to the shaft.

It is necessary now to provide a cover for the shaft, to prevent its causing injury or being injured by coming into contact with anything while it is in rapid motion, and also to add somewhat to its stiffness and not allow it to be bent at too sharp an angle. This cover is formed by first winding a single coil of iron or steel wire loosely around the outside of the shaft, allowing the latter still to move quite freely within it, and then placing on this a flexible cover of linen braid or, preferably, leather, over which at either end a ferrule is fastened. At one end a hand-piece is screwed into the ferrule, and at the other the frame which carries the driving-pulley. The spindle from the shaft, passing through the pulley, by means of a simple contrivance, is so arranged that it can move freely back and forth for a short distance, this being necessary to allow for the change in length caused by bending the shaft. A hook connected with the pulley-frame allows it to be fastened to any convenient object by means of a cord or rope, adjusting it to the length of the belt, which may be run from any adjacent counter-shaft. It will be seen therefore that the machine possesses stability, durability and great torsional strength, while at the same time having entire flexibility in all directions. At the hand-piece a clutch is provided in the spindle of the shaft, which may be thrown in and out of gear by a small projecting pin in the hand-piece, enabling the operator to stop the machine very quickly, in case of necessity, without throwing the belt-shifter on the counter-shaft. Shafts are now made from one-fourth of an inch to one and three-eighths inches in diameter, and from eight to ten feet in length, being limited in size by the weight which it is practicable to handle conveniently. One would imagine that the friction of the shaft within its cover would be very considerable, especially when bent or curved, but in practice this has not been found to be the case, and even when bent at a right angle the friction is not over fifteen per cent. greater than that occurring when the shaft is used straight.

A provision is made not only for oiling the shaft at its ends, but also within its covering for its whole length, to provide against wear and friction and insure easy working.

The uses to which this machine may be put are very numerous. In one form it is arranged as a POWER TREADLE MACHINE, and becomes available for carving, engraving, glass-cutting, wood-boring, sharpening of tools, light metal-drilling, etc., etc., a Stow Patent Treadle being used, which possesses the qualification that it is always in position ready to start. Burrs for engraving, corundum- or emery-wheels, polishing-brushes, or any tools desired, may easily be attached or removed at wish. For cleaning castings one of these shafts with the brush-tools provided is a most convenient and useful instrument, doing the work far better and more rapidly than can be done by hand.

Small breast-drills, like those generally in use about machine-shops, are exceedingly useful when worked by the flexible shaft. Our engraving shows a larger and more powerful drill, employed on a difficult piece of locomotive-work of the kind for which these machines are particularly well adapted. Stop-clutches are enclosed in the hand-pieces of the drills, which may be thrown in or out of gear by means of a trigger-piece, the forefinger resting on this trigger as the machine is held by the operator, and working it easily and effectively. For grinding operations the emery, sand or other grinding-wheel is inserted in a turned recessed plate and held firmly in position. This plate may be readily unscrewed if desired, and another plate attached in a few moments with a coarser or finer wheel. The machine may be used in the same way for wood-smoothing, sand-papering, etc., a lock-ring holding the sand-paper or cloth in place.

In working stone the flexible shaft can be employed to great advantage, the necessary tools being attached and carried to the stone without any necessity of accurately leveling up or setting the stone to any particular position. A broad circular disk of cast iron driven by a bevel-gearing has been used with great success in polishing marble, and the same arrangement with emery-wheel has been used for polishing granite. Fluted columns can be formed very rapidly by first roughing out the flutings with one tool and then finishing up with a cast-iron cylindrical tool having hemispherical ends, in connection with sand and water.

For cutting out defective boiler-tubes the machine is unexcelled. A cylindrical tool, having at one end a loose collar, is inserted in the flue to be cut off, and held in position by clamps. A short distance back of the collar a small cutting tool projects from the cylinder, which tool may be forced outward by a feather working in a slot made in the base of the cutter, the feather making an angle with the axis of the cylinder. Moving the feather backward or forward by an attached rod and thumb-screw throws the cutter in or out. A small pinion on the flexible shaft drives a spur-wheel keyed on to the cutter-bar. By operating the feed-screw the cutter is thrown out against the flue, cutting it through much more rapidly and quite as effectively as by the ordinary process, at the same time with the great advantage of destroying much less length of tube. It is stated that two-inch flues have been cut off with this machine readily in less than three minutes.

MESSRS. DAVEY, PAXMAN & CO., OF COLCHESTER, ESSEX, ENGLAND, exhibit a VERTICAL BOILER of somewhat peculiar type, that has been before the public in England for some time, meeting with considerable favor. We have already mentioned this firm's PORTABLE STEAM ENGINE, and we now desire to draw attention to their VERTICAL BOILER.

The manufacturers claim that the DAVEY-PAXMAN BOILER owes its excellence to a number of ingenious contrivances, each of which, although good in itself, would fail in being effective unless in combination with the others. Thus, water-tubes are sometimes suspended from the crown-plate to the furnace, and by this means no doubt heat is readily absorbed and steam generated more rapidly than with an ordinary boiler; but as these tubes are closed at the bottom, deposits necessarily soon occur, making their destruction inevitable. The circulation of the water also is only partial, that in the annular part of the boiler remaining comparatively inert. Sometimes, and more consistently, these tubes have been bent and let into the lower part of the fire-box, thus securing a circulation of the water in all parts of the boiler, but this is not sufficiently rapid to prevent incrustation and its evils. It is claimed that only by adopting the DAVEY-PAXMAN patented principle of tapering these tubes to certain well-defined proportions, and by using them in connection with patent Mushroom Deflectors, which wholly prevent priming, that a circulation can be obtained sufficiently rapid to prevent incrustation without producing the wet

Fig. 1.—Davey-Paxman Vertical Boiler: Davey, Paxman & Co.

bad steam usually made by boilers with brisk circulation. It is quite as important that a boiler should supply dry steam as that it be an economical generator, and one of the first essentials of economy in any steam-engine is that it should be supplied with steam of a good quality. The DAVEY-PAXMAN BOILER is claimed to be peculiar in that it combines a maximum of evaporating efficiency with the production at the same time of steam of excellent quality, not too hot nor too dry, but yet as hot and dry as required, to work with advantage.

Tubular boilers may be divided into two classes—fire tube boilers, in which the heat is wire-drawn, so to speak, through the column of water within the boiler; and water tube boilers, where the converse takes place, or the water is passed in cylindrical columns through the hottest part of the furnace. Of the former, the Cornish boiler is the most simple type, one or two flues or tubes varying from one-third to two-thirds of the whole diameter of the boiler, running through the water-space. The heat from a furnace at one end traverses these tubes, is received into a brick chamber at the opposite end, and from thence, passing through brick flues at the side, perhaps back and forth twice, it goes beneath the boiler, and finally escapes at the shaft. Although a fair proportion of the heat is absorbed by the water, a large quantity is retained by the brick-work, and much escapes by the chimney.

If the fire-tubes are increased in number and made of smaller size, we have the locomotive type of boiler, for some reasons superior to the Cornish boiler, possessing as it does much larger heating-surface. It is defective however in complete combustion, which can take place only for a short distance along the small tubes for want of oxygen, a defect improved with the use of the blast, not applied however with stationary boilers. Another defect consists in the fact that the tubes are subject to very considerable alternate expansion and contraction, from exposure, first to the heat of the furnace, and then to the cold air every time the door is opened, loosening their joints and causing leakage, a source of much trouble and expense. The cost of this type of boilers is also much more than that of the Cornish boiler.

The system of vertical tubes has been very frequently applied in vertical boilers, resulting in much greater liability to leakage, and in priming, which there seems to be no means of preventing. If these difficulties could be prevented in the vertical boiler it would largely increase its usefulness; its compactness, portability, cheapness, and general convenience making it otherwise an exceedingly desirable boiler.

To produce as nearly as possible a perfect boiler, one which should supersede all others where from one to forty horse-power is required, the following would seem to be the essential points:—

"1. An extensive heating-surface, so arranged that the water should be exposed in small cylindrical columns or otherwise to the hottest part of the furnace, with an intervention of a minimum thickness of metal, and other contrivances should be adopted to secure as far as possible the perfect absorption of the caloric.

"2. A complete and rapid circulation of the water in every part of the boiler, so that its whole volume should pass some ten or more times every minute through the most ardent part of the furnace.

"3. An entire prevention of incrustation and deposit in those parts where mischief can be done

"4. A stoppage of priming—that which renders vertical boilers so unpopular from the enormous consumption of fuel it entails.

"5. The boiler should not be subject to get out of repair.

"6. Easy access to every part for repairs.

"7. Easy of management even by inexperienced persons.

"8. Every part directly under control.

"9. Portability and no time or expense in fixing.

"10. Occupying little space.

"11. Perfect safety from explosion and its disastrous results.

"12. And as a result of these, the highest efficiency and economy."

MESSRS. DAVEY, PAXMAN & CO. claim that these are precisely the qualities that are found in the DAVEY-PAXMAN BOILER, an illustration of which and sectional view we give by Figs. 1 and 2 on pages 268 and 271.

This boiler, which is of the vertical type, consists of an outer cylindrical shell and an internal fire-box. Water-tubes extend from the crown-plate downwards to within ten or twelve inches of the fire-grate. Here these tubes are bent and tapered and inserted in the side of the fire-box, in which, as well as in the crown-plate, orifices are made for that purpose. This is an important part of the invention. The tubes conduct the water through the hottest part of the furnace; their insertion into the lower end of the fire-box allows it freely to circulate throughout the whole boiler, and the tapering of the tubes greatly increases its velocity, which again imparts a scouring action to it, and effectually prevents incrustation where it would be most dangerous. The deposit is allowed to settle in the narrowed part of the annular water-space, where the action of the fire is least, and whence it can be easily blown off and removed.

When water is placed over a fire in an ordinary vessel, and the lower part only of that vessel is exposed to the heat, ebullition takes place very slowly, because the increments of heat can only be imparted to the water through the surface so exposed, and a very small portion indeed of the molecules of the water to be heated can be exposed at one time. As these become heated they rise, and their place is supplied by colder ones from the top. This process goes on until the whole volume is raised to 212°. This being the ordinary process of ebullition, it is evident that the time required to raise a given quantity of water to the boiling-point must be in proportion to the rate at which the molecules of the whole volume are brought into contact with the heating-surface and the condition of that surface for conducting the heat. In this boiler it is claimed that the maxi-

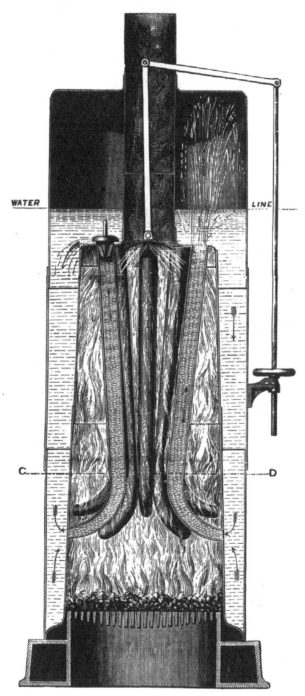

Fig. 2—Sectional Elevation of Davey-Paxman Boiler.

mum heating-surface has been gained, the larger portion of it consisting of tubes not more than one-eighth of an inch thick, placed in immediate contact with the intensest heat, and containing a large proportion of the entire volume of water. When fire is applied, the water in these tubes, immediately becoming warm, ascends, starting a circulation which increases in velocity throughout the whole boiler until ebullition ensues, when it is calculated the whole body of

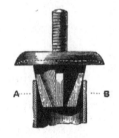

Section of Tube, with Elevation of Valve.

Section of Valve and Tube through A B.

Plan of under-side of Valve.

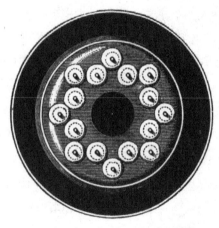

Plan of top of Tube-Plate, showing Valves.

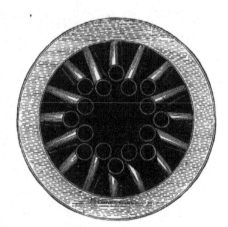

Section at C D.

water passes through these tubes in a few seconds of time. One may see the circulation by removing the manhole cover before boiling-point is reached, the water rising in fountain-like columns above the orifices in the tube-plate, and rushing with equal velocity down the annular space.

With such a circulation the boiler would be unserviceable, and indeed could not be worked without the use of the patent Mushroom Deflector. By inserting these into the mouths of the tubes, the impetus of the water is controlled, the fountains are spread out and diverted into their proper channel, the surface of the water is smoothed, and priming prevented.

A contrivance for regulating the draught, called the Baffle-Plate, is shown in the sectional view, Fig. 2, and is adjustable to the greatest nicety from the outside by a wheel within reach of the engineer. It prevents the escape of heat, throwing it back upon the tubes just where it can best be utilized. It is stated that with this contrivance a pressure of thirty pounds can be maintained for as many minutes after the fire has been extinguished. In the larger boilers two baffle-plates are employed, one just over the furnace, and the other immediately under the up-tube.

The bent taper water-tubes and deflectors can be applied even to Cornish and to Portable boilers with the best results, and for marine purposes they are invaluable.

Safety is insured in these boilers, owing to the fact that even when the water gets very low, as long as it reaches the tubes the ebullition will be such as to cause it to flow over the crown-plate and prevent trouble; and when it becomes too low for this, these tubes themselves, being very thin and exposed to the hottest fire, will serve as fusible plugs, and can do no harm except putting out the fire.

The boiler is stated to be very economical, and will burn any kind of fuel—coal, wood, peat, sawdust, or any rubbish—with advantage. It is also economical in space, prime cost and repairs, and has successfully stood many very rigid tests in England.

The earliest carriages were sleds moving on runners over the snow in cold countries, or in water-channels, to reduce the friction, in hot climates, as one sees in Madeira at the present day. This arrangement was soon improved upon by the use of large wooden rollers, which at a later date were diminished in size at the centre and at the ends, forming shafts with drums on them such as are still in use in some parts of Spain and Portugal. The invention of wheels revolving on a fixed axle followed afterwards, but in some parts of the world, at least, must have been introduced very early, as we find them on the war-chariots of the Egyptians six hundred years after the Flood, the wheels being made with a hub, and having at first only four, then six, then eight, and finally, about the close of the Assyrian or Medo-Persian Monarchy, twelve spokes. These spokes at the outer ends were mortised into a rim of wood, which may have been at first in one piece, but certainly during the Assyrian Monarchy was made in a number of pieces or felloes, and it was provided with a tire of bronze. Richly embroidered housings and trappings were furnished for the horses and cloths for the chariots. The chariot of the Romans and Greeks, about the time of the Christian era, was a four-wheeled vehicle, used only by the wealthy and extravagant, drawn by four or six horses splendidly caparisoned, and trimmed and cushioned in the most luxurious manner.

During the Dark ages the use of the carriage seems to have been almost entirely abandoned, and its revival afterwards in many respects was really a re-invention. Neither the ancient carriages nor those first used in later days had springs, although leather and steel of the best quality were both available for the purpose, and on rough roads the occupants were terribly jolted and jarred, making it necessary to their safety to hold on to bars provided for this use. In order to improve the vehicle and relieve this jolting, the body of the carriage was at first hung by leather bracing from a kind of gibbet at each end of the frame-work to which the axles were attached. This arrangement may still be found in the conveyances in some parts of this country, and the occupant, as he swings back and forth like in a ship at sea, lurches first on one side and then on the other, varying his gymnastic performances occasionally by a toss to the roof. Then springs came into use—first, wooden springs, like the spring-poles to a Sedan chair, or arranged something like the buck-board wagons of the United States. But it was in France that the first regular spring vehicles were made. The brouette was a Sedan chair on a pair of wheels, with a pair of steel springs between the axle and the body, through which the axle passed under the seat, the floor being very little elevated above the ground. So long as people traveled in their own carriages, very little improvement was made for the benefit of the horses, but when post-chaises were invented and introduced, the proprietors found it to their interest to make every effort and use every appliance which would lessen their labor. It was soon discovered that the closer the axles were placed together, the easier the pull, and the front and rear wheels were drawn as near to each other as possible, allowing only room for the door between. Steel springs, called whip-springs, were first used about the year 1750, being nearly upright like a whip-handle, and curved over at the top where the brace was attached and hung vertically. It was found, also, that the mode of hanging the body of the carriage made considerable difference. If hung leaning forward, then all the movement of the body acted as an onward weight to help the horses. The rear braces were therefore buckled up higher

Fig. 1.—Four-in-Hand Drag: Hooper & Co., London.

than those in front. The effect of this, however, was very uncomfortable to the passengers in the back seats, causing them to slide off their seats on to their knees. Nevertheless the use of the steel spring was a most important advance, increasing the speed at which traveling could be accomplished from two miles per hour to eight. The first line of stage-coaches was established between London and Edinburgh in 1754, making the distance of four hundred miles in ten days.

The whip-spring gradually changed into the C-spring for the private carriage, the perch was lengthened, the wheels set far apart, the springs curved, the braces made more nearly horizontal, and the body swung low with a deeply curved perch. Then came double springs, straight springs transverse to the axle under the frame-work, with curve-springs to hang the body above the frame-work. After this the combination telegraph-spring was introduced, the body resting on four long cross-springs pendant from four shorter side-springs bearing on the axles near the wheels, the latter moving up and down, and passing over obstacles, disturbing but little the equilibrium of the body of the carriage. Then came the "mail" suspension, on two points in front and three behind, seven springs being used instead of eight. The perch and frame-work were still retained, however, for road-work. Finally carriages were built on cross-springs without frame-work, and then a coach-builder of Lambeth, England, named Elliot, made a still further improvement by applying the elliptic-spring, which has since come into use for all kinds of carriages on account of its adaptability and cheapness.

Carriage-building, like watch-making, was at one time an art, most of the work being done by hand, and the excellence both of design and execution depending on the individual manufacturer. Now machinery is so much used that the business has become nearly stereotyped, and among those who use only first-class materials the difference is not so great. There are still, however, certain firms who stand up above the others in the estimation of the connoisseur, and hold a very high reputation for the quality of work turned out. In the carriages of the present day everything is done to make them as luxurious as possible—the seats all leather and cloth, spring cushions, the wood-work splendidly painted and highly polished, the mountings of silver or gold plated, etc., and every effort is made to reduce the cost of manufacture and allow more margin for profit. There are many varieties, from the Four-in-hand Drag, a private edition of the old public mail-coach, having inside seats for four, and outside for eight, down to the One-horse Brougham, constructed so as to reduce the draught to a minimum, and yet still give possible sitting-room inside. The lightest wheels, the smoothest springs, the shortest carriage without destroying efficiency, are the points which are aimed at, not yet however attained as perfectly as might be. For steady carriages it is desirable to have the wheels far apart, and for light and easy running, saving to the horses, and facility in turning, shortness is desirable, throwing the pairs of wheels close together. An excellent display of carriages is made at the Exhibition, and among those manufacturers represented who occupy a prominent position in the world we would mention MESSRS. HOOPER & CO., OF LONDON, ENGLAND. This firm exhibits a FOUR-IN-HAND DRAG, shown by the engraving on page 275, which displays the greatest care bestowed on its design and construction. The body, boots and under-works are so designed as to fit one another with great exactness, it being customary for MESSRS. HOOPER & CO. to have the whole design drawn out to the exact full size before any part is allowed to be commenced in construction, thus insuring perfect work. The picnic arrangements are such that a luncheon may be carried almost unperceived, instead of the owner being obliged to load his carriage up in full view, with rough and untidy hampers, baskets, boxes, and ice-pails. Wine (with arrangements to ice) can be carried in the zinc-lined mahogany wells in the hind boot, and a folding ladder is also carried here to enable ladies to mount without trouble to the outside seats, the white treads showing where to place the feet on dark nights. The various parts of the carriage and fittings form tables to enable the owner and his friends to partake of a lunch in the best and most satisfactory manner. The roof is fitted with a sun-shade as a protection from the hot rays of the sun when at races or other meetings. The springs have all been tested to carry their full load, and are so made that there is only one hole in each spring, and the ends are rolled up from the spring-plates themselves, instead of, as is sometimes done, by welding on an eye of iron to the steel. The screw-brake enables a slight or strong pressure to be put on the hind wheels, when descending hills, as may be required. The lamps are designed on the best known principles for giving a brilliant and steady light, and not being affected by their vibration in passing over rough roads.

A very handsome VIENNA PHÆTON, shown by Fig. 2, is also exhibited by this firm. This is a carriage of recent introduction into England, and partakes somewhat of the pattern of the German and Russian droskies, with many and important improvements. It is hung on high wheels, protected by carefully adjusted wings to keep off the splashing of the mud. The body is hung low, and the access is easy, with a single step. There is comfortable sitting-room for a tall person, and the carriage can easily be drawn by one horse.

Messrs. Hooper & Co. also exhibit a C-Spring Barouche, a Sefton Landau, a Miniature Brougham, and a Park Phæton, all possessing specially good qualities.

Vienna Phæton: Hooper & Co., London.

The wheels of the drag, barouche, park phæton, and landau are made of timber, all of British growth, the stocks of elm, the spokes of oak, and the felloes of ash. In the Vienna phæton and miniature brougham the wheels are of foreign timber, the stocks of elm, the spokes of hickory, and the rims of hickory. The springs are made of the finest tempered rolled and tapered British spring-steel (from Russian iron), specially manufactured for the highest quality of carriages. The axles are of British iron carefully faggoted, the surfaces of the arms and boxes being case-hardened. The axle-boxes of the miniature brougham and Vienna phæton are of wrought iron. All the cloth used is of the best English make; the moroccos are from Swiss goats, and the leather is of British manufacture. The carriages are made of timber from the following countries: ash, oak and plane-tree from England, pine from Canada, mahogany from Honduras, and hickory from the United States.

The PUTNAM MACHINE COMPANY, OF FITCHBURG, MASS., makes a large and prominent exhibit of Machine Tools, for which it claims great excellence and accuracy of workmanship, elegance of design and harmony of proportions, convenience of arrangement of parts, securing great economy in working, and a high quality of material used, which is disposed of to the best advantage for strength. Our engraving, Fig. 1, shows one of this Company's thirty by thirty-inch planers; the marked features of which are the angular feed on the head, allowing planing to be done to any angle, the method for raising and lowering the cross-beam and head by power, and the yielding arrangement by which lost motion and the momentum of the table when reversing, are received upon an elastic spring, resulting in unusual evenness of working. The machine is of great strength and power and will do the heaviest work without straining or injury to the parts. The table or platen is very thick and the occasional surface-trueing required on all machines of this kind can be done without proportionally reducing the material any more than that caused by ordinary wear on the other working parts. The shifter-dogs for varying the stroke of table are held by hand-screws, thus dispensing with the use of a wrench and rendering adjustment while the machine is in motion both easy and rapid. The vertical, angular and cross-feeds are automatic or governed by hand at will. The automatic-feed is operated by disks attached to the shaft of the driving-gear, the outer disk containing a diametrical slot having in it an adjustable lug connected by a rod to a vertical sliding rack. This rack gears into a pinion working loosely on the transverse screw-shaft and carries a pawl which gives motion to the ratchet-wheel. Variations in the feeds are effected by moving the lug from or towards the centre of the disk, thereby lengthening or shortening the stroke of the rack, and the direction is changed by passing the centre. The raising-screws for the head and beam are supported on hardened steel step-bearings, adjustable by means of set screws. The rack and all gears are cut, and work smoothly and without noise. The driving-gear shaft is massive and has four bearings, two of which extend in close proximity to each end of the rack-pinion, tending greatly to arrest vibration and give extremely smooth work, even when the machine is under heavy strain.

Fig. 2. represents this firm's PATENT BOLT-CUTTING AND NUT-TAPPING MACHINE. The frame of this machine is made entirely in one casting, and presents

Fig. 1.—Thirty by Thirty-inch Extra Heavy Iron Planer: Putnam Machine Co.

a compact, convenient and neat appearance, a very useful receptacle, accessible from the front or rear, being provided in it underneath, for tools. The machine is strongly geared and is driven by a cone with four changes of speed; it is simple in construction, easily operated, positive in action, and is not liable to get out of order. The spindle is hollow, admitting bolts of any desired length, and the dies revolve and open, allowing the bolt, which is held in the jaws of the carriage, to be withdrawn instantaneously without stopping and reversing. The dies are rigidly fixed in a solid unyielding metal ring, thus insuring perfect accuracy in the duplication of the work. They are easily and quickly adjusted or changed from one size to another without the removal of any part of the machine, and when worn can be re-sharpened without drawing the

Fig. 2.—Bolt-Cutting and Nut-Tapping Machine: Putnam Machine Co.

temper, by simply grinding their cutting-faces on an ordinary grindstone. Right- or left-hand screws can be cut, and the diameter of the threads may be made larger or smaller by means of an adjusting index. The die-head allows accurate compensation for wearage, there being sufficient range to permit each set of dies to be repeatedly re-cut and assume its original size, this in itself being a feature of great economy. The mechanical construction of the machine allows the operator complete control of it under all conditions, and the makers claim that after years of service it will perform its functions and render as reliable service as when new, the bolts at all times being cut at one operation (or once passing over) with uniform diameter, shape and lead. A special set of dies is furnished for tapping, and the tap may be removed or replaced without stopping the machine. An improved automatic oil-pump of ample capacity, serves not only to lubricate the cutting dies but also to keep them cool when undergoing severe and rapid work. The delivery-pipe is arranged to traverse with the carriage and discharge upon the tap when tapping nut-blanks, and the flow of the oil may be arrested or regulated at any time without stopping or changing the action of the pump.

Fig. 3 shows the PUTNAM MACHINE COMPANY'S PATENT GOOSENECK-DRILL, a machine of admirable design embracing in its general construction all the most essential features requisite for universal and miscellaneous use. The column is heavy, well braced, and fitted with long bearings into a large base-plate, giving ample support to the upper portion. The base-plate is planed on the upper surface and is provided with T slots, plug and bolt holes, serving as a leveling-plate, and also has a bearing for guiding boring-bars, etc. An adjustable circular drilling-table twenty-five inches in diameter, perforated with plug and bolt holes, is attached to the column by an arm, and in addition to its having a horizontally revolving motion about its own centre, it is also movable around the axis of the column with a vertical traverse of sixteen and a-half inches, thus being perfectly adjustable to any desired position. The drilling-head is adjustable, having a traverse of twenty inches, operated by rack and pinion, which adds largely to the convenience and range of the machine. The drill-spindle, which is made from hammered cast-steel, is one and three-quarters of an inch in diameter and has a run of twelve and a quarter inches without change. It has both hand and automatic variable feed, the latter with four changes, capable of being varied from one to the other instantaneously, and being operated by friction, its action is quick and without shock. The spindle has a quick return motion, and lever feeds which are very effectual and so constructed as to be available at any point of the entire traverse. The lever feed, besides being convenient for drilling holes rapidly up to three-quarters of an inch in diameter, is equally useful for small slotting and splining. The improved patent balancing device is used, by which the spindle is absolutely balanced under all conditions of usage and wear. The back-gears are conveniently arranged within working reach and are quickly operated. The bevel-gears for driving the spindles are powered to about three to one, and are encased to prevent the throwing of oil, etc., upon the operator or the work. The driving-cones have four shifts, and being back-geared, admit of eight changes of speed. Outside bearings are of improved construction, have a wide range for wear, and are susceptible of delicate adjustment, tightening to the centres and wearing to a round bearing.

Fig. 4 represents the PUTNAM MACHINE COMPANY'S IMPROVED SLOTTING AND PARING MACHINE. An important improvement in this machine consists in the method by which, by means of a simple attachment, the tool is relieved from abrasive contact on the up stroke, thus doing away with what would otherwise be a very objectionable feature of the operation. The work is moved entirely from the point of the cutting tool, preventing wear and injury to it, as well as to the article worked, and this is accomplished without any joint in the tool or in the sliding-bar, not in the least impairing their full strength. The improvement can be instantly varied to suit inside or outside work, or removed entirely if desired, without stopping the machine or using a wrench, and no alteration or

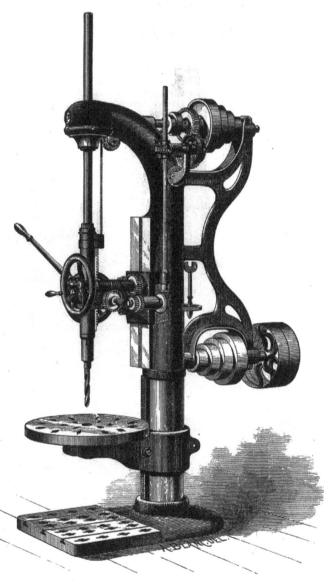

Fig. 3.—Patent Gooseneck-Drill: Putnam Machine Co.

resetting is required when the stroke is changed. The present machine has a stroke of ten inches, and a revolving-table of twenty-six inches diameter. The frame is very strong, having a cored section arched at the back and in addition stout internal ribs. A weighted lever attachment counterbalances the head which is easily adjusted for position and stroke, both operations being effected by screws. The use of elliptical eccentric gears provides a quick return motion very perfect in its action when the gears are accurately made. When the head is on the downward or cutting stroke the shortest radius of the driving gear is in contact with the longest of the driven, and in effect a small pinion drives a comparatively large gear, greatly increasing the power while diminishing the speed. On the up stroke as the only work to be performed is raising the

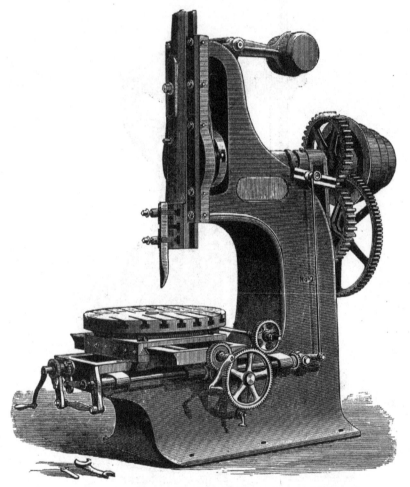

Fig. 4.—Improved Slotting and Paring Machine: Putnam Machine Co.

head already counterbalanced, and requiring little or no power, a reversal of the previous order of the gears does the operation at an increased speed, gaining in time and economizing the work at a sacrifice of the useless power. The longitudinal, transverse and circular feeds are all automatic and have ample changes of speed, while the hand-cranks, pawls, etc. are within convenient reach. The machine may be instantly stopped at any point of the stroke by means of a friction-brake on the countershaft, a great convenience in setting the work or tool and saving time. An ingenious device claimed to be original with this machine, is introduced to compensate for wear of the principal feed-screw nut, and by its use all "backlash" or lost motion in the screw or nut is

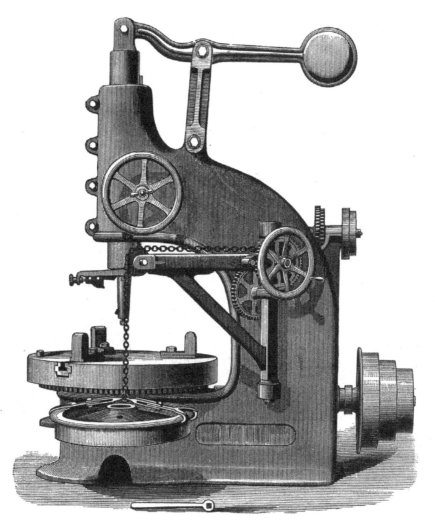

Fig. 5.—Vertical Car-Wheel Borer: Putnam Machine Co.

wholly avoided, enabling the tool to be worked with the same accuracy on the sides as upon the face.

The VERTICAL CAR-WHEEL BORER shown by Fig. 5 is of heavy construction, combining with good proportions the proper strength for the work it has to perform, and its capacity includes all sizes of wheels from fifteen to forty-eight inches in diameter. The work is held by a four-jawed chuck, the jaws of which while having independent adjustments to an accurately graduated scale on the slide, are set up or tightened on the work by means of a wrench giving a simultaneous or universal movement. The bearings upon which the chuck revolves are of the form of a double parabola with the concave faces turned in as the journal, while the seat or lower bearing is lined with Babbitt metal, producing an excellent bearing and distributing the pressure over a large area, thus, when properly lubricated, preventing contact and wear of the metal and reducing the running friction to a very small amount. These journal-bearings are surrounded by and attached to a rigid circular case, which admits of adjustment for boring either straight or tapering, without changing the vertical-line of the boring-spindle. The chuck-spindle is hollow and allows chips to fall into

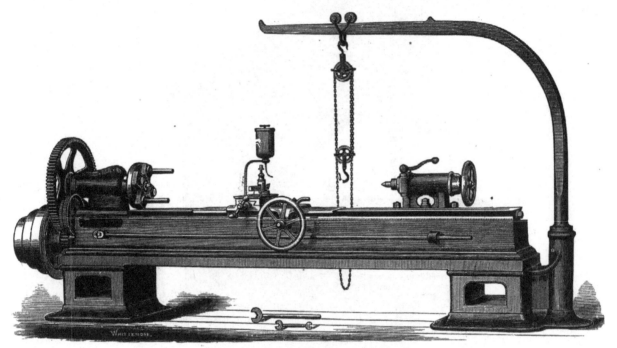

Fig. 6.—*Car-Axle Lathe: Putnam Machine Co.*

the interior of the frame from whence they may easily be removed. The boring-spindle is of large proportions, is counterbalanced, and is raised or lowered by a rack and pinion in the back, giving a very quick motion. The feed has four changes, two by belt and two by gears, and the latter admits of being changed instantaneously, independently of the former, for roughing out and finishing operations, by means of a stop-rod, while the machine is in motion. The cutter-mandrel is of steel, three and one-half inches in diameter, and has a taper bearing in the spindle, twelve inches long. An independent head for squaring the hubs of truck-wheels is quickly adjusted to, or removed from the spindle as required. A powerfully geared swing-crane is attached to the side of the machine and provided with chain- and grappling-irons for lifting and swinging wheels on and off the chuck. The driving-cone is large and has three changes of speed, and by the arrangement of the countershaft pulleys, admits of two speeds for each cone-shift without change of belt.

The CAR-AXLE LATHE of this Company, as shown by Fig. 6, is designed particularly for turning car and locomotive axles, and possesses all the requisite weight, strength and power. The bed is very heavy, having flat shears and a central feed-screw well protected from chips. The head- and draw-spindles are

Fig. 7.—*Eighteen-inch Swing Engine Lathe: Putnam Machine Co.*

of steel and are of large diameters; the former receiving the lateral thrust upon a loose anti-friction collar, and being constructed with a self-centering bearing that may be compressed to compensate for wear. The dead-spindle admits of adjustment for turning either straight or tapering, and has a self-centering binder. The feeding device is claimed to be new and admits of two changes of feed, for roughing and finishing. The machine is provided with a double set of driving-pulleys on the countershaft so that the operator may rough out and finish work at different speeds, in the same way as in the Car-Wheel Borer, without shifting belt on the driving-cone. A novel and useful wrought-iron swing-crane is attached to this lathe, socketed at the end of the bed and furnished with pulley-blocks, enabling the operator to place on the centres the heaviest axle without assistance.

This Company claims as special features of the lathes manufactured by them, the simple and effective cross-feeds, the eccentric nut on the feed-screw, the device for clamping tailstock-spindle, the binder for securing the carriage when the cross-feed is in use, and the reversing-gears for right- and left-hand screw-cutting. The Thirty and Forty-four-

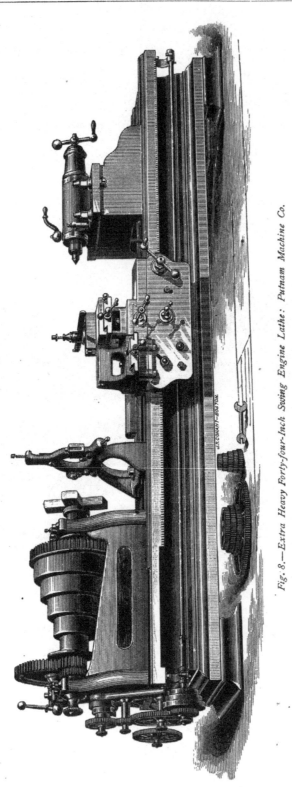

Fig. 8.—Extra Heavy Forty-four-Inch Swing Engine Lathe: *Putnam Machine Co.*

Inch Swing Engine Lathes have a compound-rest with graduated index for turning or boring tapers.

Fig. 7 shows the EIGHTEEN-INCH SWING ENGINE LATHE, with power cross-feed, and Fig. 8 the FORTY-FOUR INCH LATHE. Great care and judgment have been used in the designing and construction of these machines, and the exhibit is a very creditable one to the firm.

The PUTNAM MACHINE CO. also exhibits a Steam Engine, shown by Fig. 9, well-known as this Company's PATENT REGULATING CUT-OFF ENGINE, very simple in its construction, noiseless in its operation, and, it is claimed, unsurpassed for economy and perfection in regulation. The steam is admitted to the cylinder at boiler pressure and cut off at any point in the whole length of the stroke, as may be required by the load upon the engine, the remaining portion of the stroke being performed by the expansion of the steam thus cut off or shut up within the cylinder. This peculiarity of the cut-off, that of having a range throughout the whole length of the stroke, is considered of great importance, enabling the engine to maintain a regular speed under many circumstances of irregular load, etc., that would not be possible otherwise, and of keeping up this speed under a reduced pressure of steam, until it becomes so low as to be inadequate to the work of the whole stroke. The apparatus

Fig. 9.—Patent Regulating Cut-Off Steam-Engine: Putnam Machine Co.

for operating the steam and exhaust valves, consists of a shaft driven by gears from the engine-shaft, and by means of cams, etc., of a peculiar construction; the valves are caused to open rapidly to their full capacity, and to maintain a full opening till the time of closing, when they suddenly shut. Being balanced against steam-pressure, these valves require extremely small power to operate them, and the governor acts directly upon the mechanism and determines the time of closing. The valves are of the kind known as the "Balance Poppet," and the chest containing them is so constructed as to combine in one casting an arrangement allowing the upper and lower seats of steam- and exhaust-valves to be so near together as to leave only the requisite steam-passage between them, and thus reduce to a minimum any liability to derangement from difference in expansion of parts. Each valve is accessible and may be withdrawn by simply removing a cover.

Fig. 10 illustrates a very neat and perfect arrangement manufactured by this Company for the purpose of injecting oil into the cylinder of a steam-engine for lubricating the valves, pistons, etc. By its use the quantity of oil to be injected can be nicely regulated so that no more will be used than required. It will contain sufficient oil for several days' use, and may be set at any convenient distance from the engine and connected with the cylinder or steam-pipe by a small pipe. These oilers will save a very considerable amount of oil, which by the use of pumps, double oil-valves, hollow plug-cocks, etc. would be wasted.

Fig. 10.—Cylinder-Oiler: Putnam Machine Co.

machines exhibited by Messrs. Richards, London & Kelley, of Philadelphia, and London, England. We would like to draw attention, however, to a very ingenious Cutting-Off Saw for Hot Iron, made by this firm, and illustrated by the engraving on page 291. Circular saws have long been used for cutting

We have previously mentioned a number of off hot iron bars; but this machine possesses an important advantage in that there are no sliding surfaces, and no opportunity of derangement from iron-dust or scale. The saw-spindle is mounted on a pivoted frame, and as there is no friction or other resistance than the cutting action when the machine is working, the attendant will feel the operation of the saw as it is pressed against the iron, a matter of considerable importance, enabling him to regulate the amount of pressure and rapidity of its action, and to protect it from injury, which he cannot always do when the feed is not sensibly felt.

A number of supporting-brackets for the bars being operated on, are fitted on to a rod three or four feet long, so arranged as to be set up in the position shown, or dropped out of the way when not required, depending upon the length of the iron being cut. Adjustable gauges for determining the length of the pieces cut, are provided on the opposite side of the machine from that shown in the engraving. The belt operates from a shaft above, and is slightly tightened as the saw is pressed forward on to the iron, loosening again as it is withdrawn. This arrangement avoids unnecessary strain upon the saw-spindle when the machine is not in use, and allows the saw to be run continously, a matter of great convenience, especially when a number of workmen use the same machine, and avoiding the starting and stopping each time a piece of iron is to be cut.

Nothing at the Exhibition can be more interesting to the engineering visitor than the fine displays that are made of iron and steel, particularly of the latter material. A tremendous advance has been made in the United States in steel manufacture during the last ten or fifteen years, and our metallurgical friends from abroad have had good reason for the astonishment they have everywhere expressed at this great development. The word steel as now used is rather a vague term, being applied to certain combinations of the metal iron with carbon,—between wrought-iron on one hand, which contains little or no carbon, to cast-iron on the other, which has five to ten per cent. There is really no fixed boundary from one to the other, as the quantity of carbon in combination may be varied by almost infinitesimal amounts producing a regular grade of products, from wrought-iron up to cast, all with different qualities. The purity of the materials from which steel is made is also an important consideration, as various other substances, as tungsten, chromium, etc. will enter into the combination if present, and affect its quality and properties, more or less, depending upon their amount. A very small percentage of phosphorus will make steel brittle and unmalleable.

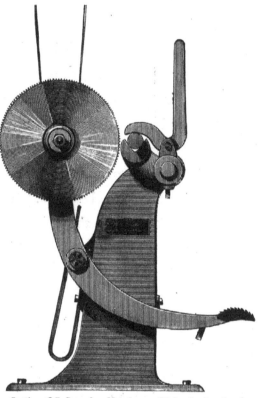

Cutting-Off Saw for Hot Iron: Richards, London & Kelley.

In the manufacture of steel various processes are employed, according to the quality of product desired, but all of them may at the present day be included under two general methods. One, which originated in Germany at an early date, is the process of cementation, in which bars of iron are treated in contact with charcoal in a closed furnace until sufficient carbon has been transferred from the charcoal to the iron by absorption, to convert it into steel. The other method includes all those processes in which the iron and carburizing materials are melted together, either in crucibles, retorts or furnaces, and subjected to a high degree of heat, until carbon in a required definite proportion has been diffused through the mass. The resulting product after being moulded into ingots, takes the name of cast-steel.

It is seen that the quality of the steel, omitting the question of impurities in the materials, depends directly upon the quantity of carbon in combination. If this is indefinite or variable in quantity the resulting product will not be uniform. The most important and most difficult point, therefore, to be attained by the steel maker, is to give each charge of his crucible or retort a definite proportion of carbon, that will make a grade of steel exactly suitable for the particular purpose that it is intended to be used. The various methods of manufacture now in use have resulted from the efforts made from time to time to secure the best and cheapest mode of doing this, depending upon the quality of steel desired, the purity of the materials employed, and the possibility perhaps of producing good results from cheap materials by the discovery of some easy method of eliminating the impurities.

The process of cementation is adapted for the production of low grades of steel only, such as the plow- and spring-steels, known as "German steel." Blistered steel is the first result of the process, and is so called, because the bars of flat-iron, after cementation is completed, have points on their surfaces that are raised or rounded like blisters. On account of the iron being thus converted, without change of form and without correction of any defects of quality or workmanship that may exist in the bar as it is placed in the furnace, such steel may be regarded as a crude and heterogeneous combination of the iron and carbon. By fagoting, welding and hammering, it may be refined and improved in quality; but to obtain a homogeneous steel with carbon diffused uniformly and combined in a relative proportion previously determined, the steel-maker must resort to some one of the processes employed in manufacturing cast-steel, such as the crucible process, Bessemer's, Berard's or Siemens-Martin open hearth process. All fine steels are made in crucibles, the other processes referred to being used for the production of such special and commoner qualities, as rails for the tracks of railways, shafts, car-axles, plates for boilers, ships, etc. MESSRS. HUSSEY, HOWE & CO., OF PITTSBURGH, PENNSYLVANIA, make at the Exhibition, a very fine display of cast-steel, the product of their works, a portion of which we have endeavored to illustrate as faithfully as possible by the engraving on page 294. Cast steel only, is produced at these works, and the whole product is melted in crucibles. Formerly, coke furnaces were used, but have now been abandoned, the melting being accomplished better and more economically in the "Siemens' Regenerative Gas Furnace," six of which are in operation, having an aggregate capacity for melting daily thirty to forty tons. The muck-bar iron, intended for the common kinds of steel, is also manufactured at the works, and for this purpose sixteen puddling-furnaces are kept in operation. For the finer qualities of steel, refined American blooms and Swedish and Norway bar-irons are used in large quantities. The process of manufacturing the steel may be described briefly, as follows:—The bar-iron is cut into conveniently small pieces for charging the crucibles, each grade or quality being kept separate until wanted for use. One hundred and forty-four crucibles are charged at the same time, with the mixtures for the kind of steel intended to be produced. The mixtures composing the charges are made up of the iron selected, the carburizing materials, chemicals to eliminate any impurities in the iron, and some black oxide of manganese to render more accurate the proportions of the combination of iron and carbon. The cap or cover is then adjusted on the crucible, and the steel-melter commences his work. He lifts the charged crucible with a pair of tongs and lowers it into the already glowing Siemens' furnace. After all of the crucibles have been placed in position on the hearths of the furnaces, the doors are closed, and the temperature is gradually raised for the first two hours. During this time the cap has become softened and sealed to the crucible, excluding air and confining generated gases, and the pieces of iron in the crucible have become thoroughly heated, with opened pores and softened fibres, and the molecules of carbon are penetrating from the surface towards the centre. The reaction is precisely analogous to the changes which occur in the process of cementation.

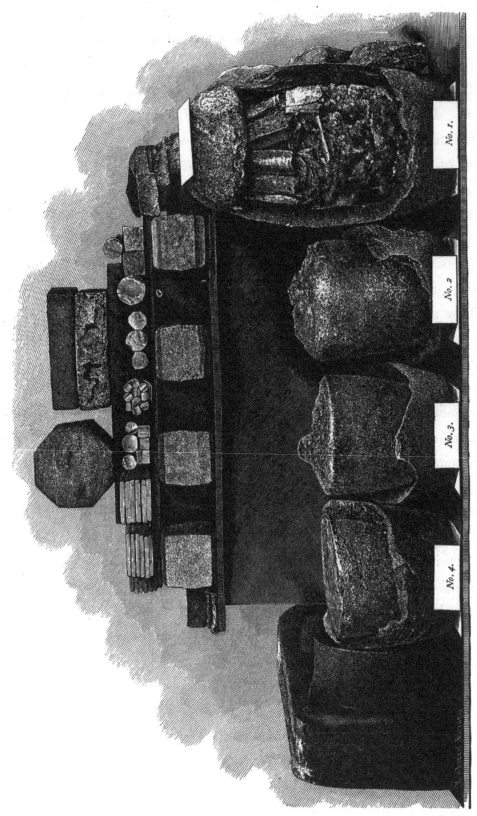

Steel Exhibit: Hussey, Howe & Co., Pittsburgh, Pennsylvania.

The operation goes on, however, at a much greater temperature, and before the carbon molecules have penetrated to the centre of the pieces of iron, during the two hours' heating, the surfaces first carburized will have commenced to trickle down in drops, charged with the absorbed carbon. During the third hour the temperature of the furnace is forced to a white heat, ranging from twenty-five hundred to three thousand degrees, in order to produce that complete fusion of the whole charge necessary to accomplish the desired reactions and results among its materials, and to incorporate the carbon with the iron so thoroughly as to give to all its particles the peculiar and remarkable characteristics of steel.

After the three hours' heating, the crucibles are lifted from the furnace, one or two at a time; the cap is removed, and a workman skims off with an iron rod the flux that rests on the melted steel. The steel-melter then takes up the crucible with long-handled tongs, and steadying the handles on his thigh, slowly turns the crucible over the mouth of a cast-iron mould and "teems" out the contents in a small stream, so that it will fall into the mould without touching its sides. In a little while after the casting is finished, the moulds are knocked apart and the ingot of cast-steel is passed over to the inspector.

Among the specimens of the product of these works, shown in our engraving, are four crucibles, each having been charged with mixtures similar in quantity, and illustrating different stages of the melting operation while in the furnace. Crucible No. 1 remained in the furnace one hour and a half. The arrangement of the charge of iron can be observed, as the greater part of it, the upper portion of the mass, remains unmelted. The next crucible, No. 2, was taken out at the expiration of two hours, with its charge in a semi-fluid condition. Crucible No. 3 was heated for two hours and a half, showing the operation yet incomplete. Ebullition, caused by escaping gases, has taken place while the melted charge was cooling down, and a projected bubble of the metal has been solidified on the surface. Crucible No. 4, after being in the furnace three hours and exposed to its highest heat, shows a homogeneous mass with complete combination of the carbon and iron, ready to be moulded into an ingot of cast-steel. Before the steel is well melted and completed, it shows a state of ebullition, but after it is finished and ready for pouring, its surface becomes clear, and the metal rests in the pot without motion.

After removal from the moulds the ingots are first inspected by an expert, to determine approximately the carburization, quality and suitableness for special uses in other manufactures. When thus assorted and marked, they are ready to be reheated and hammered at a welding heat, under ponderous steam-hammers, to close up any pores or cells that may have been formed in casting. Such as are intended for the manufacture of edge tools, drills, chisels, taps, dies, turning-tools and all similar articles requiring a fine quality of steel, are afterwards hammered at lower temperatures, into flat, square or octagon-shaped bars, commercially known as tool-steel. Other ingots, which the inspector has passed as adaptable to the purpose, are rolled into sheets and bars for the manufacture of plows, shovels, saws, springs, swords, gun-barrels, steam-boilers, fire-boxes and the whole range of uses to which steel is applicable. Any desired quality of steel can be supplied to meet the wants of the consumer. Bars can be hammered to sizes varying from eight inches to one-eighth of an inch square, and sheets can be rolled seven feet wide and eighteen feet long.

MESSRS. HUSSEY, HOWE & CO., also make a special kind of steel which the peculiarities of soil in our Western prairies has made an indispensable article in the manufacture of prairie-plows. One or more plates of heated wrought-iron are inserted in a cast-iron mould that has grooved sides to receive them. Melted steel is then poured in and welds the plates so as to form an ingot of steel with laminations of iron. These ingots are rolled into plates that may be hardened without danger of breaking, and are used in making mould-boards and other parts of plows, and also in the construction of safes and vaults.

After the steel is finished under the hammer or at the rolls, a final inspection is given to every bar and sheet for the purpose of detecting any imperfections or flaws that may have escaped observation at first or have developed in subsequent stages of the manufacture. The unsound bars and sheets are condemned and returned to be re-melted, while all that pass inspection are placed in store ready for the market, entering into use in all the mechanic and industrial arts of the world.

The PRATT & WHITNEY COMPANY, OF HARTFORD, CONN., makes a large exhibit of excellent and very interesting machinery, remarkable for qualities of exactness, accuracy and adaptation to the purposes intended. The parts of these machines are conveniently arranged, the metal well distributed for proper strength, and all the working portions accurately fitted, insuring durability and ease in running. In fitting, all sliding surfaces are scraped to a bearing, no emery being employed, particular attention is paid to the form of the teeth in racks and gears, and all actuating screws are of steel. The nuts, screws and wrenches are case-hardened, and all journals run in cast-iron boxes lined with the best Babbitt-metal, compressed after being cast into shells. The shells are in some cases plugged with Babbitt-metal instead of presenting a face wholly of that material. Patent friction clutches, operating with ease and certainty, and without jar or strain, are used for reversing the motion on counter-shafts, and for the instantaneous starting and stopping of machines. The spindles of lathes, milling-machines, drills, screw-machines, etc., are of steel.

The foot-stock spindles of the lathes are held by a clamp that embraces the spindle firmly on all sides, insuring exactness of centre; and the cones of all the lathes are turned and finished inside as well as outside, to balance them and secure steadiness at high rates of speed. The screw-cutting lathes are provided with independent worm-gear and friction-feed, driven by gearing that may be instantly thrown out to allow the spindle to run free; and both the turning and the screw-cutting lathes, with weighted or gibbed carriage, have unusual length of bearing in the carriage and very large spindles. The gibbed carriage-lathes are furnished with Pratt's patent tool-post, an arrangement that may be accurately adjusted vertically without loosening the tool in the post, and yet possesses perfect rigidity.

The gibbed carriages are all gibbed their whole length on the outside of the bed, and Slate's taper attachment is furnished with lathes, if desired. This attachment while it does not interfere with any of the ordinary functions of the turning or screw-cutting lathe, has positive advantages which at once commend themselves, it being a great convenience for turning tapers either outside or inside. It does not require the placing of lathe centres out of line, and the varying length of the pieces to be turned does not affect the taper, the degree of which may be accurately adjusted by a graduated scale on the attachment, without any preliminary or trial cuts being necessary. It can readily be released from operation without removing any part of it from the lathe.

Our Fig. 1., on page 298, represents the PRATT & WHITNEY COMPANY'S TEN-INCH PILLAR SHAPER; or, as it sometimes called, COMPOUND PLANER, an exceedingly useful and convenient tool for a variety of work, being particularly adapted to die-work, and in very many cases a substitute for the ordinary and more expensive shaping-machine. The machine stands on a hollow column, the base measuring thirty-six by twenty-seven inches. The stroke may be graduated to any point within its extreme limit; the cutter-slide has a quick return, and the cross-feed is automatic and adjustable. All the slides and bearings are fitted perfectly by scraping, the nuts, screws and wrenches are hardened, and the whole is of excellent and accurate construction throughout. This machine is provided with the "Newell" patent vise, which has a circular graduated base, and hardened steel jaws, ground true, one of which swivels, to present either a straight or V-shaped side, the latter for holding circular work. The extreme length of stroke of this machine is ten inches; the traverse of bed, sixteen inches; the distance between table-top and bottom of slide, eleven and a half inches, and the weight seventeen hundred pounds. In addition to the shapers, intended for light work, this Company also builds power-planers, with beds from five to thirty feet in length, and capable of doing work from sixteen inches square to forty-eight inches wide and high. The size designation of the tool denotes the width and height that it will plane. Thus, a "Sixteen-Inch Planer" receives work sixteen inches wide and high, limited in length only by the capacity of the bed and traverse of the table. The length of the planer-beds is so much greater than that of the tables, that the latter cannot tip and unseat themselves, even when the rack is run out of gear with the pinion. The gears and racks are strong and accurately cut, and the shafts are unusually large. All of the sizes but the sixteen-inch planer have automatic

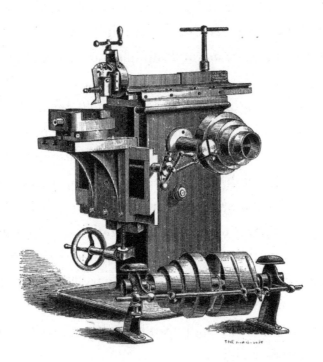

Fig. 1.—Ten-Inch Pillar-Shaper: The Pratt & Whitney Co.

angular and vertical-feed, as well as self-acting horizontal-feed. The belts have a speed of fifty to sixty feet, to one foot movement of the table. A patented shipping movement, and movable dog, allows the work to be run either way from under the cutting tool, without changing the stops.

We illustrate by Fig. 2 the PRATT & WHITNEY COMPANY'S SIXTEEN-INCH PLANER of which we have just spoken. This machine has a five-foot bed, and planes sixteen inches in height and width, and three feet two inches in length,

Fig. 2.—Sixteen-Inch Planer: The Pratt & Whitney Co.

the table being three feet six inches long. The horizontal-feed is automatic, and the planer has the patent shipping-movement with movable dog arrangement previously mentioned. The machine is run with two one-inch belts, the weight with counter-shaft is sixteen hundred and fifty pounds, and the speed of counter-shaft is two hundred and seventy revolutions per minute.

The PRATT & WHITNEY COMPANY make a number of Boring and Drilling Machines; a horizontal boring-mill of several sizes, the largest of which has a length of bed of sixteen feet, the greatest distance between centres of eleven feet five inches, and a weight of eight thousand two hundred pounds. From among the upright drills we select this firm's No. 2 FOUR-SPINDLE DRILL, as shown by Fig. 3. The spindles of all the gang-drills run in gun-metal boxes, are split and furnished with a nut to compensate for wear, and they may be run at the highest rate of speed of which the machine is capable without the slightest danger of binding in the boxes, the longitudinal expansion not being checked by a fixed collet, but being allowed freedom without impairing the accuracy of the machine. In this machine the table is counterbalanced, and is actuated either by an adjustable hand lever or by the foot. A stop-rod, which is not shown in the engraving, prevents the table from dropping while being adjusted. The spindles are of steel and the machine will drill holes up to five-eighths of an inch in diameter. When heavier work is to be done, as counter-boring or facing, the two central spindles may be geared to any speed desired. A gauge is provided for determining the depth of hole to be drilled, obviating the necessity of frequent withdrawals and measurements, and saving much time and labor. The distance between table and spindle is seven to twenty-four inches, diameter of spindle one and a half inches, and weight with counter-shaft, seventeen hundred and twenty-five pounds.

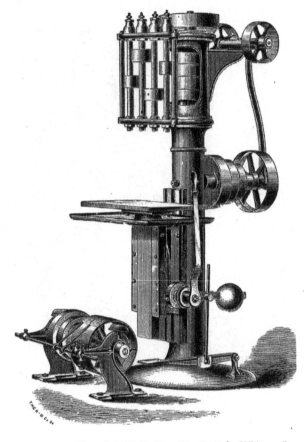

Fig. 3.—No. 2 Four-Spindle Drill: The Pratt & Whitney Co.

Fig. 4, on page 301, illustrates this firm's No. 2 HAND-MILLING MACHINE, an exceedingly convenient and useful tool, adapted to a great variety of work. The cone has four grades, of three, five, seven and nine inches diameter, with two and a quarter inches face, and the spindle is of steel, running in cast-iron boxes lined with Babbitt-metal. The head is mounted on a hollow column of convenient height, in the interior of which is a convenient receptacle for cutters, wrenches, etc. The range of speed adapts the machine to the use of cutters, from one-quarter inch to two and a half inches in diameter. The table has a vertical adjustment of fifteen inches, and the upper slide a vertical movement of two and a half inches, a horizontal adjustment, toward or from the column, of five inches, and a transverse movement of five and a half inches. The weight with counter-shaft is nine hundred and ten pounds. A number of other milling-machines are turned out by this firm; a small hand-machine, very nice for light work, also index milling-machines, double-face milling-machines and large power-machines. The No. 2 Power-Milling-Machine is in extensive use in armories and sewing-machine establishments, both in this country and Europe, being well adapted for general work, and appears to meet with great favor.

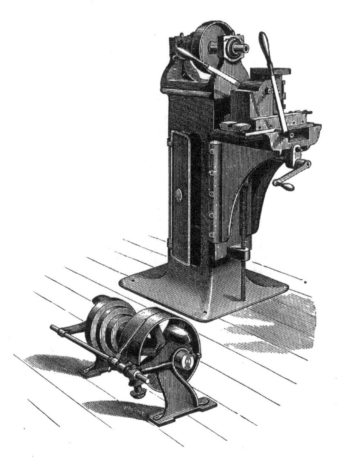

Fig. 4.—No. 2 Hand-Milling Machine: The Pratt & Whitney Co.

Fig. 5, on page 302, shows a tool made by the PRATT & WHITNEY CO., very convenient for shops and manufactories where rotary-cutters of any style are used for gear-cutting, milling, slotting, etc.,—viz., the Cutter-Grinder. The spindle-head and the cutter-holder and guide, are sustained by a columnar support with a broad base, and the platen to which the holder and guide are attached, may be adjusted in height to suit the diameter of the cutter to be operated upon. The guide rests against the tooth that is being ground; thus, gauging the work perfectly, even though there may be irregularity in the size of the teeth. The machine is adapted to cutters of all sizes and styles of teeth, whether straight, beveled, or spiral, and either small grind-stones or emery-wheels may be attached to the spindles. The weight of this machine, with counter-shaft, is four hundred and twenty-five pounds. The cone has two grades, and the speed of counter-shaft is four hundred and eighty revolutions per minute.

The PRATT & WHITNEY COMPANY make Revolving Head Screw Machines, either with or without wire feed, back gears or automatic feeding apparatus, that are claimed to be unsurpassed for ease of manipulation and perfection of results, and have attracted a large amount of attention at the Exhibition. Our engraving, Fig. 6, on page 303, shows the No. 1 SCREW-MACHINE WITH WIRE-FEED. The turret or revolving-head, is usually made to receive six distinct tools, including gauge-stop, but may be made for eight if so desired. Every operation required in the production of screws from wire or rods, except slotting the heads, may be performed successively on these machines without removal of the work from the chuck, stoppage of the machine or change of tools. The turrets are self-rotating and self-fastening, those of Nos. 0 and 1, being of steel, and of Nos. 2, 3 and 4, of cast iron with a hardened steel indexing-ring. The tools are held firmly in the turret by means of a clamp, which, by the partial turning of a binding-screw, embraces the diameter of the tool-shanks, holding them securely. These machines are furnished with case-hardened wrenches, oil-tank, dripper, and Pratt's patent reversing counter-shaft. The usual tools for the turret are box-tools, die-holders and dies, cutting-off tool, and stop-gauge. The die-holder is a chuck and clutch combined, and a split-

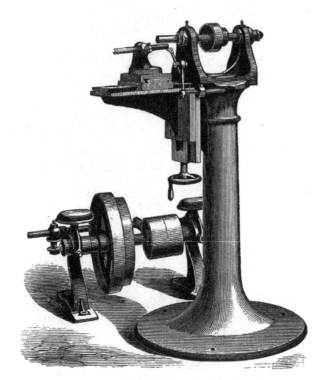

Fig. 5.—Cutter-Grinder: The Pratt & Whitney Co.

collet is furnished for delicate adjustment of the thread of die. The holder can receive a tap in place of a die if so desired. The shank of the holder revolves in a sheath or sleeve, which has its ends formed into right- and left-hand clutches, which engage with pins or projections on the shank and head of the die-holder; thus, allowing the die, or tap, to remain stationary at the instant of reversing speeds, and to be backed without jar, or danger of being

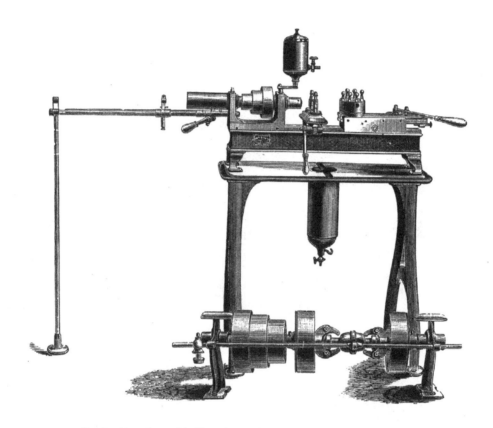

Fig. 6.—No. 1 Screw-Machine with Wire-Feed: The Pratt & Whitney Co.

broken, while the die may cut close up to a shoulder, or the tap be sent in to an exact and unvarying distance. The box-tool will receive from two to five cutters (including one in the hollow shank) of a form adapted to the work to be done. This tool has an adjustable back-guide, or bearing, set on an angle of 45°, to present its support always towards the centre of varying diameters of screw or stud. The cutting-off tool is a cutter dove-tailed into a holder so as to be readily adjusted, and the cutter may be sharpened by grinding, without changing its form.

The Parkhurst patent wire-feed attachment, which is shown in connection with the No. 1 screw-machine in Fig. 6, is a simple and efficient device, not liable to derangement, operated by the movement of a hand-lever, and adapted to rods up to seven-eighths of an inch in diameter. It feeds the wire forward to a length regulated by an adjustable gauge-stop held in the turret, and the same movement that brings the wire forward closes the jaws of the chuck,

Fig. 7.—Two-Spindle Profiling-Machine: The Pratt & Whitney Co.

holding the wire firmly, while the reverse movement opens the jaws to receive another length. These movements are performed instantaneously, without stopping the machine, so that the use of the attachment effects great saving of time. The No. 1 machine will make screws from one-sixteenth to five-sixteenths of an inch in diameter, and the No. 4 machine will operate rods up to one and a half inches in diameter. Other intermediate sizes are made.

Fig. 7, on page 304, illustrates the PRATT & WHITNEY COMPANY'S TWO-SPINDLE PROFILING-MACHINE, a well-known and highly valued tool for cutting irregular forms. Many improvements have been introduced by these manufacturers, one of which, the Parkhurst patent device for cutting formers without reversing the fixtures, will be fully appreciated by those who have had experience in the usual way of making and adjusting the forming pattern. This improvement may be furnished with the machine, or not, however, as desired. To produce the forming pattern with this device, the operator secures the model piece in the place and position afterwards to be occupied by the work to be machined, and the piece from which the pattern is to be made in the place which it is to permanently hold. The guide-pin is put into the spindle which usually carries the cutter, and the cutter into the guide-pin spindle. The former is then made to follow the outlines of the model-piece, and the latter cuts out the forming-pattern in the exact position it will retain when in use. Now disconnecting the gearing upon the spindles, reversing the relative positions of the guide-pin and cutter, and smoothing

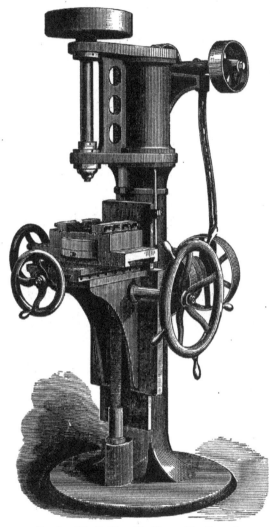

Fig. 8.—Die-Sinking Machine: The Pratt & Whitney Co.

the edge of the forming-pattern if it should be necessary, and the machine is ready for work. The gearing for moving the table and cross-slide is adjustable by means of double-gears, set to prevent back-lash by two independent adjusting-screws, and also by a double rack adjusted in the same manner. This arrangement is indispensable to secure perfect accuracy in cutting irregular forms, especially in turning corners. The cutter will profile or surface work to the extreme limit of the table area.

Fig. 8, on page 305, shows the PRATT & WHITNEY COMPANY'S DIE-SINKING MACHINE, a tool that operates in a similar manner to the profiling-machine, the spindle being stationary, and the work to be operated on, being held in a vise, which may be moved in all directions horizontally, by compound slides on the table of the machine, and may be also elevated or depressed by a vertical movement of the platen or table. The cutter may be of any suitable size or form, and revolves with the spindle, which is driven by a belt, giving much smoother action than is possible with gears. A pattern or forming-piece may be used to guide the work or it may be controlled wholly by the operator. The machine is very strongly built, insuring smooth work, free from chatter-marks, and is well adapted to a variety of work, being particularly useful in forming and finishing recesses of either circular, annular or irregular shape, and for recessing dies for the drop-press. The various movements of the vise and platen are independent of each other and all under full and perfect control of the operator.

The visitor to Machinery Hall notices in the Belgian Department some curious-looking pieces of Machinery, built up of cast- and wrought-iron, with some exceedingly heavy forgings, resembling some sort of mammoth drilling and dredging-apparatus. This is the exhibit of M. JULIEN DEBY, C. E., OF BRUSSELS, to whom we are indebted for the data and information from which this article is prepared, and comprises the tools used in the KIND-CHAUDRON PROCESS FOR SINKING AND TUBING MINING SHAFTS.

The question of sinking deep shafts has always been considered a very serious one, not only where the miner has reason to anticipate trouble from water or from loose or flowing water-soaked material, but also when he is obliged to pass through strata of great hardness. In the latter case, he usually resorts to hydraulic or steam drilling-machinery, with the use of suitable explosives, until the main water-levels are reached, when, in addition, powerful and expensive pumping apparatus must be employed, which is often required to perform much greater duty than could possibly have been anticipated or provided for, very much increasing the expenses, and only resulting perhaps in the end in failure and disappointment. In the former case, if some one of the usual methods of mining is employed, the process is at the best, somewhat problematical, always one of great difficulty, and the resulting shaft sure to be more or less leaky, and a great and continuous expense to the company prosecuting the mining operations. The locations in which such difficulties as we have described exist, called by the miners "bad ground," occur in numerous instances all over the world, deterring the owners from working, otherwise exceedingly valuable mining property.

In the year 1849, M. Kind, a European engineer, who had already attained great reputation as a well-borer, conceived the idea of sinking shafts mechanically, through any kind of strata, on the largest scale, starting out with the supposition that the only thing necessary, was to be provided with sufficiently powerful and weighty tools. Professor Combes, had previously, in 1844, made the same suggestion, but had never put it to a practical test. M. Kind, however, designed and patented apparatus for the purpose, and about the year 1850, undertook to sink three shafts through water-bearing strata, by his method, two of these being situated at Stiring Wendel, in the Department of the Moselle, and the third in the Valley of the Ruhr, in Westphalia. His attempts failed, partly owing to the inefficiency of his tools, but more especially to the impossibility of making any kind of wooden casing or tubing sufficiently tight at the horizontal joints, or strong enough to resist the tremendous outside pressure to which it was submitted. He used staves up to twelve inches in thickness, carefully banded together by means of iron hoops, and yet in every instance they eventually gave way and caused the loss of the shaft. An additional external coating of twelve inches of concrete did not cause any improvement, and even the use of boiler-plate tubing was found unsuccessful.

M. Mulot, the well known engineer of the Grenelle artesian well, in Paris, also made an attempt to sink a colliery-shaft by mechanical means, in 1849, but failed signally, through the imperfection of his tubing.

M. Kind's and all kindred processes were now looked upon among mining engineers on the continent, as impossibilities in the way of success, and a great invention would probably have been lost, had it not been that M. J. Chaudron, an eminent Belgian engineer had his attention drawn to the subject, when it had been nearly abandoned by its first promoters.

M. Chaudron soon modified most of the details in the construction of the boring tools, and replaced the inefficient wooden and sheet-iron casings by one formed of a series of superimposed, heavy, cast-iron rings, which he found it was a practical necessity to employ in a single piece, not in sections, and with the flanges carefully planed on the surfaces of contact. He also discovered that it was essential to test every ring as to its resistance, by means of external hydraulic pressure, many rings being found defective from imperfections in the castings. The thickness required for the tubing at various depths below the surface of the ground, was determined theoretically by formulæ, and every care taken to meet all the points of the problem. M. Chaudron's labors, however, were not limited to the consideration of the actual boring of the shaft and the subsequent introduction into it of a solid cast-iron column, but he added to his already ingenious apparatus, the only rational and practical method of sinking vertically and at one time, such an unwieldy mass as the assemblage of cast-iron rings, the total weight of which often exceeded several hundred tons. This he did by putting a false bottom on to his huge cast-iron tube and floating the whole system on the water in the shaft. He perfected this process of shaft-sinking still further by introducing beneath the false bottom a stuffing-box, or, as he called it, a "moss-box," filled with moss, and of the same diameter as the outer tubing, constituting, when the case has finally reached the bottom of the shaft, a perfectly water-tight joint, shutting out permanently from the mine below, all water around the outer walls of the shaft, either stagnant or flowing.

M. Chaudron having perfected his plans, soon undertook a contract for sinking two shafts in Belgium, for a company that had spent immense sums of money in vainly attempting to reach the coal seams. The first of these had to go about four hundred and five feet, and the second three hundred and fifty feet. They were bored in two successive operations, the first bore of small diameter, being afterwards enlarged to its required permanent dimensions. The first shaft was sunk with a diameter of four and a half feet to its whole depth in one hundred and twenty-one working days. It was then widened to fourteen feet diameter for a depth of three hundred and fourteen feet, in seven months additional time, during which period, fully two months of stoppages occurred from accidental causes. The cast-iron tubing had a total height of two hundred and four feet, and weighed two hundred and fifty-five tons, the whole of the tools and machinery employed not costing more than fourteen thousand dollars, and the total expense of shaft, when completed, being about forty-four thousand dollars. The second shaft, intended as a ventilating shaft, was sunk to a depth of three hundred and twenty-five feet, with a bore of four and a half feet, and widened afterwards to seven feet for a depth of three hundred and fifteen feet.

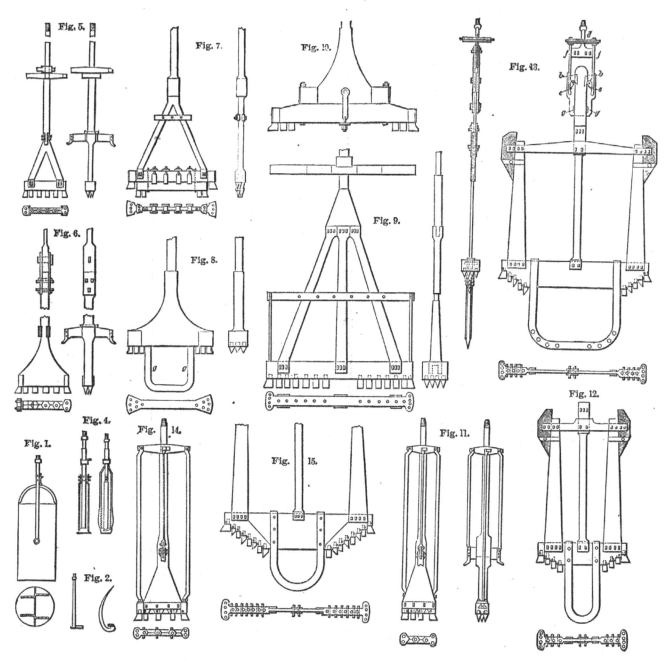

The Kind-Chaudron Process for Sinking and Tubing Mining-Shafts: *Julien Deby, C. E., Brussels, Belgium.*

occupying in its construction less than seven months, and costing about thirteen thousand dollars.

M. Chaudron next undertook two shafts of L'Hopital, in the Department of the Moselle, a locality in which over four millions of dollars had been spent within a short period of time previously, in unsuccessful attempts to reach the

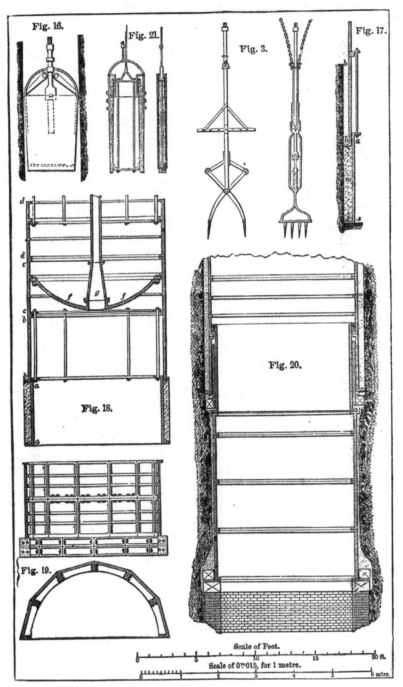

The Kind-Chaudron Process for Sinking and Tubing Mining-Shafts: Julian Deby, C.E., Brussels, Belgium.

coal lying below. The shafts had to be bored through the tough but highly water-bearing sand-stones of the Vosges, and through various strata of the new red sand-stone, including its basal conglomerates of exceeding hardness.

One of the shafts was to be four hundred and eighty feet deep, and twelve feet in diameter, the other of the same depth, and five and a half feet in diameter. The larger shaft took about two and a half years for its completion, and the smaller one about three years, the difficulties that had to be contended with being much greater than had been before experienced, and the ground of a very much harder character. All obstacles were however finally overcome, and the work finished at a cost of thirty thousand dollars less than the original estimates. The large shaft cost eighty-eight thousand dollars, and the small one fifty-one thousand two hundred dollars, the amount of iron in the tubing of the former being six hundred and thirty-five tons, and of the latter, two hundred and fifty-eight tons.

The success of the KIND-CHAUDRON METHOD was now fully established, and its employment spread rapidly throughout Belgium, France and Germany, the most incredulous of its former adversaries admitting its triumph. The total number of shafts already sunk by this means, or under way, now amounts to forty-three, with an aggregate length of the borings of nearly ten thousand feet. M. DEBY states that every shaft sunk in Europe by the CHAUDRON PROCESS, without a single exception, has been a success in every way. All have kept water-tight and have resisted external pressure and are to this day giving the greatest satisfaction to their owners.

We shall now proceed to describe the principal tools used in boring mining-shafts by the KIND-CHAUDRON PROCESS, of which we furnish illustrations on pages 309 and 310. Trepans are those tools which are used to disintegrate the rock or earth by concussion. They are attached to the extremity of a series of wooden connecting-rods which can be added on to, in order to increase the length as the depth of the work increases, having iron armatures and screw-ends, the series of rods being suspended to the end of a balance or striking-beam and put into motion by means of a single acting or bull engine, worked by hand. The shafts are bored in two and occasionally in three successive operations. The first bore is made by a small trepan, generally about four and a half feet in diameter, through which the debris is extracted until the final completion of the shaft. This first bore is then widened by means of a large trepan. The small trepan is formed of two distinct parts, the blade and its stem. The first is made of a solid block of forged iron, into the lower portion of which are inserted a number of steel or of chilled teeth of a wedge-like shape, held in place by conical keys. The stem is attached to the blade by another set of strong keys, and to the suspension appliances by means of a sliding box, the latter being a very important part of the apparatus, as without it the violent vibrations transmitted by the concussions of the trepan on hard rock, would inevitably rupture the connecting-rods at every blow. The weight of the small trepan is variable, depending upon the kind of work which it is to perform. That on exhibition weighs fifteen tons. In the trepans first constructed by M. Kind, the upper portion of the central stem was threaded to receive a screw which united to the slide, but this arrangement gave much trouble and soon got out of repair, necessitating an adaptation which is now used, consisting of two plates permanently keyed to the stem and replacing the male portion of the older model. The large trepan, employed for widening the bore made by the small trepan, consists also of a ponderous forged iron blade, united to the central stem by three arms strongly keyed. At the two extremities of the blade are teeth, and a U-shaped guide, of the diameter of the small bore, projects below, in the central space between the two sets of teeth. The weight of this tool is about twenty-five tons. The sand-buckets are large, plate-iron cylinders, with valve bottoms, admitting of the dumping of their contents, and they are used to dredge out the dirt and slush from the bottom of the shaft as the work progresses. The tools employed in case of accidents or special emergencies, consist of a safety-hook, a grappling forceps of very ingenious construction, and the fanchere or holding nippers.

The whole apparatus made use of in sinking and tubing a mining-shaft by this process, is operated by two engines, one to raise the trepans during the act of striking, the other to work a capstan which is used in the lifting and lowering of the various tools and of the tubing.

In our illustrations on pages 309 and 310, the various tools are shown as follows:—Fig. 1 is a sand-bucket or dredging-apparatus; Fig. 2 shows the safety-hook for lifting the trepans and their connecting-rods in case of rupture of these last; Fig. 3, the grappling-hook for extracting blocks of rock, detached teeth from the trepans, etc., from the bottom of the shaft; and Fig. 4, the fanchere replacing the safety-hook in the event of a rupture of the main stem, or that of one of the rods below the prominent collar at its head.

Fig. 5 shows a small trepan, used at L'Hopital for the first bore of four and a half feet; Fig. 6, a small, massive trepan for the same purpose, but in hard rock; and Fig. 7, a widening trepan with a double blade, used in the air-shaft for a diameter of eight and two-tenths feet.

Fig. 8 is a large trepan for hard ground, in which g, g, is a central guide, occupying the bore previously made by the smaller tool and maintaining the apparatus in a central position.

Fig. 9 is a large trepan for boring diameters of from thirteen and a half to fourteen feet; and Fig. 10, also a large trepan, made by adding a blade to trepan shown by Fig. 7.

Fig. 11 is a new form of trepan proposed by M. Kind for diameters of two and three-tenths to three and three-tenths feet, and Fig. 12 shows a trepan for a first widening of the shaft to eight and two-tenths feet. Fig. 13 being one for shafts of thirteen and eight-tenths feet in diameter, which may follow the previous one. It has teeth arranged on an incline so as to direct the debris of rock to the centre.

Fig. 14 is a small trepan for bores of five feet diameter in soft ground, and Fig. 15 shows a large trepan for widening this bore in the same kind of ground.

Fig. 16 shows a kibble, or vessel for receiving debris, intended to be suspended in the shaft during the work of widening.

Fig. 17 is a vertical section of the moss-box, used as a fitting to the tubing of shaft No. 2 of L'Hopital; a, a, being the internal cylinder, carrying a flange at the bottom, forming the wall of the moss-box. This cylinder is suspended by means of six screw-bolts, which allow of its gliding on them as guides during compression. b, b, is the first section of the tubing, which carries an outer flange and forms the other wall of the moss-box; s, s, sheet-iron segments, which press on the moss and prevent exclusive vertical compression of the same, and m, the moss contained in the joint before compression.

Fig. 18 is an assemblage of the parts which constitute the lower end of the tubing. This portion alone is lowered to the surface of the water before the series of rings of the tubing are adapted successively to it. a, a, is the internal wall of the moss-box; b, b, the first section of the tubing, forming the outer wall of the moss-box; c, c, the second section of the tubing, which carries the false bottom, and eventually floats the whole column, and d, d, the third section, with the suspension flanges which attach to the guide rods for the maintenance of the system in a vertical position while sinking. $f, f,$ is the false bottom, and g, a central pipe connected at its lower end with the false bottom at its centre, and carried to the top in successive lengths along with the outer tubing; water being allowed to penetrate by means of suitable cocks inserted at various heights in this tube, permits the gradual and simultaneous lowering of the whole casing independent of its weight. When the casing has reached bottom, and the moss-box has closed by compression, the water is pumped out of the shaft, and the false bottom and central tube extracted, after which the permanent foundations are established. Before the water is removed from the shaft, however, a coating of concrete is introduced between the tubing and the outer walls of the shaft, and permitted to harden. If the work has been properly done, the shaft is now found to be perfectly tight in all of its parts.

Fig. 19 shows the foundation for the tubing as used at L'Hopital, and Fig. 20 that for the shaft of Sainte Barbe.

Fig. 21 shows the special ladle used for the introduction of the concrete, it being furnished with a movable bottom connected to a piston-rod in such a way that pressure on the latter will cause the evacuation of the contents of the ladle.

It is stated by M. DEBY, that under the best conditions, the cost of sinking and tubing a shaft by this process may be stated at about four hundred and fifty dollars, gold, per yard, on an average diameter of twelve feet, the amount having never in the worst cases exceeded about seven hundred and thirty dollars, gold, as a maximum. He states that for a diameter of fifteen feet, it would be safe to estimate on a minimum cost of seven hundred and thirty dollars, gold, per yard, and not to exceed eleven hundred dollars, gold, as a maximum. When shifting-sand, gravel, loose clay, or quicksand occurs, extra expense will always ensue from the necessity of a certain amount of protective or temple tubing.

Hereafter, the boring of a mining-shaft through even the most highly water-bearing strata, need no longer be feared by any mining engineer who thoroughly understands the KIND-CHAUDRON PROCESS, and a debt of gratitude is due M. Chaudron by the whole profession for this important contribution to the science of mining technology.

M. DEBY sums up the advantages of the CHAUDRON PROCESS over the older methods of mining, as follows:—

1. The water from all the water-levels situated above the bottom of the tubing is isolated and kept permanently excluded from the shaft as well as from the workings below it.

2. The solidity and durability of the shaft is very great, and much superior to that obtained by any other means.

3. The cost of sinking a shaft through water-bearing or caving strata is reduced to a minimum.

4. A great saving in time is realized.

5. The possibility, not to say certainty, is obtained of traversing, without much difficulty, any number of successive water-levels and any kind of water-bearing strata without having recourse to any pumping machinery whatever.

6. The absence of all danger and inconvenience to the miner during the operation is complete, and contrasts with the perils and discomfiture attendant on the ordinary mode of sinking shafts below the water-level.

7. Safe and reliable preliminary estimates as to the cost of sinking a mining-shaft through unpropitious ground are attainable only by the adoption of the CHAUDRON PROCESS.

MESSRS. CHAMBERS, BROTHERS & CO., OF PHILADELPHIA, exhibit a brick-making machine which possesses particular merit on account of its great simplicity, there being no expensive details requiring continual renewal, and an entire absence of links, joints, pistons or nicely fitted surfaces to be worked in contact with the clay, so often found in machines of this class. In the operation of this machine, which is represented by the engraving on page 316, the crude clay is taken directly from the bank, and after being mixed upon the dumping floor when necessary, with a little loam, sand or coal, and the requisite amount of water added to reduce it to a proper consistency, it is fed through a hopper directly to a tempering chamber. This chamber is of a cylindrical form, and a horizontal revolving shaft passes through it, having tempering knives attached to it, which project radially and are arranged spirally around it. These knives thoroughly mix and temper the clay, moving it forward at the same time until it is delivered in one homogeneous mass to the expressing screw, which is attached to the end of the tempering shaft and revolves in a cone-shaped case. The screw-case is conical in form and the inside is fluted, this arrangement preventing the clay from simply revolving with the motion of the screw, and causing it to become, as it were, a nut on the screw, which, since the latter cannot move backward, must go forward, and is therefore forced into the case. The case changes from a circular to a rectangular section, and reduces the dense cylindrical column of clay to a rectangular shape, the material being forced well into the corners, so that, as it is pushed out from the finishing die, which is of chilled cast-iron, it issues in the form of a continuous rectangular bar with firm, well defined corners and of the exact proportions for cutting apart into bricks. MR. CHAMBERS, a member of the firm, has recently invented a most ingenious device for cutting this continuous bar of clay into proper lengths for bricks, and has, he believes, practically overcome the most serious objections which have heretofore been urged against this machine, to wit: the slight irregularity in the length of the bricks and their rough ends. When the bar of clay issues from the die, it is received upon an endless bed, made up of jointed plates, each the length of a brick. Between these plates are narrow spaces, to permit the passage of a continuous steel blade, which is held by being passed spirally around a large cylinder. The projection of this blade from the cylinder gradually increases, its pitch being the length of a brick, so that when the cylinder revolves the blade, it will cut the bar of clay into uniform lengths proper for bricks. The velocity of the movement of the clay and the speed of revolution of the spiral are very ingeniously governed so as to insure this uniformity of lengths in the bricks. Simple and effective arrangements are made to prevent any damage from the accidental presence of large stones in the clay. After the bricks are cut off they are traveled on an endless

Archimedean Brick-Making Machine: Chambers, Brothers & Co.

belt under a dusting machine, to put them in better condition for handling and to improve their appearance when burnt. They are then removed from this belt by hand and pass through the usual processes of stacking, drying and burning, when the finished brick is obtained ready for the use of the builder. The machine throughout is a marvel of simplicity and adaptation of mechanical means to the manufacture of a material naturally antagonistic to machinery. Its continuous rotary movement enables it to be run at a very high speed with satisfactory results and without undue wear.

Messrs. Chambers, Brothers & Co. also exhibit some excellent Folding

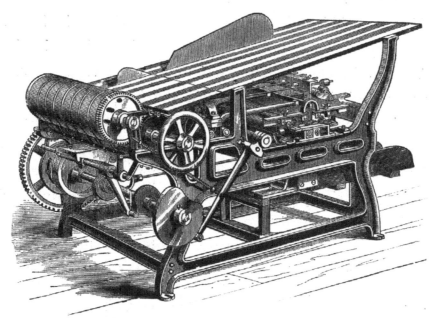

News-Folding and Pasting Machine: Chambers, Brothers & Co.

Machines, one of which is illustrated above. The merits of these machines, which consist of a newspaper-folder, paster and trimmer; a book folder; and a periodical folder, paster and coverer, are well known among the book-binders and newspaper-men throughout the country, and they represent not only fine specimens of design and workmanship, but at the same time their performance is such as to fully indorse their high claim in regard to genuine labor-saving qualities. The various book-folding machines manufactured by this firm are very extensively used in our large binderies, and may be seen, each daily performing the average work of five hand-folders, doing it in a much better manner and at very considerably less cost. The folding, pasting and covering machine exhibited is particularly noticeable for the almost lifelike way in which it performs its work, and has attracted a large amount of attention from visitors. A sheet of paper intended for a pamphlet is laid on the table of the machine and is first taken up by a pair of rollers, then carried on back and forth from one part of the machine to the other on endless tapes, being folded and pasted, and pasted and folded, over and over again until reduced to the proper size. A cover is then fed in at one end of the machine and attached to the folded paper, and finally, the once plain sheet is dropped out in the form of a neat thirty-two page pamphlet with the leaves all pasted together at the back and supplied with a well secured cover. Another machine exhibited among these, an adjustable 12mo folder, although comparatively simple in its construction, performs a variety of work formerly considered impossible to be done on one machine. This machine will fold plain sixteens, cut a twenty-four, and out or inset it, as may be desired, with equal facility, and it is also so arranged as to throw out all imperfectly printed sheets and place in the packing trough only those that are perfect and fit for binding. The rotary newspaper folding machine is in use by nearly all of the principal daily and weekly papers having large editions, and among the publishers of large eight-paged papers, the folder, paster and trimmer is recognized as a most important machine. It delivers the paper neatly pasted at the back, with trimmed edges, so that all the pages may be read without the annoyance of cutting or turning a large uncut sheet, thus very much increasing the value of the inner pages for advertisements.

The Government of France appreciating the efforts of the United States in the creditable display which it made at the Paris Exposition, of 1867, has reciprocated at this present Exhibition, and, through its Ministry of Public Works, has had collected and placed on exhibition a series of models, charts and drawings, so selected as to convey to the visitor some idea of the works of the "Ponts et Chaussées" and the "Mines," during the past few years. The great railway companies of France were also solicited by the Government, and have promptly given their co-operation towards the furtherance of the exhibit. No space having been assigned by the Centennial Commission in the Exhibition Buildings for this special display, the French Department of Public Works, has had erected, at its own expense, a special fire-proof structure for this purpose, from plans presented by M. de'Dartein, engineer "de Ponts et Chaussées, professor of architecture at the Polytechnic School and at the "Ecole des Ponts et Chaussées.' M. Lavoinne, engineer "des Ponts et Chaussées," and M. d'Hervilly, conducteur, have been delegated in charge of the exhibit. We owe special thanks to M. Lavoinne for the courtesies he has extended and facilities he has given to us in the study of the plans and models, and in the preparation of this article.

Catalogues in French and English have been published and distributed with great liberality, giving full descriptions of the exhibits and furnishing material of great value to the engineer. The display has been divided into sections, illustrating works in the different departments, of roads, railways, internal navigation, maritime improvements, light-houses, water supply of towns and canals, miscellaneous subjects, and charts, and various objects in relation to mines. It is only possible for us, however, even with a liberal allotment of space at our command, to draw attention to a few of the more prominent examples of the list.

The lines of communication in France are divided into terrestrial, fluvial and maritime. Of these, the terrestrial lines comprise roads and railways, roads being subdivided into national and departmental; and fluvial communications embrace rivers and canals. The map displayed at the Exhibition shows these different divisions by a very excellent conventional arrangement. National roads are represented by a dark-brown line, departmental roads by a narrower line of the same color, railways by a white line, navigable rivers and canals by a blue line. The arrangement of the mountains shows the summits of the watersheds, the large rivers occupying the intermediate valleys, and the principal routes of maritime navigation are given in gold lines. Various depths of the sea, from one hundred, two hundred, and five hundred to one thousand metres, are marked by red lines, and white circles show the luminous ranges of the different light-houses. Chief towns of departments are represented by gilt hollow-top buttons, chief towns of "arrondissements" by black dots, and seaports by flat gilt buttons of two sizes, the larger ones indicating the great ports.

Taking up the subject of "roads," we would mention first the SWING BRIDGE AT BREST, shown by the engraving on page 320, which was represented at the Exhibition by a model of the bridge on a scale of one-fiftieth, a model of the centre and its mechanism, on a scale of one-tenth, and a perspective view of the commercial port. This bridge provides a communication between the towns of Brest and Recouvrance, separated by the Penfield. It is at a height above high tide of about sixty-four feet, and about five hundred and seventy-one feet in length between faces of abutments. The two portions of the iron superstructure spanning this distance, are supported, each, by a circular pier nearly thirty-five feet in diameter at the top, about which they revolve, the distance between these piers being such as to give a clear passage of nearly three hundred and forty-eight feet when the bridge is open. The trusses, which are two in number, in each part, are twenty-five and three-tenths feet deep at the piers, and four and six-tenths feet at the centre extremities. The roadway is

Fig. 1.—Swing Bridge, at Brest.

sixteen and four-tenths feet wide and the footways three and six-tenths feet. Each truss is composed of upper and lower members, T-shaped in section, with upright braces between, and cross-braces in the panels. Diagonal and lateral ties and struts are placed between the trusses on each panel, so as to give ample stiffness sideways, and advantage is also taken of the flooring to increase this stiffness, by making it in two courses of timber, overlaid so as to cross all the joints and make an inflexible surface. In the rear portion of each of the two parts of the Bridge is arranged a counterpoise weight to preserve the equilibrium of the part on its pier. The most difficult portion of the structure to design, was that over the piers where the action of the dead weight is the greatest, and where the proper arrangements for rotation in opening and closing the draw must be made. The weight of the framework being placed on a bed of friction-rollers, it was essential that it should be distributed as equally as possible on all of them. To effect this, the division of the uprights in the girder has been so regulated as to make them correspond with the circle of the centre line of friction-rollers, therefore, giving four principal points of support, extra columns being placed here in the trusses for additional strength. These columns have four strong sets of diagonal stays binding them together, and they are further connected by a circular drum or tower of boiler-plate having a platform at top and bottom, and serving to distribute the load. The friction-tables or centres are the same in principle as the French railway turn-tables, having a diameter of twenty-nine and a half feet, and being subjected each to a load of over six hundred and sixty tons of two thousand pounds. There are fifty rollers to a table, with an average diameter to each of about one and six-tenths feet, and a length of nearly two feet. A rack is provided on a fixed circle into which a pinion on the movable-table works, the latter being operated by a wheel and second pinion. The latter is fixed to a vertical shaft reaching to the floor of the bridge and arranged with a cross-head for capstan-bars to be worked by hand-power. To insure the stability of the superstructure during the traffic, two sets of bolts are arranged in the forward end of each pair of trusses to fasten them together, and in the rear there are two sets of levers, similar to the jaws of a vise, in a horizontal position, which seize a piece of cast-iron let into the solid abutment, the axes of the levers being fastened to the framework of the bridge. To relieve the rollers of the great weight on them, two jack screws are fixed on the forward side of each centre, which are supported on plates bedded in the crown of the pier. Each is intended to raise one of the girders, and they are operated by means of a connection with a jointed capstan which can be lowered and kept under the flooring of the bridge. It is stated that in perfectly calm weather, with two men on each portion of the bridge, it can be opened or shut in fifteen minutes at the most. This does not compare very favorably with our American swing bridges, as such a length of time would be considered entirely out of the question in this country.

The foundations of the stone-work are all on solid rock, the facings of the piers being in dressed free-stone, and those of the abutments in dressed quarry-stone, with chain courses of dressed ashlar at the angles. Hydraulic pumps are arranged to lift the spans off of the friction rollers and beds, should any repairs be necessary. The bridge was constructed from plans of M. Oudry, and cost about four hundred and twenty-four thousand dollars, the total weight of iron being about two million six hundred and forty-six thousand pounds.

Fig. 2.—*Bridge of Saint-Sauveur, on the Gave, Pau.*

The next bridge we desire to notice is that of SAINT-SAUVEUR, ON THE GAVE, PAU, shown above, represented by a fine model of one twenty-fifth full size. It consists of a single semicircular arch of one hundred and thirty-seven and eight-tenths feet span, with a length between approaches of two hundred and seventeen and two-tenths feet, and a height of roadway above low water of about two hundred and fifteen feet. It affords accommodation for a carriageway of fourteen and eight-tenths feet, and two foot-ways of two and eight-tenths feet, each, the total breadth between outer faces of balustrades being twenty, and thirty-four hundredths feet. The arch rests on abutments of solid rock, the crowns, brackets and plinths being of dressed stone, the main body of the vaulting of rag-stone, rubble and Vassy cement, and the spandrils of calcareous rubbles and fat lime, mixed with a tenth of Vassy cement. It presents, therefore, quite a bold specimen of this method of construction. The arrangement of centre and false-works was somewhat peculiar, a timber pier having been erected in the centre of the stream, and of the arch opening, as shown by our engraving, carrying a platform nearly forty feet wide, level with the springing of the arch, and formed of cross-beams on longitudinal girders, supported by the banks and pier, and additionally by inclined struts. This platform was built up to the lower part of the centre in two stages of twenty-six and a quarter, and nineteen and seven-tenths feet, respectively, and the centering consisted of four ribs, each composed of six beams, mutually supporting each other, and framed with courses of horizontal ties and queen posts. The centres were removed by means of jack-screws, and the settling observed, was about sixteen-thousandths of a foot. The work was commenced in 1860, by order of the Emperor Napoleon III, and the bridge was opened for traffic on the 30th of June, 1861. It was planned and executed by M. M. Schérer and Marx, engineers in chief, and Bruniquel, resident engineer des "Ponts et Chaussées." The total cost being a little less than sixty-four thousand dollars, of which the sum of about twenty-four thousand two hundred dollars was for provisional works.

Under the head of Railways, the great iron viaduct of Busseau-d'Ahun, on the Montluçon and Limoges line, over the Creuse, for a double set of rails, forms a prominent exhibit, and is represented by a model of a pier and a portion of the flooring on a scale of one twenty-fifth, and general drawings. It consists of six bays, four of them with spans of one hundred and sixty-four feet, each, one of one hundred and forty-eight and a half feet, and one of one hundred and thirty-five and three-tenths feet, making a total length, including approaches of about eleven hundred and eleven feet. The two abutments are of masonry, and the five piers consist of open cast-iron frame-work, anchored on to bases of masonry, the height from low water mark to rails being about one hundred and eighty-five feet. A pyramidal arrangement has been adopted in the piers, and each pier is formed of eight hollow columns in two pier frames, the total height being made up of seven stages in addition to an annular foot-plate, bound together at each stage by horizontal braces and diagonal stays, and the whole capped by a crown and hinged capital supporting the superstructure. The columns have a uniform exterior diameter throughout of about fourteen inches, the thickness of metal decreasing from the base upwards for the various stages, the successive lengths being joined together by bolts. They are so arranged that the plane of each stage is formed of three squares in juxtaposition, making a rectangle at the top, of about six and a half by nineteen and a half feet, and at the bottom, of about eleven by thirty-three and a half feet. The eight columns of each stage are braced by pairs of diagonal stays, in the form of a Saint Andrew's cross, six of these connecting the columns of one frame and four uniting the two frames to each other. There are also ten horizontal braces, six of them parallel to the axis of the pier in plan, and four perpendicular to it, these being all placed at the points of junction of the several stages. The cross-bars are double, and formed of U-iron. The braces are made of two T-irons, arranged crosswise. The horizontal cross-bars making the diagonals of the three squares formed at each stage by the ten cross-braces, are also of T-iron. The foot-plates of the columns are strongly anchored to the masonry by iron rods built in and keyed to anchor-plates, the rods being from two and a half to four inches in diameter, and from twelve to eighteen feet long. The foot-plates give a maximum pressure on the masonry of about six hundred and eighty-three pounds per square inch, and they rest on three courses of Volvic lava, over two feet thick, and more than twenty-five feet square in area, thus reducing the pressure on the masonry below, to about one hundred and fourteen pounds per square inch. The superstructure consists of four girders, about fourteen and a half feet deep, placed about six and a half feet apart, centre to centre, and built continuously from abutment to abutment. They are of wrought-iron lattice construction, formed of plate, angle and T-sections, additional stiffness being obtained by vertical stiffners of double T-iron and rolled plates at each standard. The girders are braced together, vertically, by a series of diagonal stays of

U-iron, placed about thirteen feet apart, and laterally by diagonal stays, above and below, between the two central girders, in squares about six and a half feet each. Friction-rollers are introduced under the girders on the small piers and on the abutments. Each girder has only one bearing upon the heads of the piers, thus avoiding unequal distribution of weight between the standards of the two pier frames during the passage of trains. A hinged capital is placed between each girder and the crown of the pier.

The total cost of this structure was about three hundred and three thousand dollars, and the work was executed under M. M. Thiriou, engineer in chief "des Ponts et Chaussées" and director of the central lines of the Orleans railway, Nordling, engineer in chief, and Geoffroy, resident engineer of the same line.

The Bridge VIADUCT OF THE POINT-DU-JOUR, AT PARIS, AUTEUIL, shown by engraving on page 326, for the passage of the Circular Railway over the Seine, also attracts particular attention from visitors, owing to the exceedingly handsome and complete models exhibited; there being one of the entire structure, on a scale of one-hundredth, and also one of an abutment arch, and one of two arches of the viaduct proper, on a scale of one-twenty-fifth. The location has required the construction of heavy works, comprising a street bridge over the river and a viaduct at a higher grade for the railway, consisting of a central portion, supported by the bridge below, and two long approaches. The different parts of the structure are designated by different names, the total length being divided as follows: Viaduct of Auteuil, from the station of that name, to the Versailles road, amounting to about thirty-five hundred and twenty feet; Viaduct of the Point-du-Jour, from the Versailles road to the quay of the Seine (right bank), making about five hundred and eight feet; Bridge Viaduct over the Seine, of nearly eight hundred feet; and the Viaduct of Javel, from the quay of the Seine (left bank), to the commencement of the embankment; the whole footing up to a total of about five thousand two hundred and eighteen feet.

In reference to the Bridge Viaduct—this work consists of two perfectly distinct parts; a bridge or lower stage, intended for foot-passengers and carriages following the military road; and a viaduct or upper stage, occupying only the central part of the bridge, and belonging to the Circular Railway. The bridge or lower stage is composed of five elliptical arches, of about ninety-nine feet span, and twenty-nine and a half feet rise, having a width between heads of about one hundred and two feet, made up of two parapets of about one and seven-tenths feet each, two footways of seven and four-tenths feet, two roadways of twenty-three and eight-tenths, and a central part of about thirty-six and two tenths feet, the latter being covered by the railway viaduct, and forming a footway. Three arches or vaults are built parallel to the stream, in the spandrels over each pier, having fifteen feet span, and one and eight-tenths feet thickness at the key. They are constructed of rough millstone grit and cement, except in the central part when merely supporting the footway, where they are of perforated brick and only seventy-two hundredths of a foot thick. The abutments of these arches are also pierced by eight series of bays parallel to the axis of the bridge, and a vaulted subway for water-pipes is built under each footway, about five feet wide and two feet high, covered with a brick arch two and a half inches thick. The construction in all cases is sufficiently heavy to allow a steam roller of thirty-two tons to pass over the lateral ways, and the whole result of the lightening is such as to diminish the weight sustained

Fig. 3.—Viaduct of the Point-du-Jour, at Paris-Auteuil.

by each pier by three and three-tenths tons, and that on the concrete by about twenty-three pounds per square inch.

In regard to the viaduct proper, or upper stage, the quays of Auteuil and Javel are covered by segmental arches of about sixty-five and a half feet span, and eight and seven-tenths feet rise, there being between them thirty-one semicircular arches of fifteen and three-quarters feet span, one and a half feet thickness at key, and twenty-nine and a half feet width between heads. The thickness of piers at springing is three and thirty-seven hundredths feet, and the piers and abutments are pierced by two bays of seven and thirty-eight hundredths feet span, and fourteen feet height to crown, so as to afford a passage for pedestrians on the central footway. The spandrels are built with small brick vaults, in order to decrease their weight.

The foundation soil consists of chalk, covered in the bed of the river, and on the left bank by gravel, but on the right bank, by layers of clay, mixed with peat and mud. The right bank foundations are on piles, penetrating the chalk to a depth of at least six and a half feet, and covered by a layer of concrete, from five to six feet in thickness. The piers are founded on beton laid on the chalk-bed, previously cleared by dredging, and the foundations of the left bank are constructed on the gravel by means of a solid mass of beton, laid dry in an enclosure of piles and sheet piling. The beton for the piers was run into timber caissons without bottoms, put together on the spot, that is, on the movable scaffolding constructed for their immersion, their dimensions being about one hundred and thirty-one feet in length, by about twenty-nine and a half feet in breadth and twenty-six feet in depth. The centres of the arches were removed by means of sand-boxes, and the subsidence of the arches varied from three-eighths to half an inch. These arches are in rough millstone grit and cement mortar. The total cost amounted to about five hundred and seventy-three thousand dollars.

The Viaduct of the Point-du-Jour, consists of twenty-six semicircular arches of sixteen and three-tenths feet span, about two feet thickness of crown, and forty-nine and two-tenths feet breadth between heads, this breadth being necessary on account of the location of a passenger station at this point. The piers are three and thirty-seven hundredths feet thick at the springing, and are pierced by four openings of six and six-tenths feet span, and fourteen feet height to crown, for foot passengers. A portion of the foundations, those for twelve piers, rest immediately on the gravel, but the others had to pass through a soil about twenty-five feet thick, composed of recent alluvial deposits, with alternate layers of clay, mud and peat, requiring a height of pier of nearly sixty-six feet to the arches. An arrangement was therefore adopted of dispensing with the alternate foundations, and supporting the piers of these on ogival arches resting on the solid foundations of the piers of even rank on either side, the construction being all below the level of the boulevard and not apparent to the eye.

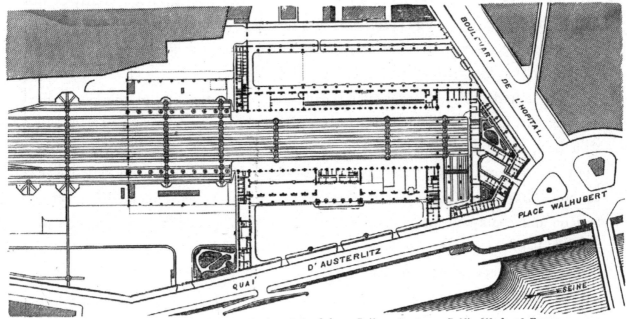

Fig. 4.—Plan of the Paris Terminal Station of the Orleans Railway Company—Public Works of France.

The socles and angles of the piers, the cornices, parapets and the roussoirs of the heads are in dressed stone, the balance of the work being in ashlar or rubble. The pressure on the foundation soil is about sixty-eight pounds per square inch, that on the concrete about seventy-five pounds, increasing in the ogival arches to one hundred and forty-two pounds per square inch. The cost of this portion was a little over one hundred and eight thousand dollars, and the entire work was constructed in 1864 and 1865, M. M. Bassompierre-Sewrin, engineer in chief, and de Villiers du Terrage, resident engineer, "des Ponts et Chaussées."

In connection with railways we would also draw attention to the PARIS TERMINAL STATION OF THE ORLEANS RAILWAY COMPANY, represented by a set of drawings on scales of five-thousandths and two-hundredths. The erection of this building was commenced in 1865, to replace the old station which had become far too contracted in accommodations to suffice for the greatly increased business of the road, and it occupied in completion about four years and a half, being opened for traffic in July, 1869. The former station was situated between the Boulevard de l'Hôpital and the Rue de la Gare. For the new station all of the ground was obtained between this street and the quay d'Austerlitz, the entire space being bounded by the Boulevard de l'Hôpital on one side and by the quay on the other, and the superficial area covered by the old station, which amounted to only four hundred and forty-one thousand three hundred and twenty-four square feet, was now increased to one million seventy-six thousand four hundred square feet. The new station has been designed on a very similar plan to the old one, only occupying a larger area. The company's offices are placed in front at the angle of the boulevard and the quay, while in the centre are the tracks covered by a roof, the departure platform with the required rooms and offices being on the left, and the arrival on the right. That portion of the station for the departure of passengers is the most important part of a terminus, and in the present building it is arranged with a pavilion in the front of its centre, containing a vast hall with ticket offices, and with wings on the sides, each nearly two hundred feet long, one containing the waiting-rooms and the other the luggage and registration office. A portico constructed in cast- and wrought-iron, a little over twenty feet in width, extends the whole length of the side. From the two extremities of this side, rectangular buildings extend out as far as the quay, which are devoted to the various accessories, such as refreshment-rooms, post-office, telegraph, station-masters' apartments, etc. The arrival side is of the same length as the departure, but the breadth of the building is less and the construction more simple. It contains only a waiting-room and a hall for the delivery of baggage, the portion shown in the figure as projecting on the court being used for porters' rooms, the octroi or city dues, the goods department and for various other adjuncts. The custom in Paris of levying city import dues on certain classes of goods, necessitates very complete arrangements for the rapid and effective examination of baggage before it can be delivered. It is taken from the cars and all placed on long platforms, the custom house officers standing on one side and the traveler on the other, the latter, when he finds his trunk, patiently waiting, keys in hand, ready to open it for examination when requested.

The arrival platform is about six hundred and twenty-three feet long and one hundred and thirty-one feet wide, one-third of its length being roofed over, and the remainder bordered with an awning. The distance between the walls of the arrival and departure portions of the station is about one hundred and sixty-nine feet, and this is occupied by eight lines of railway and two footways. A roof of a single span covers this whole space, extending out three hundred and twenty-eight feet beyond the extremity of the goods department, so as to protect the cars housed on the intermediate tracks. Fig. 4, on page 328, shows a plan of the

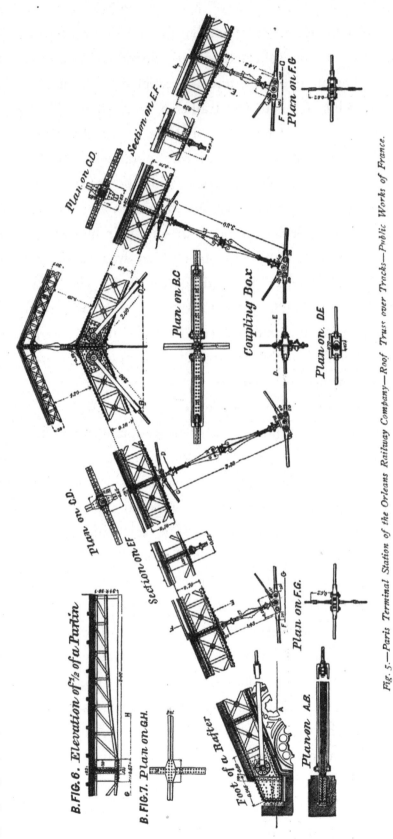

Fig. 5.—Paris Terminal Station of the Orleans Railway Company—Roof Truss over Tracks—Public Works of France.

station on a scale of 0.0004, Fig. 5, on page 330, Fig. 6 on this page, and Fig. 7, on page 332 give details of roof on scales of 0.00125, 0.025 and 0.05.

The exceptional span and vast dimensions of this roof make it one of special interest to the engineer. Its total length is about nine hundred and eighteen feet, and its span about one hundred and sixty-nine feet, the trusses being nearly thirty-three feet apart, with deep-built purlins between them. The form of truss used is that generally called by us the French triangular system, but there known as the Polonceau system. It consists of wrought-iron tie-rods, cast-iron struts, and built principal rafters, composed of cross-

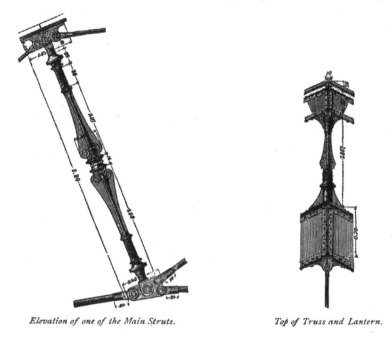

Elevation of one of the Main Struts.   *Top of Truss and Lantern.*

Fig. 6.—*Paris Terminal Station of the Orleans Railway—Details of Roof—Public Works of France.*

pieces in flat and angle irons, united by rivets. Where the trusses rest on the walls there are ornamental cast-iron brackets. The purlins are made with cross-bars similar to the ribs, the lower flange being curved, giving greater depth at the centre than at the ends, and thereby increasing the rigidity, while at the same time lessening the monotony of the long lines to the eye. The total weight of iron in the roof is about fourteen hundred and thirty-three tons of two thousand pounds each, and the dimensions of the parts have been so proportioned that the stress does not anywhere exceed eighty-five hundred and thirty-three pounds per square inch, not even on the tie-rods. The area of ground occupied by actual buildings, amounts to 131,859 square feet, that covered by roofs and awnings independent of the buildings, is 295,170 square feet, and that belonging to the station, but not covered, is 346,299 square feet. The foundations were very unfavorable, necessitating a depth to suitable bottom of twenty-six to thirty-six feet, and owing to the proximity of the Seine, and the permeable nature of the strata, a large amount of pumping was required. The roof had to be raised without interference with the use of the station for the daily traffic, and this was accomplished by means of a traveling scaffold of the same breadth as the roof. The total sum expended was $3,600,000, of which $1,670,000 was for the ground, $328,000 for the company's offices, $1,380,000 for the station proper, and $222,000 for accessory works. The roof cost about one dollar and thirty cents per square foot covered, the departure buildings, nine dollars and sixty-six cents, and the arrival building six dollars and ten cents per square foot. The work was executed under M. Sevène, as chief engineer, and M. Louis Renaud, as principal architect.

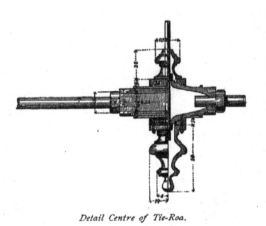

Fig. 7.—*Paris Terminal Station of the Orleans Railway—Details of Roof—Public Works of France.*

Detail Centre of Tie-Rod.

Great attention has been bestowed in France upon the improvement of internal navigation. Many of the rivers, if left in their natural state, could be used for water traffic during only a very small portion of the year. The navigation between Paris and Auxerre, following the Yonne, between Auxerre and Montereau, for a distance of about seventy-five miles, and the Seine between Montereau and Paris for about sixty-one miles, during more than three centuries was intermittent, and for eight or nine months in the year was carried on by means of "éclusées" of the upper Yonne, which are temporary risings of the water, caused from time to time by the regular and successive closing and opening of navigable passages and barrages on its course, the bodies of water so created, carrying the various boats, rafts, etc., in the river, along with them. This has been very much improved of late years by the construction of a number of movable barrages, twelve on the Seine with locks, and twenty-five on the Yonne, twenty-two of the latter having locks, and three coming under the head of derivations or cuttings. These improvements have given a continuous system of navigation, which has been in operation between Paris and Laroche, since the month of September, 1871, and as far as the Auxerre, since September, 1874, furnishing a minimum depth of water of five and a quarter feet.

The river Yonne, in its course between Auxerre and Montereau, is divided into two parts. The first from Auxerre to Laroche of about seventeen miles in length, varies from two hundred to two hundred and sixty feet in breath, and, with an average fall of about three and a half feet to the mile, discharges about four hundred and sixty cubic feet at low water, and from ten to eighteen thousand cubic feet, when the river is high. Before the present arrangement, when the éclusées were depended upon for the navigation, the

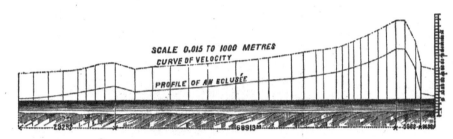

Fig. 8.—*Form and Velocity of an éclusée. Public Works of France.*

number made per year was about eighty-five. The movable dams were opened successively at suitable intervals, about twice per week in summer, the effect of each éclusée being experienced for about four hours. The regulations which governed the management of the dams, were very strict and had to be closely followed. Each dam was closed as soon as the water had fallen to a certain point in order to collect water for the next éclusée. The closing was, of course, followed by a subsidence of the water behind the previous éclusée, below its lowest natural level, which lasted for about a day, and was felt for a long distance down the river, rendering the passage of the smallest barge impossible over the sand-bars in the channel, and requiring all ascending boats to be empty or nearly so. The velocity of the éclusée was governed by various circumstances; the volume of water let free, the rate of descent, the windings and other obstructions of the river, the force and direction of the wind, etc., etc., and a violent head-wind has been known to retard an éclusée two hours between Auxerre and Laroche. Notes on the velocity of éclusées have been taken by M. Chanoine at Laroche, our Fig. 8 giving the results of one of his observations. This velocity varied from point to point, being greatest at the highest point of the wave, and a boat on top of the éclusée was impelled actually at a faster rate than the mean movement, allowing it to make short stops on its downward passage, while at the same time continuing all the way by the same éclusée. The second part of the river Yonne, from Laroche to Montereau, a distance of about fifty-seven miles, has a breadth of bed of from two hundred and sixty to three hundred and twenty-five feet, and an average descent of one and eight-tenths feet per mile, discharging some six hundred

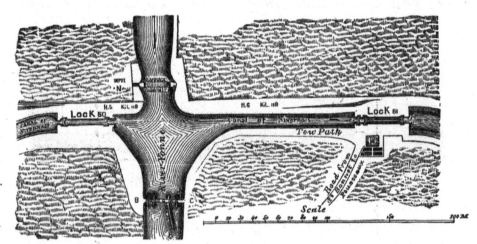

Fig. 9.—*The Basseville Barrage at the crossing of the Yonne River by the Nivernais Canal. Public Works of France.*

cubic feet at low water, and twenty-five to thirty-nine thousand feet at great floods.

Twenty-two of the locks on the Yonne are about thirty-five feet in width, by three hundred and fifteen feet working length, and will receive at one time six canal-boats, or two wood-rafts coupled. Two of the remaining three locks are only about twenty-seven feet wide, but six hundred feet working length and have the same capacity, while the third, that of Chaînette at Auxerre, is twenty-seven feet wide by only a little over three hundred feet working length, and will receive but three boats, or two rafts. All of the lock-gates are of wood, each being worked by a circular rack, and pinion with a winch.

Three of the movable barrages between Auxerre and Montereau, viz., those of Chaînette, Épineau, and Port-Renard are old, on the system of Poirée, having a permanent or fixed weir in masonry, and a passage closed by trestles or fermettes and paddles or needles, hence called needle-barrages, or sometimes *barrages à fermettes*. In order to describe the construction of these, we illustrate here the crossing of the Yonne River by the Nivernais Canal at Basseville, where is the first needle-barrage ever erected. Fig. 9 shows a general plan of the location, giving the position of the barrage, and Fig. 10 is a sectional elevation and plan of the barrage itself, half of it being closed and half open. Fig. 11 is a transverse section, parallel with the current, on a much larger scale, and Fig. 12 gives elevations of the left and right abut-

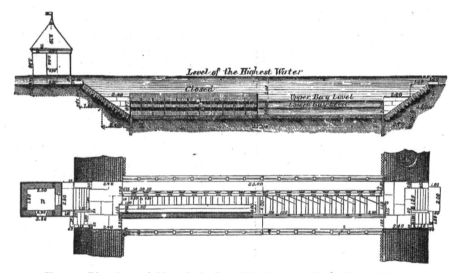

Fig. 10.—Elevation and Plan of the Basseville Barrage   Public Works of France.

ments, looking each way from the centre of the stream. The arrangement of the masonry, with the opening for the water, is clearly shown. All the way across between the abutments, the masonry of the bottom of this opening is very carefully built, a cross section being shown in Fig. 11, and a succession of iron frames, or fermettes, are placed on this bed with their planes vertical and parallel to the current of the stream, each capable of revolving to a horizontal position about an axis at its base, in bearings firmly fixed to the masonry. These fermettes are united at their tops by bars with jaws or catches at their extremities. Fig. 13 shows a plan and elevation of the details for uniting the adjacent fermettes, and Fig. 14 is the bar which joins them together. Fig. 15 is a plan and transverse section of the gudgeons and bearings, the gudgeons themselves being shown by Fig. 16.

The fermettes being placed upright in position, and connected by the bars,

Fig. 11.—Transverse Section of the Basseville Barrage. Public Works of France.

a vertical screen composed of wooden battens or needles about three inches square in section, and seven and a half feet long, is rested against them on the up-stream side as shown in Fig. 11, the lower ends of these needles bear-

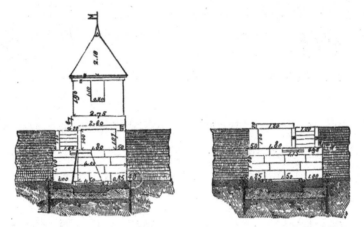

Fig. 12.—The Abutments of the Basseville Barrage. Public Works of France.

ing against a sill, and the upper ends against the connecting bars of the fermettes. The fermettes support on top a temporary foot-bridge used by the lock-keeper in working the barrage. When the barrage is to be opened, the attendant and his assistant first remove the needles, one by one, and afterwards the planking of the bridge, the connecting bars, etc., and the fermettes are successively turned on the axes at their bases into a horizontal position in the recess back of the sill provided in the bed for them. When the barrage is to be raised, the attendant grapples for the fermettes under the water, raising one after the other, reconstructing his foot-bridge as he proceeds, and then replacing the needles. The time necessary to open the barrage is about one-half minute per lineal yard, and for closing it, one and a half minutes;

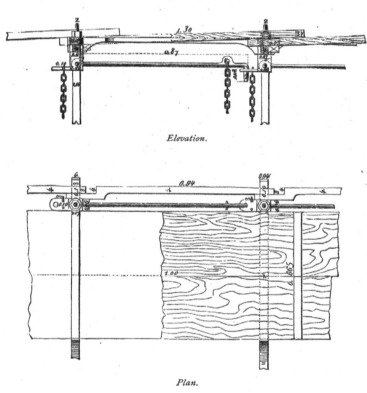

Fig. 13.—Details for uniting adjacent fermettes. Public Works of France.

the entire opening and lowering of the water taking forty-five seconds per yard.

The twenty-two new barrages have a passage closed, with movable shutters, on what is designated as Chanoîne's system, worked by a bar furnished with cams, and by the aid of a boat. Thirteen have the weir surmounted by movable shutters, and worked from a foot-bridge sustained by a system of fermettes, and in six there are trestles and paddles with a foot-bridge about eight-tenths of a foot above the head-water level. One barrage, that of l'Ile-Brûlée, near Auxerre, is fitted with large shutters on Girard's system.

The three derivations or cuttings, those of Gurgy, Joigny, and Courlon, of about three and two-tenths, two and two-tenths, and two and six-tenths miles in length respectively, are for the purpose of avoiding abrupt bends in the river, shortening the distance between Montereau and Auxerre by about seven miles. A stop-gate at the head of each keeps out the flood-water. These derivations have a width at bottom of fifty-two and a half feet, a depth below the normal level of water of about six feet, and a width at surface of about seventy feet, the slope of banks being three horizontally to two in heighth. The tow-paths are from thirteen to nineteen and a half feet wide, and are

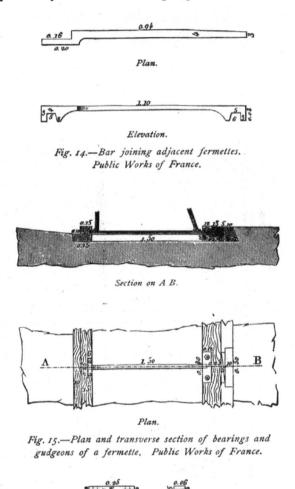

Fig. 14.—Bar joining adjacent fermettes. Public Works of France.

Fig. 15.—Plan and transverse section of bearings and gudgeons of a fermette. Public Works of France.

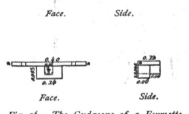

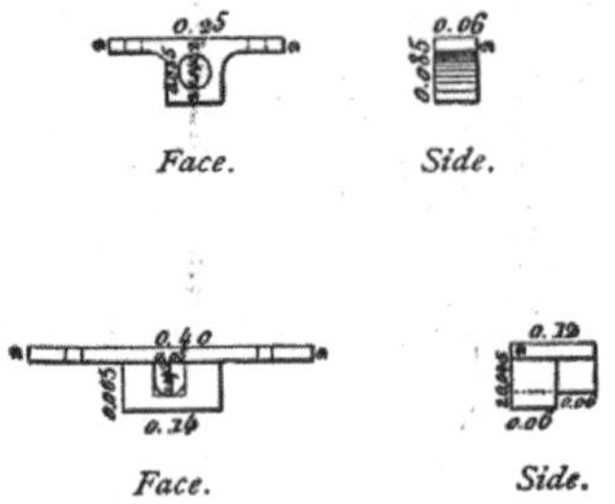

Fig. 16.—The Gudgeons of a Fermette. Public Works of France.

placed a little over one and a half feet above the highest known inundations.

The twelve barrages on the Seine, below Montereau, are also on Chanoine's system. The barrage at Melun has retained the trestle and paddle system for the weir, the only instance, however, in which this has been done. The barrages on upper Seine, above Paris, consist of two parts, a navigable passage from one hundred and thirty to one hundred and eighty feet in breadth, and a weir from one hundred and ninety-six to two hundred and thirty feet in breadth, both provided with movable shutters and separated by a pier, a lock being generally connected with the arrangement. The locks have a chamber of about thirty-nine by five hundred and ninety feet, sufficient to receive twelve canal-boats, or four rafts of wood. The lower sill of the lock is placed at least five and a quarter feet below the head-water of the lower barrage, and the cappings of the barrages and locks are one and three-tenths feet above the upper water level. These movable shutter-barrages are well illustrated at the Exhibition by a number of drawings, and by three models on a scale of one tenth.

Fig. 17.—Barrage of the Port à l'Anglais. Public Works of France.

The navigable passage and weir have different levels, the sill of the former being about two feet below low water-mark, and the shutters nearly ten feet in height and level with the water, while the weir is finished off about one and six-tenths feet above low water-mark, and the shutters are six and a half feet in height. The floors or platforms of the passage and weir are laid in beton, ran in an enclosure of sheet-piling, and paved with stone, a wooden sill, or ledge, being furnished, extending the whole length of the flooring, and well fastened down by disc anchors, its upper face giving a support to the base of the shutters, and the lower part carrying the sockets of the chevalets. To provide against possible subsidence of the beton under the movable parts of the weir, cross-pieces and longitudinal beams are employed, laid on the two lines of piles comprising the enclosure, a third row of intermediate piles being also driven.

The shutters are made of stout frame-work of uprights and cross pieces,

Fig. 18.—*Barrage of the Port à l'Anglais. Public Works of France.*

differing in size, as before-mentioned, for weir and passage. They are set with an intervening space of from two to six inches between them, depending on the discharge of the river at low water, their lower edges bearing for a part of their thickness, on the sill. Each shutter is furnished with a wrought-iron chevalet and a counterfort. The chevalet is a frame in the form of a trapezoid, the bottom terminating in two gudgeons received by sockets let into the sill, on which the chevalet can turn when it is being lowered on to the floor with the shutter, and the upper part also having two gudgeons connected with the shutter at its axis of rotation by two collars bolted to the frame-work. This axis is placed at about five-twelfths of the height of the shutter, varying from this, in some cases, depending on whether the shutters are self-acting, as on the weirs, or otherwise, as on the passages. The action of the shutters, to raise and lower themselves, depends upon the position of

Fig. 19.—Barrage of the Port à l'Anglais. Public Works of France.

the axis of rotation. The upper gudgeons have stops or catches to limit the turning of the shutters to an inclination of fifteen degrees, and to avoid too great strain on the breech in lowering. A counterpoise is attached near the bottom of the shutter to balance the weight of the upper part when the breech is submerged. The counterfort is a wrought-iron bar, hinged on to the upper part of the chevalet and inclined out so as to support the shutter and the weight of the head-water, the lower end abutting against a cast-iron shoe or catch, strongly embedded in the flooring. The rear of the shoe is an inclined plane, and it has a bent guide-bar to one side of it with a passage between. When it is desired to lower the shutter, the counterfort is moved sideways into this passage, along which it slides as soon as it leaves the front of the shoe, the chevalet turning on its base and falling with the shutter on to the flooring. The bent guide-bar draws the counterfort over as it slides, until it is directly

Fig. 20.—*Barrage of the Port à l'Anglais. Public Works of France.*

back of the shoe, and when the shutter is raised again, it moves on a groove up to the inclined back of the shoe, and over it, until caught in its original position. A broad iron handle on the base of the shutter allows the keeper to attach a rope by a hook, and by a small winch in a boat, he effects the raising, moving successively from shutter to shutter. The fall of the water assists the operator to a considerable extent. The disengagement of the counterforts from the front of the shoes, for lowering, is effected by means of a bar with projecting cams, running under the feet of the counterforts for the whole length of the shutters, resting on rollers, and worked by a rack and wheel-gearing with a crab winch on the pier. By adjusting the distance between the cams, in relation to the distance between counterforts, it may be so arranged that the shutters will lower in succession, one by one, or two by two, etc., as desired. A passage can be opened by the winch at the rate of one second per foot run, and closed by a boat at twenty-three seconds per foot.

The striking simplicity of the ingenious system of self-acting shutters for the weirs first led to some isolated experiments in a single barrage. But its application, on a larger scale, to the twelve weirs on the Seine, between Montereau and Paris, showed that some grave miscalculations had been made. They lowered themselves too quickly, and did not rise until after a reduction of the upper or head-water by about three feet, stopping navigation and causing the barges to ground. Foot-bridges were therefore constructed above each weir, built of iron trestles after the Poirée system, the trestles being hinged at the bottom, and movable on a horizontal axis, perpendicular to the axis of the weir. Two lines of bars connect the tops of the trestles, forming rails for a truck, which carries a winch with two chains, one for the head, and the other for the breech of a shutter. A planking is laid between the bars for a foot-walk. This arrangement allows of the necessary operations being performed with facility and without danger. If a flood occur, the trestles are lowered behind a frame or protection, nearly level with the platform, turning on their hinged bases, and the planking, bars and winch are placed in safety on shore. The new system has proved a complete success. At night the keepers are warned of any rise in the upper water of the barrage by a bell, actuated by a float, and telegraphic communication between the various barrages allows full information to be given in advance of the condition of the weir, precluding all possibility of surprise.

The total cost of the work on the Yonne, between Auxerre and Montereau, has been about three million three hundred and sixty-five thousand dollars, and the Seine, between Montereau and Paris, about two million eight hundred and seventy-one thousand dollars; the interest on the cost, including the annual expenses of management, amounting to about two-thirds of a cent per ton, per mile for the Yonne, and nearly one-third of a cent per ton, per mile for the Seine.

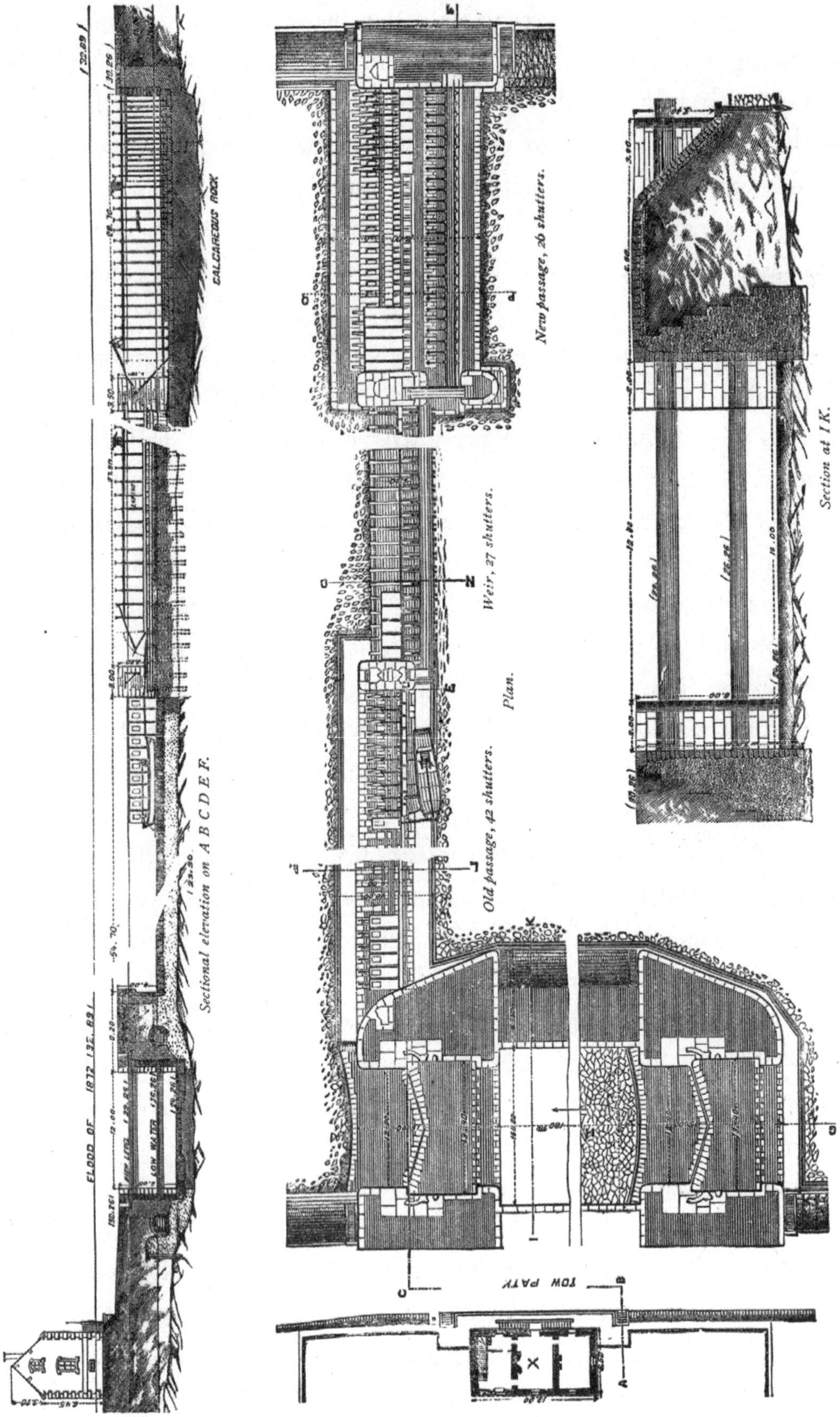

Fig. 21.—Barrage at Port à l'Anglais. Public Works of France.

In consequence of the new arrangements made for establishing an uninterrupted communication by the Seine to Paris, it became essential to lower by about three and a quarter feet, the tail-sill of the lock at Port à l'Anglais, a point about two and a half miles above Paris. This necessitated the construction, on the left-side of the weir, of a new navigable passage about ninety-four feet between abutments, reducing the width of the weir to about one hundred and twenty-four feet, and leaving the former passage of one hundred and seventy-nine and a half feet in its primitive state. This new passage is closed by twenty-six movable shutters on the Chanoine system, the sill being

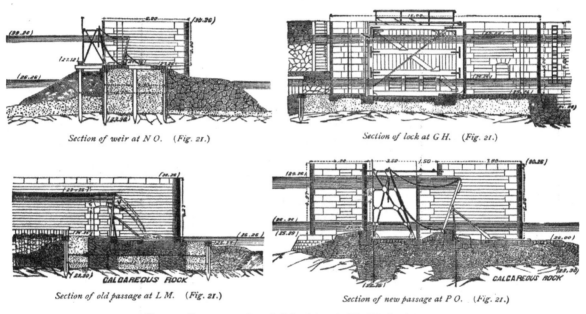

Section of weir at N O. (Fig. 21.)

Section of lock at G H. (Fig. 21.)

Section of old passage at L M. (Fig. 21.)

Section of new passage at P O. (Fig. 21.)

Fig. 22.—Barrage at Port à l'Anglais. Public Works of France.

two and three-tenths feet below that of the old passage, and requiring therefore shutters that much longer.

We give here illustrations of this barrage which are interesting as showing very clearly the general arrangement of the Chanoine system. Fig. 17 represents a general view of the dam, looking up-stream, the opening in the weir made to furnish a channel for the traffic during the rebuilding of the lock being shown on the right. On the left is seen the lock in process of reconstruction. Fig. 18 is a view from the shore on the right, on the upper side of the dam, the new passage being in the foreground, with its iron fermettes and gates, next to it is the weir, then beyond that, the old passage,

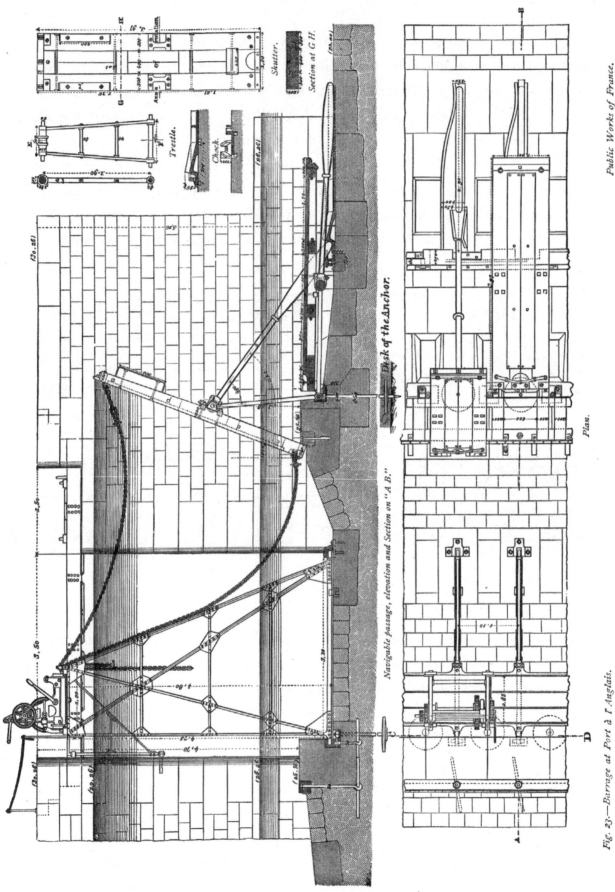

Fig. 23.—Barrage at Port à l'Anglais. Public Works of France.

the gates of which are lowered, and finally, in the distance, is seen the lock and also the lockman's house. Fig. 19 is a view from below, looking up-stream, and gives a near view of the new passage with its shutters and the supports for them. Fig. 20 gives a still nearer view of a portion of the new passage, the shutters and fermettes being shown in various positions.

The pressure of the body of water required to be sustained at Port à l'Anglais was so great that it was found necessary to somewhat modify the design adopted by M. Chanoine for movable shutter barrages. In the first place the breadth of each shutter was reduced by nearly eight inches, the intervening space between adjacent shutters, of about four inches, being retained, and the frame-work was simpli-

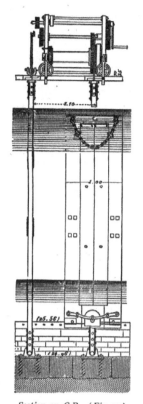

Section on CD. (Fig. 23.)
Fig. 24.—Barrage at Port à l'Anglais.
Public Works of France.

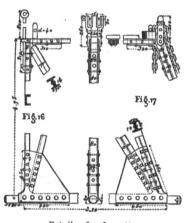

Details of a fermette.

Bearings.

Fig. 25.—Barrage at Port à l'Anglais.
Public Works of France.

fied. Then, again, the inclination of the shutters from the perpendicular, originally eight degrees, was increased to twenty, so as to diminish the strain tending to tear up the sill of the platform. When the shutter is lowered, it is made to bear upon four rests, and it is raised by two cleats fixed to the uprights, making it perfectly supported under all circumstances, and obviating entirely any risk or danger of dislocation of the frame. In M. Chanoine's original design the axis of rotation of the shutter was placed at five-twelfths of the total height. This, for the new shutter at Port à l'Anglais, would have given to the breech a height of nearly five and a quarter feet, but five and three-quarters feet was adopted, making the axis of rotation only about six inches below the axis of the figure and preventing spontaneous rotation, an inconvenience sometimes occurring with the former shutters when the tail-water was too high and the fall of the barrage unusually small. The introduction of small self-acting valves, called butterfly valves, in the upper part of the shutter, between the uprights, is removing this objection. These valves open of their own accord before the shutter comes to a balance and can easily be closed by the keeper, by means of a gaff, from a boat behind the barrage. Finally, for raising the large shutters of the new passage, a foot-bridge on Poirée's system, with a traveling winch, has been adopted in place of a boat. The trestles of the foot-bridge are about fifteen and a half feet high, ten feet wide at base, and four feet at the top, constructed of wrought iron. We give in Figs. 21 and 22, a plan and sectional elevation of this barrage, with also cross sections at various points, which show very clearly the general form and mode of operation of the construction at Port à l'Anglais. Figs. 23, 24 and 25, show plan, elevation, and details of a portion of the navigable passage, giving the fermettes, the shutters, etc.' fully illustrating the special points of the Chanoine system, as we have been describing it for this barrage, and generally for others on the Seine and Yonne.

We have already mentioned that the weir of the barrage of l'Ile-Brûlée is fitted with large shutters or sluice-gates on a system invented by M. Girard. We will now explain more in detail the special features comprised in M. Girard's system. It consists essentially of a series of wooden sluice-gates, movable around a horizontal axis, and turning into a cast-iron groove or case, let into the top of a masonry weir, when in a horizontal position; these gates being worked by hydraulic presses fixed on to the lower slope of the weir and solidly anchored into the masonry. To the piston rod of each of these presses is affixed a strong cross-head, supported and guided by slide bars, and having attached to it three connecting rods, jointed to another cross-head which is hinged on to the middle of the movable sluice-gate. The presses are operated by hydraulic machinery, constructed on the abutment of the barrage, consisting of a turbine with vertical axis, a double-action forcing pump for water and an air pump, on the same piston, both worked by the turbine, and a generator or reservoir for compressed air. The pumps and generator are connected with each other and with the presses by a series of copper tubes, and by means of three-way cocks, the water may be forced either into the generator, or the presses, or be discharged into a waste-pipe. The entire working of the sluices is performed by the simple action of these cocks. Putting each press into communication either with the pumps or with the reservoir under sufficient pressure, causes an upward movement of the piston and the sluice rises. On the other hand, if the discharge cock is opened, the

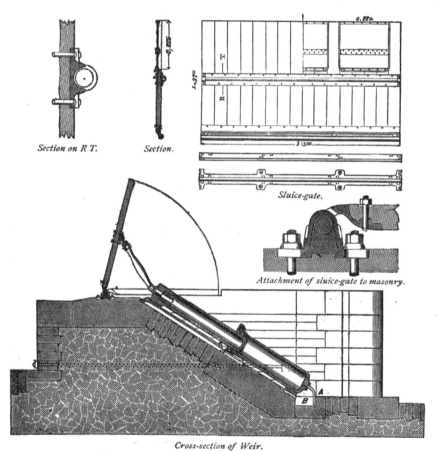

Fig. 26.—Weir of the Barrage of l'Ile-Brûlée. Public Works of France.

water escapes under the pressure of the head-water on the sluice-gates, the pump is emptied and the gate falls to its horizontal position. The reservoir of compressed air regulates the action of the pumps, and also gives an accumulation of power to raise the sluices when there is not sufficient head of water to work the turbine.

Fig. 26 gives a cross-section of the weir, showing the hydraulic press with its piston and connection with sluice-gate. It also gives a plan, elevation and section of one of the sluice-gates and details of its attachments to the masonry and to the cross-head. Fig. 27 shows details of cross-head, the guides or slide-bars for the same, and a butterfly-valve, the use of which will be presently explained.

The details of the hydraulic machinery, pumps, turbine, working-valves, etc., are shown in Fig. 28.

The weir of l'Ile-Brûlée is eighty-two feet in length, and the level of the sill is about six and a half feet below the head-water, the fall being six feet. There are seven sluice-gates, each about eleven and a half feet wide by six and a half feet high, giving an obstruction of about one foot four inches when raised, and lying horizontal when lowered. Each sluice-gate has an hydraulic

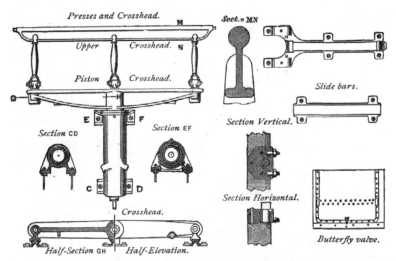

Fig. 27.—Weir of the Barrage of l'Ile-Brûlée. Public Works of France.

press to raise it. These presses are of cast-iron, about one foot four inches in exterior diameter, and have a thickness of metal of one and a half inches. The piston is also of cast-iron with a bronze casing. It has a diameter of about one foot and a packing of hot pressed leather, making a joint more and more water-tight as the pressure on it is stronger. The copper connecting pipes are about one inch in diameter, there being one for each press.

To protect the presses from the frost, they were placed entirely below the down-stream level, but as this arrangement interfered very much with inspection and repairs, stone walls were built in between each press, having vertical grooves in them as shown in Fig. 26, into which plank partitions may be slid, forming a small coffer-dam for each press. These can be pumped out and any necessary examinations and repairs made as desired.

The turbine is about four feet in diameter, and a crank at the upper end

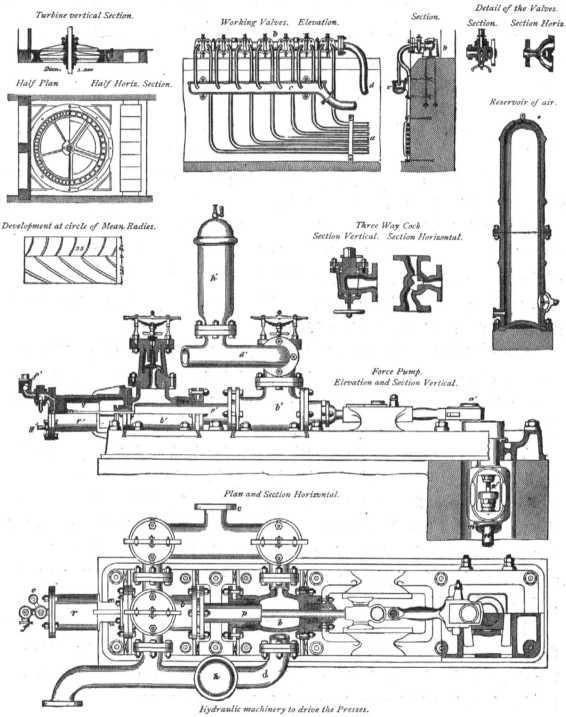

Fig. 28.—Weir of the Barrage of l'Ile-Brûlée. Public Works of France.

of its axis connects directly with the pumps which drive back the water and air into the generator, under a pressure which may be raised up to twenty-five or thirty atmospheres. All of the machinery was tested under a pressure of thirty-five atmospheres, and the dimensions of the various parts of the weir were calculated throughout by M. Girard, so that a pressure of more than twenty-five atmospheres would never be exceeded, this being the amount necessary to raise the shutters against the complete fall of six feet.

The machinery works with precision and rapidity, the opening of a large sluice-valve occupying less than a minute. The amount of pressure required to raise the water can be diminished by fitting the large sluices with small butterfly-valves, as shown in Figs. 26 and 27; these valves, three in number, being placed in the upper part of the shutter, with their axis of rotation at one-third their height. The experiment has been made upon two of the shutters with success. When the shutter is lowered, they incline in the direction of the stream which covers them, and, being of thin sheet-iron, offer no resistance to the pressure of the water, considerably reducing the strain upon the shutter at the starting point in raising.

When the shutters are raised, they may be retained in position by a pressure of seven to eight atmospheres, and it is not necessary to keep the presses in communication with the reservoir; but being perfectly tight, the distribution cocks are closed and they remain as they are placed.

This system of barrage costs about one hundred and twenty-two dollars per lineal foot, or about one hundred and eighty-three dollars with the stone-work, considerably more than the trestle and paddle weir, which only amounts to about seventy-three dollars per foot, including everything, in this part of the Yonne, where the foundation is not difficult. This excessive cost, therefore, operates against the hydraulic system, notwithstanding its superiority in convenience of working.

Between the years 1855 and 1867, fourteen barrages were constructed for the purpose of improving the navigation of the lower Marne, and of this number eleven had weirs fitted with movable wickets, on the system of the late M. Desfontaines, in addition to the usual lock and navigable passage. The excellence of this system has been confirmed by the test of some eighteen years' experience, and the complete model at the exhibition gives the visitor a very clear idea of the construction. M. Desfontaines' system consists in the use of drum shutters, which are moved by the application of the power resulting from the difference of head between the upper and lower water levels. This difference of water level, at its maximum, is generally about six and a half feet, being at Joinville, however, seven feet. A permanent stone weir is built to half the height and the movable shutters occupy the other half. An iron plate sluice or valve about six and a half feet in height by five feet in breadth, having a horizontal fixed axis at half its height, is placed with its axis at the top of the masonry weir and parallel with its direction, a recess in the form of a quadrant being made in the masonry, on the up-stream side, for the lower half of the plate, so that it may turn from a vertical plane to a horizontal, the upper part revolving in a direction down-stream. The latter thus forms a movable wicket, while the lower part constitutes a counter-wicket, and it is on this that the pressure of the water is made to act. By closing the ends of the masonry recess, close to the wicket, by two vertical cast-iron diaphragm plates, and covering the top on a level with the axis of rotation of the wicket by a horizontal plate, a closed box is formed, in the shape of a quadrant of a cylinder, divided by the counter-wicket into two sectors of varying dimensions, depending on its position. The wicket plate is bent slightly out of plane on the under side next to its axis, and then back again about a foot below, thus making the depth of the upper sector or space never less than a certain amount, generally about a foot. The plate works freely in the drum, there being a small space around its three sides for clearance, except when it is in a vertical position, when this space is closed perfectly water-tight by an India-rubber band. A series of these drums, placed end to end for the whole length of the weir, forms the complete arrangement. Drums formed entirely of iron were used in the first two barrages that were built. They presented on the head and tail sides, horizontal flanges, which were bolted to the wales of the two rows of piling. For the other barrages, however, they were constructed in the stone-work, as here described, a cavity of convenient section being formed in the whole length of the weir, and divided into divisions of about five feet by cast-iron diaphragms let into the masonry about three inches.

Now, by putting the space of the upper sector or head compartment of the counter-wicket in connection with the head or upper water level, while the tail compartment, more or less empty, is placed in connection with the lower or tail-water, it is evident that the counter-wicket may descend, notwithstanding the static and dynamic pressure which tends to retain the wicket in a horizontal position. The counter-wicket, presenting a larger surface and being placed lower, has a preponderating force acting on it, causing it to lower and the wicket to rise. When the wicket is up, it may be lowered by simply intercepting the communication between the upper or head-water, and the upper compartment of the drum; but the operation may be much facilitated by placing the lower compartment in communication with the upper water level and the upper compartment with the lower or tail level, thus giving to the apparatus a double action. These facilities are provided for, by having two openings in each division between the drums, one corresponding to the head compartment and the other to the tail compartment, so as to connect those of each series together. In the abutment of the barrage is a culvert, con-

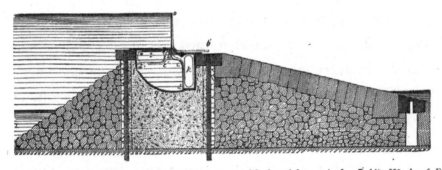

Fig. 29.—Cross-section of Weir on Desfontaines' System, with the wickets raised. Public Works of France.

necting at one end with the head-water and at the other with the tail-water, having an opening, rectangular in cross-section and divided by a horizontal plate into two parts, which communicate respectively with the upper and lower compartments of the end drum. By means of two valves and a single movement, transmitted from the upper to the lower valve by a balance beam, like that of a steam engine, all the operations desired may be accomplished. One movement of these valves causes the water to enter the head compartment of the first drum, and leave the lower compartment. Its action on the counter-wicket causes the wicket to rise, and the water, after filling the upper compartment, passes into the second drum with similar effect; and so on from drum to drum, wicket after wicket successively rising into position, until all are up. This manner of working is one of the most important points of the system. If the water were admitted into all of the drums at once, the leakage around the three free sides of the counter-wicket would be so great as to probably defeat the purpose intended, but going as it does into each drum successively, it successively transforms each into a water-tight compartment and insures the desired result. If, when the wickets are up, it is required to lower them, the movement of the valves is simply reversed, and the head-water enters the lower compartment, while what is in the head compartment passes out to the tail level and the wickets are successively lowered until all are down.

Our engraving, Fig. 29, shows a cross-section of one of these weirs. $f\ g\ h$ is the curved recess in the masonry; $b\ a\ c\ d$ the upright position, and $b'\ a\ c'\ d'$ the horizontal position for the wicket and counter-wicket; l and k the openings from drum to drum. Fig. 30 gives plan and elevation of one of the wickets and its drum.

It is necessary sometimes, owing to variations in the volume of water in the river, to modify the overflow at the weir, and to accomplish this, M. Desfontaines had fitted to the tail face of the wickets at four of the barrages, a jointed crutch, similar to the counterfort previously mentioned in the other systems, and so arranged that when the wicket is inclined at say an angle of forty-five degrees, the crutch, sliding down on top of the weir, is stopped by a catch or sill of angle iron and sustains the wicket in this position.

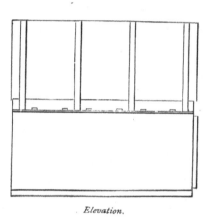

Elevation.

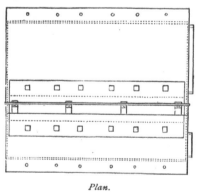

Plan.

Fig. 30.—Wicket and Drum: Desfontaines' system. Public Works of France.

When the wickets are to be lowered all the way down, it is not necessary to unfix the props laterally, by main strength, under the pressure of water; but the wickets are first raised to their full height and then the steps or catches being removed, they are lowered to the horizontal. In the latest barrage constructed no crutches were used, the shutters being always, with the exception of two or three, entirely up or entirely down; but they can be raised or lowered at pleasure. If valves are made at both pier and abutment, one-half the wickets can be moved from one end and one-half from the other; and the water being always let into the culverts from one extremity of the weir, at the pier for instance, the abutment valve must be worked in a contrary direction.

The operator on the abutment, by a simple movement of a handle commanding the valve, at the same time watching a hydrometric scale, admits water to the drums in just such quantity as he desires, raising or lowering only a single wicket or as many more as he wishes. If he continues the operation, the wickets as they are successively raised, form a line with almost mathematical precision until one-half of them are up. If all are to be raised, the pier and abutment valves can be worked simultaneously, and the operation at the Joinville barrage, where there is a width of about two hundred and seven feet, is accomplished in two minutes, forming a really beautiful sight, as the surging mass of water, which rushes to the gradually contracting outlet of the weir, is driven back behind the wickets.

The closing of the navigable passage in these barrages is effected, for ten of them, by swinging shutters, on the same principle as used on the upper Seine, except that the shutters were never raised by a boat, but from a footbridge supported by Poirée trestles. For the eleventh and last barrage constructed, trestles were employed in preference to swinging shutters.

M. Desfontaines has laid it down as a principle, that the width of navigable passages ought not to exceed the requirements of navigation, the expenses being thereby reduced, and the serious difficulties attending the care and operation of these passages, very much lessened. For this reason, an opening of eighty-two feet only has been given to the navigable passages of the Marne.

The idea of utilizing the head of water as a motive power, as here explained, was originally derived from Holland, where from time immemorial, fan-gates have been employed to close the irrigation canals. The French author states that the suggestion probably passed from Holland to America, about the year 1818, as the principle was then employed on the Lehigh river navigation system, in Pennsylvania, to gates not with vertical quoin-posts, but with horizontal axes. Afterwards taken up in France, it now returns to us in the present exhibition as the "American barrage perfected under French auspices;" and most beautifully has the principle been carried out, with all the perfection of theory for which the French engineers are noted.

Under the head of Maritime Works, there are exhibited a number of beautiful models and drawings, illustrating the Lock of the Port of Dunkerque, the Basin of the Citadel of the Port of Havre, the Port and Lock-gates of Saint Nazaire, the Port of Marseilles, etc., etc., all of which are exceedingly interesting to the engineer.

The Port of Marseilles, which in 1844 only possessed its old natural harbor, covering about seventy-two acres, and having about eighty-nine hundred feet of quay, now comprises a series of exterior basins all reclaimed from the sea, and giving a surface of perfectly sheltered water of some three hundred and thirty-six acres, with a development of quay of about forty-one thousand four hundred feet, of which about twenty-eight thousand can be utilized for loading and discharging cargo. Vast warehouses, capable of receiving one hundred and thirty thousand tons of goods, have been erected around the basins, and large spaces of ground adjoining, add to their value. Five graving docks have also been constructed, giving all necessary facilities for examination and repairs of vessels of every class. One of these, about four hundred and sixty-five feet in length, will, under certain conditions, receive the largest vessel, and the Great Eastern, six hundred and fifty-six feet in length, can enter the port without inconvenience. In the course of the year some twenty thousand vessels, with a measurement of about five million of tons, enter and leave this port. The works have not yet been entirely completed, but when finished, the total length of quays will be about fifty-six thousand five hundred feet, of which forty-one thousand will be available for loading and unloading purposes. The depth of water in the old basin is from twenty to twenty-three feet below low tide, and in the new basins the minimum depth, with a few exceptions, amounts to twenty-three feet, being, for an area of about one hundred and twenty-five acres, over twenty-nine and a half feet. Along the exterior dike, which is now about ten thousand feet long, the water ranges from thirty-six to fifty and sixty-five feet in depth.

There are three passages, that of the old port, and the southern and northern passages of the new basins, having depths of water below the lowest tides, of about twenty-five, thirty and fifty feet respectively.

In the design of the plans of the exterior basins, the principal feature consists in the construction along the shore of a series of basins, separated from each other by interior moles or causeways, with strong foundations, protected on the side next the sea by a dike parallel to the coast, and leaving between it and the ends of the moles a channel sufficiently large for easy communication between all of these basins. This arrangement allows the capacity of the port to be always extended to the increasing requirements of commerce by adding new moles in succession to those existing, and protecting them by a corresponding extension of the exterior dike.

In the department of Light-houses and Beacons, a large volume, with map, shows their condition on the coasts of France on the 1st of January, 1876, when there were three hundred and seventy-nine light-houses, not including those of Algeria. A number of plans and models are exhibited, from which we select two for illustration.

Fig. 21.—Light-house of New-Caledonia. Public Works of France.

The Light-house of New-Caledonia, represented by a model on a scale of one-twenty-fifth, was erected on the islet of Amède, thirteen miles from Port-de-France, in New-Caledonia, in 1865. It is constructed entirely of iron and is founded on a solid mass of beton, in which are sunk cast-iron foot-plates to support the uprights. The tower is about one hundred and forty-seven and a half feet high from the level of the ground to the platform at the top, and the lantern, arranged for a fixed light, is one hundred and sixty-four feet above the highest tides.

In designing the structure, the following special conditions were kept strictly in view:—

1. "To render the skeleton of the edifice entirely independent of the exterior covering; to shelter it from the sea-fogs, which rapidly cause oxida-

tion; to facilitate inspection and proper keeping, and to reduce as much as possible the extent of surface exposed to humidity."

2. "To manage the process of construction, in such a way, that the tower should be erected without scaffolding at the bottom, and without clinching a single rivet on the place."

"Stress was also laid upon the necessity of the parts being of such dimensions as to obviate any difficulties of embarkation, of stowage on board, and of placing in position."

The frame-work of the structure consists of sixteen great uprights or standards of fifteen panels, each, in height, each panel being of simple **T** irons put together and riveted so as to ensure perfect solidity and resistance to change of form, contingent upon the greatest shocks or forces that could be foreseen. The panels are bolted, one upon the other, and cross-braces, inside as well as outside, also bolted, keep the standards in position. On the last cross-braces and on the exterior standards, sheets of iron are fitted to form a covering, the joints being well protected by wrought-iron plat-bands fixed by bolts.

Fig. 31.—Light-house of La Banche.
Public Works of France.

The platform required for the external service of the lantern is corbelled out on cast-iron brackets attached to the standards. The chambers are surrounded by a brick wall placed about two inches from the sheet-iron covering of tower, and the partitions are also of brick. The floors are of concrete and the ceiling of stone-work on iron beams.

The cost of iron work, including putting together and taking to pieces in Paris, amounted to about forty-five thousand eight hundred dollars.

The Light-house of la Banche is situated west-south-west of the embouchure of the Loire, nearly six miles from the nearest land, eight miles from Pouliguen and fifteen miles from Saint-Nazaire, the only ports in which materials could be prepared and embarked. It was commenced in 1861, and first lighted on August 15th, 1865.

The bank of le Turc, on which the light-house is located, is only about a foot or so above the surface of the water at ebb tide, and is a part of the great plateau of la Banche, which constitutes the most serious danger in entering the Loire, having at low tide only from three to sixteen feet of water, and being thickly studded with projecting rocks.

Great difficulty was experienced in obtaining foundations. What appeared to be rock at the site selected, was found, on commencing the work, to be only an enormous mass of calcareous stones which had been broken off from the plateau by the sea and cemented together by sand and mud. The location had, therefore, to be shifted to another site more difficult of access by the boats used in the construction, and exposed to the full force of the sea. Even here the rock was not uniformly solid, but changed in parts to a mass of calcareous materials of the consistence of muddy sand. The greatest care had therefore to be exercised to clean out this material from the solid rock. This was effected at each tide by means of coffer-dams hastily constructed, most frequently of the seaweed close at hand, and then packed with a mortar-bath of Portland cement to a depth, in some cases, of over six feet below the lowest ebb-tides. A dike had also to be built, over three hundred feet in length, with a line of rails on it, to convey materials from the spot where vessels could approach, to the foot of the light-house.

The light-house, as built, is a stone tower resting on a solid stone foundation about eight and a half feet in thickness, the total height being eighty-seven feet, and the focal plane of the lighting apparatus sixty-nine and six-tenths feet above the highest tides. The tower includes a cellar, vestibule, kitchen, two keepers' rooms and a service room, the threshold of the entrance door being only six and a half feet above the highest tides. As it is on the north side, however, and sheltered from the waves of the offing, the sea rarely reaches it. The basement window is protected by a thick copper shutter fitting into a bronze frame. The stairways are of cast-iron, and the floors are formed of strong girders of plate and angle iron filled in with brick-work. This system of flooring gives additional support to the hollow part of the tower, which is also sustained by three wrought-iron bands embedded in the

Fig. 32.—Iron Viaduct over the Valley of the Osse. Public Works of France.

masonry at different levels. The stone-work is of granite, and the total cost was a little less than seventy-five thousand dollars.

There is also on exhibition a large collection of photographs, in a series

Fig. 33.—Aqueduct du Gard. Public Works of France.

of twenty-two volumes, containing views of the principal works constructed on the different lines of communication in France.

We present on page 361 an engraving, Fig. 32, of the Iron Viaduct over the Osse, on the railway between Agen and Tarbes, the line crossing the valley on a grade of about a hundred and thirty-two feet to the mile, and at a height above the ground of from fifty-eight to seventy-two feet. There are seven spans, the intermediate ones being one hundred and twenty-six feet each, and the end spans ninety-four and a half feet, making a total length of eight hundred and nineteen feet. The abutments are of masonry and the piers of double cast-iron tubular columns, each about five feet seven inches in diameter and filled with béton, the interior mass giving stability to resist both the action of the wind and the vibrations caused by passing trains. The columns are formed of a series of rings of cast-iron, bolted together in the usual way, making a continuous cylinder, fixed to a solid foundation, and receiving on top the weight of the superstructure, on cap plates provided with guides and expansion rollers.

The superstructure, which is inclined on the grade of the road, is constructed of wrought-iron. It has been stated by an authority that the conditions in this structure are such that a horizontal component of force, due to the inclination of the plane of the rails, produces a tendency to overthrow the piers, rendering it necessary to sink them deep, make them very strong and bed them firmly to resist it. The author is not sufficiently familiar with the details of the structure to say whether this is the case or not, but would state that the question depends entirely upon whether the trusses have a horizontal bearing on the piers. If they are constructed with an inclined bearing then there will be a horizontal thrust, but if with a *horizontal* bearing, which is the only proper way to build them, the piers will receive only a vertical load, and there is no tendency to horizontal motion in the truss as a mass, not taken up by itself, any more than if the railway track were perfectly level.

The total cost of the viaduct was about one hundred and twelve thousand dollars.

A part of the collection of subjects as shown by these photographic views is reproduced in a work entitled: the Public Works of France, published in Paris under the auspices of the Ministry of Public Works, the first numbers being on exhibition, and we are indebted to the editor, M. Rothschild for two engravings taken from his publication.

The first is a view of the Aqueduct du Gard, Fig. 33, a work of the

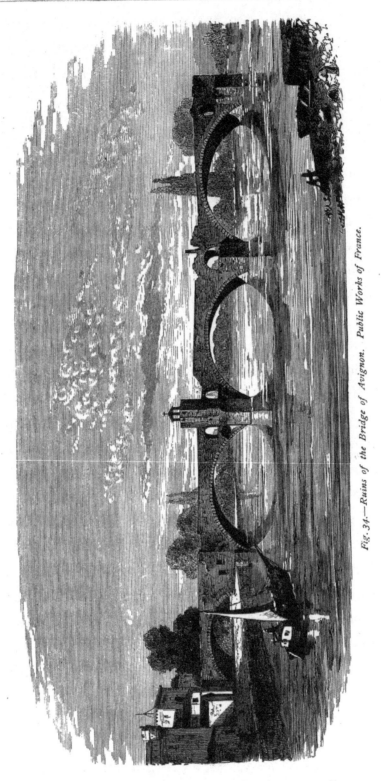

Fig. 34.—Ruins of the Bridge of Avignon. Public Works of France.

Gallo-Roman period, constructed over the Gardon, a tributary of the Rhône for the purpose of conveying water to Nîmes from the springs of Eure and Airau. This aqueduct was built in the same manner as other Roman masonry, of hewn stones laid dry, without mortar, the voussoirs of the arches not being bonded together as in modern work, but each arch formed of a series of separate parts, of one stone each in width, and placed in juxtaposition but not connected with each other.

The total height of the aqueduct is about one hundred and fifty-eight feet, there being two tiers of larger arches surmounted by a third tier of smaller ones, the lower one, four hundred and sixty-six feet in length, the second, seven hundred and ninety-four feet, and the third or upper one, eight hundred and ninety-six feet. Each arch is formed of four annular, parallel and independent rings of hewn stone, laid dry. The extremities of this beautiful work were broken down about the commencement of the fifth century by the barbarians who besieged Nîmes. In 1743 some portions of the structure were strengthened and repaired, and the lower piers were lengthened out and an arch roadway bridge erected on them. The projecting corbels that may be seen within the arches, at a certain height above the springing, were designed to support the centre during construction. It is well known that the Romans did not generally make use of centres for the lower part of the arch, employing, however, in some cases, where necessary, a sort of mortise and tenon between the voussoirs to prevent their sliding during construction. The absence of bonding simplifies the arrangement of the centres, dispensing with the necessity of bolstering. With one rib in each head plane and one in each plane of separation between consecutive parts of the arch, the requirements of the construction were satisfied, each voussoir filling the space comprised between two consecutive ribs and resting upon them at its two extremities. At the place of juxtaposition of the arches, on the intrados, a regular space would be left which could be readily filled up by a flag curb, projecting out from the regular face, and forming as it were, a skeleton of the arch, producing a very fine effect.

Our second engraving, Fig. 34, is a view of the ruins of the bridge of Avignon. This bridge was composed of eighteen arches. Its construction was commenced in 1176 and completed in 1188, being under the direction of Saint Benezet, who, it is said, was a shepherd, and that when not twelve years of age repeated revelations from heaven directing him to leave his vocation and undertake this work.

Some of the arches of the bridge were destroyed in 1385, during the contentions of the popes, and in 1602 three others fell from neglect in making repairs. In 1670, a very unusual ice-freshet in the river destroyed a number of piers, the one with the chapel of St. Nicholas on it, however, as seen in the engraving, being still left standing. This bridge was of the same age as Old London Bridge.

In conclusion we would remark that the exhibit, as a whole, does the utmost credit to the French nation, and we cannot too warmly express our thanks for the opportunity she has afforded this country for studying her internal improvements.

The STILES & PARKER PRESS COMPANY, OF MIDDLETOWN, CONNECTICUT, make a fine display, in Machinery Hall, of Drop-Presses, improved Punching- and Trimming-Presses, cutting, stamping and forging dies, etc., etc. The old processes of forming manufactured articles—by hand-forging, by planing, filing, drilling, etc.—have now become entirely too slow and expensive, and American ingenuity has superseded them by Power-Presses, which do the work almost automatically, and with great thoroughness, precision and rapidity. In the form of power-press as made twenty years ago, the method of stopping was very imperfect, being accomplished merely by running off the belt, and perfect control of the machine by the operator was out of the question. The punch could not always be stopped on the up stroke, and before being brought to rest it might, by repeated strokes, destroy work of great value. A high rate of speed was inadmissible, not giving the operator time between strokes to put his work in position. In the improvement of this Press the automatic stop was the first invention of importance, insuring always a disconnection of the wheel from the shaft at a certain point after one revolution, and thereby giving to the operator complete control of his machine. There remained, however, still another defect in the press, owing to the fact that the punch could not be adjusted with sufficient accuracy to insure perfect execution. The invention of an eccentric adjustment by Mr. N. C. Stiles obviated this difficulty, and power-presses are now among the most efficient labor-saving machines of the present day. The Drop has always shared to some extent the efficiency and valuable labor-saving qualities possessed by the Press, and important improvements have been made in it from the time when it was simply a weight attached to a rope over a pulley, until now, when the Power-Drop, as manufactured by this firm, is a machine eminently adapted to act in conjunction with the improved Power-Press. The Friction Roll Drop, as improved by Mr. Stiles, is a machine perfectly adjustable to the requirements of the labor it has to perform, and entirely

Fig. 1.—Double-Acting Drawing-Press : Stiles & Parker Press Company.

under the control of the operator, capable of giving a uniform, an occasionally varied, or a constantly varied blow, according to the nature of the work to be done.

The branches of industry in which these machines prove themselves invaluable are very numerous. They are largely used in the manufacture of tin-, silver- and Britannia-ware, buttons, jewelry, cutlery, musical instruments, locks, nails, sewing-machines, etc., etc., and above all in the making of watches. There is hardly a portion of a watch which is not all or in part made by one or the other of these machines or by both. Every species of punching, cutting, trimming, drawing, shaping, stamping, and forging comes within the range of their united capacity.

In Fig. 1 we show a new DRAWING-PRESS made by this firm for cutting and drawing sheet-metal into hollow-ware. It is a machine which, while possessing great power, and solidity and durability of construction, requires but small space, and is light in weight compared with other machines for the same purpose. In making articles of moderate size, it cuts the blank and forms the article at one operation. The lower slide, C, of the machine, is carried up by

Fig. 2.—Copper Jelly-Mould: Stiles & Parker Press Company.

the cam, A, on the main shaft, which is so made that when the slide is raised to a proper height it cuts the blank and holds it firmly. The upper slide or plunger, E, then comes down and forces it into shape. The upper slide is actuated by the yoke above, which in its turn is operated by the pitmans, D, D, and the cranks, B, B, on the main shaft. The shaft is of steel, and the pitmans and yoke of best wrought iron. These sustain the strain, thus relieving the cast-iron frame and allowing greater lightness. The Stiles Patent Eccentric Adjustments are placed on the shaft on which is the roll in the lower slide, and on the crank-pins operating the pitmans carrying the upper slide, and by these adjustment can be made easily and with great accuracy to the various thicknesses of metal used.

Fig. 2 shows a COPPER JELLY-MOULD in two stages of manufacture, the first being the form in which it comes from the drawing-press, and the second its finished state after it has been put under a drop to give it sharp and well-defined corners and edges.

Fig. 3 gives the shape on a scale one-half full size of two articles made by the drawing-press. The upper line represents a cross-section of a brass clock case for the Seth Thomas Clock Company, drawn up in one operation. The method heretofore used for making these has been to spin them in three operations, requiring them to be twice annealed. In the drawing process fifteen are made per minute, and the brass is not hardened as it is by stamping or spinning. The lower line on this figure shows the cross-section of a Britannia coffee-pot, also made in one operation, two at a time, of No. 23 gauge iron.

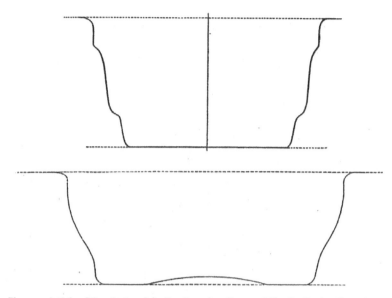

Fig. 3.—Articles Manufactured by the Drawing-Press: Stiles & Parker Press Company.

These have heretofore been stamped in a drop, one at a time, and each one requiring thirteen operations. Now, the drawing-press makes twenty per minute.

Fig. 4 represents the STILES & PARKER ADJUSTABLE BED-PRESS, a machine which, being set on an incline, facilitates very much the removal of work not passing through the die, thus enabling the operator to produce finished work, cut, formed and stamped in the same time as would usually be required to cut a plain blank. The adjustable feature allows this press to be set level and used as an ordinary press if desirable, which also obviates the difficulty generally experienced in setting dies on an inclined press, and allows the press to be set level for that purpose, and then moved back to the incline.

Fig. 5 shows the FRICTION ROLL DROP made by this firm, which is one of that class where the hammer is raised by a stiff belt or board passing up between two friction-rolls. A trip-pin, actuated by the hammer, allows the operator to obtain any height of blow desired, automatically, and the upright rod on which this is placed connects with both hand and foot levers. Fig. 6 shows the arrangement of clamps that hold up the hammer, the belt or board passing up between them. They are so arranged that as the hammer ascends they open freely of themselves, but on descending they will close and hold the hammer up. A pressure on the foot-treadle opens them, and the hammer falls. The back roll and clamp are made adjustable to different thicknesses of board or belt. If one blow is wanted from the height of trip-pin as set, the treadle is pressed upon, and the pressure removed as soon as the blow is given. If the foot is kept on the treadle, the blows will be repeated for the same height until the pressure is removed. If a blow is desired of less height than the trip-pin is set for, it may be given to any height wanted, by working the hand or foot lever, as the clamps will hold the hammer at any required point below the collar, when the hand or foot lever is brought into action, and the next blow can be given from a state of rest. This is a feature peculiar to this drop. A gentle pressure upon the treadle will allow the hammer to descend slowly, but it will stop and remain suspended at any point, as soon as the pressure is removed.

Fig. 4.—*Adjustable Bed Press: Stiles & Parker Press Company.*

In the Main Building, south of the central aisle and east of the transept, are situated the exhibits of American Chemicals, a display which exceeds in variety and quantity, anything of like character from abroad, a circumstance that is probably due to some extent to the difficulty of transporting articles of this kind so great a distance without injuring their appearance for exhibition purposes. The chemicals shown by France and Germany consist mainly of Aniline colors and other coal tar products, forming, it is true, an exceedingly interesting class, of great use in the arts, and those from England are principally the Soda and Potash salts, of which she is a very large manufacturer. In the foreign exhibits a point of special interest is made of what might be regarded as curiosities in the way of chemicals, and France also, has a very fine collection of chemical apparatus, very little of anything of this sort being exhibited from the United States.

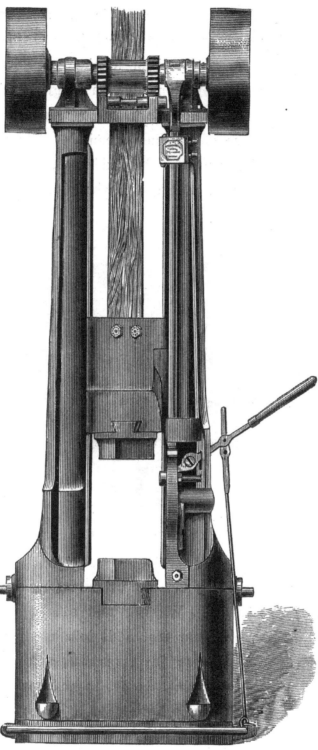

Fig. 5.—*The Hotchkiss and N. C. Stiles' Friction Roll-Drop: Stiles & Parker Press Company.*

In the American display it is doubtful if any chemicals of importance that could be exhibited, are omitted, whether used in the arts or in medicine, and it is extremely interesting to note the great artistic taste that has been shown in their arrangement, so as to exhibit them to the best advantage. Here for instance, is a massive formation of alum, tons in weight, made in imitation of the recesses of an Arctic cave with its icy stalactites and stalagmites, and there are feathery creations of some of the finer chemicals. We may mention, as a prominent feature in this department of the Exhibition, the collection of POWERS & WEIGHTMAN, OF PHILADELPHIA, a firm with whose predecessors, in 1818, the history of chemical manufacture in this country commenced. They have erected a handsome pavilion, covering an area of twenty-five by thirty-seven feet, and decorated in the polychromatic style, with plate-glass on the sides, and an entrance at each end to the interior where the exhibits are placed. The engraving on page 373 will give the reader a fair idea of this display, although it is impossible with a wood-cut to do full justice to it, to supply the brilliant colors of the specimens and the beautiful forms of the crystalline shapes, with their lustre and bright flashes of light so attractive to the eye in chemical products. We are told that to place the display in position has cost the firm over ten thousand dollars, while the value of the goods must amount to from three to five times that sum.

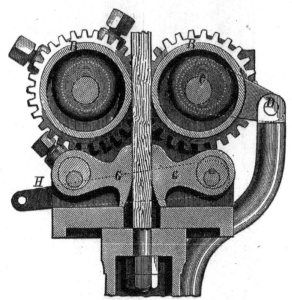

Fig. 6.—*Friction Roll-Drop—Section showing Clamps: Stiles & Parker Press Company.*

Taking a glance at the engraving, we see in that portion of the pavilion to the extreme right, in an elevated position, a crown supported by four uprights, all of crystals of tartaric acid, the finished product of argols or the deposits in wine-casks, and a material which in its present handsome state it is difficult to realize forms an ingredient of that domestic article of every-day use—the baking-powder. Below this, on either side, are inverted vases, that to the right containing a column of blocks of gallic acid, and the one on the left,

MECHANICS AND SCIENCE.

Department of Chemicals: Powers & Weightman, Philadelphia.

a porous-looking, dark pillar of tannic acid, both valuable in the arts. Arranged between these are a number of tall but shapely bottles, in which are liquids of innocent transparency as of water, but concealing the elements of deadly, though—in the hands of the skilled operator—controllable power. These are the liquid acids. Near the tartaric acid is a Gothic window, its outside border blue with crystals of sulphate of copper—commercially known as blue vitriol—and its inner lights transparent with those of citric acid.

On the other side of the entrance may be seen the rugged forms of masses of alum with their well-defined crystals, the darker object between, covered with an inverted vase, being a structure of intersecting arches of blue vitriol crystals; and near these, also around and below them, extending almost to the left of the pavilion, are large collections of various salts, the blue showing their copper origin, and the green their base of iron. Here are seen also the pink, candy-like crystals of sulphate of manganese, preparations of mercury in red and white powders, and scale preparations of iron that glisten with varying hue in the ever-changing light. The tall white pile, near the left, is formed of blocks of sulphate of morphia—a very valuable mass; and to the extreme left is an inverted vase of opium, the dark loaves of Lethe (the crude material from which the morphia is extracted), coming from Persia. In the next corner is another pile of opium from Turkey.

On the other side of the pavilion and not visible to us, are large white monumental masses of the sulphates of quinia, cinchonia, and quinidia, all bitter to the taste, but the pocket companion of the western pioneer in his contests with malaria. Near them are specimens of the classes of bark from which these alkaloids are extracted. A very prominent feature in this collection is the exhibition of the crude materials with the derived products, argols with tartaric acid, lemon-, lime- and sour orange-juices with citric acid, nux-vomica with strichnia, etc.

Hemispheres of crystallized nitrate of ammonia, from which dentists prepare nitrous oxide, and of caffeine, the active principle of coffee and tea, also form beautiful objects in the collection. The caffeine looks like tufts of down that a mere breath would blow away, and upon nearer view it shows the delicate needle crystals peculiar to this chemical. It would be impossible for us in our limited space to even enumerate, much less describe, the two hundred and seventy preparations forming this collection, and embracing chemicals for medicine, photography and the arts in general, but we would like to make special mention of citric acid, of which this firm is the only manufacturer in this country, and has established the industry at great expense. The juice for its manufacture, which is brought principally from the cultivated plantations of Sicily, but also from Mexico, Jamaica and the native sour orange groves of Florida, is prepared by pressing the fruit in a similar way as apples are pressed for cider. The liquid is then strained to remove the pulp, and being put into a copper boiler, it is boiled, with occasional skimming and stirring, until its bulk is reduced to one-sixth its original quantity. After cooling, it is ready for shipment to the manufacturers. Citric acid has become of great use in medicine and the arts.

Handsome exhibits are made of chemical manufactures by many other firms—pharmaceutical preparations, liquor essences, flavors, miniature icebergs of sugar of lead, tall bottles of glycerine, perfumery, paints, etc., etc., all testifying in the strongest manner to the great reputation which this country, and Philadelphia especially, has made in this branch of manufacture.

THE
HISTORY
OF THE
EXHIBITION
BY
JOS. M. WILSON.

INDEX

TO

HISTORICAL INTRODUCTION.

BY

JOSEPH M. WILSON, C. E.

INTERNATIONAL EXHIBITIONS.—Early origin, xvii; French Exhibition of 1798, xvii; Those of 1802, 1806, 1819, 1823, 1827, etc., 1844, xviii; Paris Exhibition of 1849, xix; Exhibitions of other nations, xix.

FIRST INTERNATIONAL EXHIBITION.—That of London, 1851, xix; Its great success due largely to energy of Prince Albert, xix; Selection of plan, xx; Opening day, xxi; The appearance of the Exhibition, xxiii; Description of building, xxiii; United States representation, xxv; Results due to Exhibition of 1851, xxv; Sydenham Crystal Palace, xxv; Advantages gained by England, xxxvii.

DUBLIN EXHIBITION—Of 1853, xxvi; Description of buildings, xxvi; Exhibits, xxvi.

NEW YORK EXHIBITION—Of 1853, xxvi; Organization, xxvi; Want of success, xxvii; Description of building, xxvii; Decorations, xxviii; Proper rules for decoration, xxviii; Exhibits, xxx.

PARIS EXHIBITION—Of 1855, xxxi; Reasons for having it, xxxi; Preliminary arrangements, xxxi; Contracts, progress of work and opening ceremonies, xxxii; Description of buildings, xxxiii; Prominence of Fine Art department, xxxiv; Number of exhibitors, xxxiv; Closing ceremonies and distribution of awards, xxxiv; Exhibits, xxxv; Good results of these exhibitions, xxxv.

MANCHESTER EXHIBITION—Of 1857, xxxv; Description of building, xxxvi; Art treasures, xxxvi.

ITALIAN EXHIBITION at Florence, in 1861, xxxvii.

LONDON EXHIBITION—Of 1862, its incorporation, xxxviii; Selection of site and designs for building, xxxix; Description of building, xl; Decorations, xli; Area, domes, cost, xlii; Opening ceremonies, loss in death of Prince Albert, xlii; Display from United States, xliii; Other exhibits, xliv; Closing ceremonies, xlv.

PARIS EXHIBITION—Of 1867, xlv; Interest of Napoleon III in the development of his Empire, xlv; Arrangement of details for holding the same, xlvi; Plan of Exhibition, site and classification, xlvii; Beautiful appearance of the grounds and description of buildings, xlviii; Opening ceremonies, xlviii; Tour of the Exhibition, xlix; Department of Fine Arts, l; Liberal Arts, l; The United States exhibit in musical and surgical instruments, li; Furniture and glass, lii; Pottery, textile fabrics, silver and bronze, Milton Shield, liii; Furniture and carvings, articles for clothing, lv; Lace and embroidery, Cashmere shawls, Goldsmiths' work and jewelry, lvii; Raw and manufactured materials, United States mineral exhibit, lviii; Iron-work, furs and machinery, lix; Philadelphia exhibits in machinery, lx; Railway exhibits, Grant locomotive, civil engineering and architectural exhibits lx; Food exhibits, agricultural annex, lxi; Distribution of prizes, lxi; Number of awards, pecuniary success of the Exhibition, lxii.

MINOR EXHIBITIONS of Dublin, Leeds, Copenhagen and Moscow, lxiii; Annual exhibitions of England, lxiii.

AUSTRIA DECIDES ON AN EXHIBITION, lxiii; Active measures for an exhibition in Vienna in 1873, lxiv; Great interest taken by foreign governments, lxiv; Selection of site, lxiv; Vast extent of ground enclosed, arrangement of classification, grouping of buildings, lxvi; Description of the Palace of Industry, lxvii; Interior decorations, lxviii; Exterior effect, arrangements of countries, Gallery of Fine Arts, Machinery Hall, lxix; Agricultural Hall, lxx; Opening ceremonies, lxx; Review of exhibits, lxxi; Machinery, lxxiii; American exhibits, lxxiv; Pottery and porcelain, lxxv; Glassware, mosaics, furniture, carpets, lxxvii; Album cover in enamel painting, lxxviii; Awards, closing of Exhibition, lxxix; Total cost, pecuniary loss, but great indirect benefits, lxxix.

GREAT EXHIBITIONS IN THE UNITED STATES, Franklin Institute, lxxix; American Institute of New York, Exhibitions of Cincinnati and Chicago, lxxx.

INDEX

TO

HISTORY OF THE EXHIBITION.

INTRODUCTORY.—Discovery of Delaware Bay, settlement of its shores, lxxiii; Grant of Pennsylvania to William Penn, he takes possession of the country, lxxiv; Laying out of Philadelphia, lxxxv; Prosperity of the city, its active interest in colonial affairs, first Continental Congress meets here in Carpenters' Hall, lxxxvi; Declaration of Independence, old Independence Bell, adoption of the Constitution of the United States, lxxxvi; Reasons for celebrating the Centennial Anniversary by an International Exhibition, and for holding it in Philadelphia, lxxxvi.

PRELIMINARY PROCEEDINGS.—Initiatory move taken by the Franklin Institute, lxxxvi; Request to City Councils to memorialize Congress on the subject, lxxxvii; Favorable action of Councils, and also of the Legislature of Pennsylvania, lxxxviii; Joint committees visit Washington and present a memorial to Congress, lxxxviii; Act of Congress, xc.

THE CENTENNIAL COMMISSION.—Appointment of commissioners, first meeting, expenses defrayed by the City of Philadelphia, xci; Election of president and arrangement of committees, selection of site, xci; Second session of commission, and consideration of measures for obtaining funds, xci; Application to Congress for charter for a "Centennial Board of Finance" and passage of act for the same, xcii; Powers and duties of the "Centennial Board of Finance," xciv; Third session, small amount of work accomplished, design for seal, xciv; Committee on plans instructed to advertise for designs, but could do nothing for want of funds, xciv; Appropriation by Philadelphia for incidental expenses, discouraging prospects, xciv.

FURTHER PROGRESS OF THE WORK.—The Citizens' Centennial Finance Committee and its object, xcv; Large subscriptions and assurance of success, xcv; Imposing mass-meeting of Feb. 22d, 1873, xcv; Action taken in other states, xcv; Appointment of Blake and Pettit to go to Vienna, xcv; Organization of Women's Centennial Committee, and election of Mrs. E. D. Gillespie as president, xcv; Full completion of organization of Centennial Board of Finance, xcvi; Issue of invitations for preliminary designs for buildings, xcvi; Election of Director-General, xcvii.

IMPORTANT ADVANCES.—Formal transfer of the grounds to the Commission, xcvii; Preliminary plans received, second competition and awards, xcviii; Report of Committee on Plans, xcix; Approval of plan for Main Exhibition Building, c; Design for Memorial Building, c; Selection of architects for Main Building, c.

DIFFICULTIES.—Interpretation of the Secretary of State concerning the International features of the Exhibition, cii; Necessity of additional legislation, cii; Influence of the financial panic of 1873, cii; Unsuccessful effort to obtain pecuniary aid from the Government, cii; Amount of present subscriptions, cii; Absolute necessity for economy, cost of adopted plans for Main Building precluding their use, ciii; Successive modifications of plan result in successive reductions of cost, ciii; Still further modifications, ciii; Plan of Mr. Pettit, cvi; Bids received on the various plans, cvi; Adoption of Mr. Pettit's plan, cvi; The architects for the Main Building decline to execute the work and their contract closed, cvi.

ACTUAL WORK AT LAST.—Arrangements with Messrs. Pettit and Wilson, modification of plan for Main Building, its cost, cvi; Memorial Building, Mr. Schwarzmann's plan adopted, contract made for the same, its cost, cvi; Plans prepared for Machinery Hall, contract for the same, its cost, cvii; Conservatory Building, contract and cost, cvii; Agricultural Building, contract and cost, cvii; Area covered by these buildings, cvii; Rapid progress of the work, Government Building, offices, special buildings, etc., laying out of grounds, bridges over Landsdowne and Belmont ravines, cvii; Organization of Water Department to the Exhibition, cviii; Transportation facilities, cix.

CLASSIFICATION AND ORGANIZATION.—Arrangement of general classification, cix; More unsuccessful efforts to obtain an appropriation from Congress, cx; Women's Centennial Executive Committee and special Women's Building, cx; Organization of Bureaus, cx; System of awards adopted, cxi; A final and successful effort to obtain aid from Congress, cxi.

DAWN OF THE CENTENNIAL YEAR.—Celebrations and rejoicings, foreign representatives arriving, goods begin to appear on the grounds, cxi; Mildness of the winter, and rapid completion of the buildings and preparation of grounds, cxi.

OPENING DAY.—Decorations, great enthusiasm, assembling of the multitudes, cxii; Arrangements for the opening exercises, cxiii; Entrance of the distinguished visitors, the official ceremonies begin, cxiv; Wagner's Centennial Inauguration March, prayer, cxiv; Whittier's Hymn, presentation of the buildings to the Centennial Commission, Lanier's Cantata, cxv; Presentation of the Exhibition to the President of the United States, who declares it open to the world, cxvii; Hallelujah Chorus, cxvii; A procession formed, it passes through the Main Building and on to Machinery Hall, where the President and the Emperor of Brazil start the great Corliss Engine, cxvii; End of the first day, cxvii.

THE EXHIBITION.—Those pleasant days of May and June, the varying crowds, the Krupp Guns, the Corliss Engine, the pleasures and benefits of the Exhibition, cxvii; The hot days of July, the celebration of the Fourth, the latter days of August, with the great crowds of September, the Horticultural grounds and the first of October, an intimation of the end, cxx.

THE MAIN BUILDING.—Entrance at Belmont Auenue, offices of Centennial Commission and Board of Finance, cxx; Bartholdi Fountain, exterior appearance of Main Building, restrictions in preparing the design, cxx; Location, area, details of plan, cxxii; Further details, foundations, columns, roofs, entrances, cxxii; Framing, floors, gas, water, cxxiv; Drainage, painting and decorations, interior effects, cxxv; The rich Exhibits and the effect they produce, cxxvi.

MACHINERY HALL.—Its location and area, the boiler-houses, points governing the design, cxxvi; Further points, description of plan, entrances, floor area cxxvii; Hydraulic Annex, water-fall, foundations of building, walls, frame-work, flooring, roof, ventilation, cxxix; Gas, water service, painting, cxxi; Boiler-houses, Corliss boiler-house, the hour of noon, plantation songs from the negroes, cxxxi; The lake, the chimes, Col. Lienard's relief-plan of Paris, cxxxii.

THE GOVERNMENT BUILDING.—By whom constructed and for what purpose, cxxxii; Description of building, departments of Government exhibited within it, cxxxiii; The Post-Office Department, the Agricultural Department, cxxxiii; Department of the Interior, cxxxiv; Smithsonian Institute, War and Treasury Departments, cxxxv; Trois Frères Restaurant, Post Hospital of Medical Department, cxxxv; Signal Service, cxxxv.

HORTICULTURAL HALL.—Fountain Avenue and its lovely flowers, cxxxv; Exhibit of English Rhododendrons, cxxxvi; Description of Horticultural Hall, collection of plants, Miss Foley's Fountain, contiguous grounds, Cacti exhibit, cxxxvii.

TOUR THROUGH THE GROUNDS.—The Philadelphia City Building, the Bible Society, the Milk Dairy Association, cxxxix; The Brazilian and German Government Buildings, the Moorish villa, cxxxix; Landsdowne ravine, the marine band, the hunter's camp, the bridges over the ravine, cxxxix; Details of the bridges, cxl; the Singer Sewing-Machine Company, the Pennsylvania educational exhibit, cxli.

THE DAY CLOSES.—The warning of the fog-horn, the numerous cheap routes to and fro connecting with the city, our choice, the paddle boats, on the Schuylkill, cxli; The Sudreau restaurant, the way to the river, the Worthington pumps, cxli; Up to the Falls and back, the lovely river, Laurel Hill, cxliii.

THE ART GALLERY.—A memorial of the Centennial year, the bronze horses from Vienna, Wolf's "Dead Lioness," Mead's group of "The Navy," cxliii; Description of the Memorial Building, cxliv; Further description, cxlv; The Art Annex and its plan, cxlv.

SUNDRY EXHIBITS ON THE GROUNDS.—The French Government exhibit, the Vienna Bakery, the police barracks, cxlvii; The Empire Transportation Company, the Bankers' exhibit, cxlvii; The Photographic Building, the Carriage Annex cxlvii.

THE SWEDISH SCHOOL-HOUSE.—So thoroughly natural in its appearance, cxlvii; Its mode of construction, the Swedish Commissioner of Education, cxlviii; Education in Sweden, the building before us a country school-house, size and arrangement of the rooms, cxlix; the appliances provided and methods of study, the various branches taught, cxlix; Further concerning the course of instruction, what Sweden owes to her schools, cl.

OTHER SPECIAL EXHIBITS.—The Japanese Bazaar, the building and grounds, cl; The little fountain, the bronze cranes and pigs, cli; The curious garden, the "Japs" and their wares, clii; The Department of Public Comfort, the Judges' Hall, the Pennsylvania Railroad Ticket-Office, clii; The Cook Tourist Agency, the Palestine encampment, the Centennial Photographic Association, clii.

AN EXCURSION.—A trip on the narrow-gauge railroad with the "Emma," the lake with its fountain, the lovely flowers, the Women's Pavilion, the New Jersey Building, the Southern Restaurant, the Agricultural Building, the American Restaurant, cliv; the land of wind-mills, the State Buildings, a fairy scene, cliv; Back to our starting-point, all in fifteen minutes, clv.

THE WOMEN'S PAVILION.—All by women, the organization of the Women's Centennial Committee, how its work began, clv; What was done by the women, clvi; Their aid in the presentation of the International feature of the Exhibition, clvii; Their entertainments and tea parties, clvii; The Washington Assembly, the Fête Champêtre, the International Assemblies, clviii; The work in Rhode Island and other localities, clix; The efforts made towards a special exhibit of women's work, clix; The amount of subscriptions raised by the women, the cost of Wagner's Centennial March defrayed by them, clix; Voluntary labors, the building for their special use, the exhibits in it, clx.

THE KINDERGARDEN.—A genuine Frœbel Kindergarden, all due to the women; the opening exercises, the pretty songs, clay-model day, clxi; Marching and other exercises, all to song, lunch, play songs, clxii; The metallic rings, studies from nature, closing exercises, clxiii.

ADJACENT STRUCTURES.—The New Jersey Building, the Southern Restaurant, the Kansas Building, Virginia, the New England Log House, clxiii; The farmer's home of a century ago, clxiv; An Algerian booth, clxiv.

AGRICULTURAL HALL.—Description of the Building, clxiv; The exhibits, the Pomological Annex, clxv; The Wagon Annex, the Brewer's Building, the Butter and Cheese Factory, clxv; The Tea- and Coffee-Press Building, the single rail elevated railway, the German Restaurant, clxv.

THE STATE AND NATIONAL BUILDINGS.—The Centennial Fire Patrol, the Ohio building, the Indiana building, clxvii; Illinois, Wisconsin, Michigan, New Hampshire, Connecticut, clxvii; Massachusetts, Delaware, Maryland, Tennessee, Iowa, and Missouri, clxvii; The purpose of these buildings, Rhode Island, Mississippi, clxvii; The Hungarian Wine Pavilion, the California building, the New York building, clxviii; The St. George's House, the barracks and workmen's quarters, clxviii; A Kettle-drum, the interior of St. George's House, clxviii; The Japanese Dwelling, the Spanish Government buildings, West Virginia, Arkansas, the Canadian Log-House, clxix; The Fountain of the Catholic Total Abstinence Union, clxx; The Vermont building, the Turkish Café, the Jerusalem and Bethlehem Bazaars, the Pennsylvania building, clxxi.

STILL ANOTHER DAY.—West of Machinery Hall, the Saw-Mill Annex, the Glass-Works, the Pennsylvania Railroad exhibit, the "John Bull," clxxi; The Krohnke Silver Reduction Process, clxxii; The Campbell Press Company, Gas Machines, Castings, the State of Nevada, clxxiii; The Shoe and Leather building, its description, the exhibits in it, clxxiii; The New England Granite Company, back to the Plaza of the Bartholdi Fountain, clxxiii.

THE AWARDS.—What the Judges have been doing, the announcement of awards, clxxiv; The ceremonies of the occasion, the total number of exhibitors and the number of awards, clxxv.

WHAT THE EXHIBITION DEMONSTRATES.—Iron manufacture, its development, immense coal exhibits, clxxv; The great iron-ore deposits of the United States, America up to the age in iron manufacture, clxxvi; Steel exhibits, the perfection of the Bessemer process in the United States, the value of a protective tariff, clxxvii; Swedish exhibits in iron and steel, clxxvii; Display from the United States in Ceramic and Glass Wares, clxxvii; Wonderful development of the industry, clxxviii; Ceramic wares from abroad, Great Britain, France, Sweden, clxxix; Porcelain and Pottery from Japan, Glass exhibit from United States, clxxx; Chemical products, textile fabrics, clxxx; America able to cope with foreign competition, oil cloths, raw cottons, clxxxi; Wool and silk fabrics, carpets, clxxxi; Jewelry, watches and silverware, the consternation among Swiss manufacturers, clxxxi; Paper, stationery, and book-making, printing presses, clxxxi; Hardware, the prominent display from the United States, clxxxi; Edge-tools, the American ax, carpenters', agricultural and laborers' tools, cutlery, fire- and burglar-proof safes, clxxxii; Railway plant, clxxxii; Hydraulic motors, transmitters and pneumatic apparatus, clxxxiii; Machinery, great prominence of the United States, clxxxiii; Sewing, knitting and embroidering machines, clxxxiii; Electric and telegraphic apparatus, clxxxiii; Civil Engineering exhibit, clxxxiv; Agricultural machinery, live-stock exhibitions, clxxxiv.

CLOSING CEREMONIES.—Farewell dinner to the Foreign Commissioners, the 10th of November, a rainy day, the ceremonies confined to Judges' Hall, clxxxv; Dignitaries present, Wagners' Inauguration March, prayer, addresses from the various officers, clxxxv; Musical selections, the hymn "America," the flag of the "Bon Homme Richard," the last act of the drama, clxxxv; Old Hundred and the chimes, clxxxvi; Total number of visitors, total receipts for admissions, number of buildings on the grounds, passengers carried on the narrow-gauge railway, clxxxvi.

HISTORICAL INTRODUCTION

INTERNATIONAL EXHIBITIONS.

EXHIBITIONS undoubtedly date back to a very remote period, even the Olympic games of the Greeks might be classed as such, and the ancient periodical fairs for the display and sale of natural and industrial products, some of them continuing to the present day, although not properly speaking, true expositions or intended for such, yet gave great encouragement to the arts and manufactures of their time.

After Europe began to recover from the blight of the Dark Ages, the arts of civilization and luxury, centering and developing in Italy, rapidly found their way into France, a country already prepared for them by its ancient Roman education; and from being the recipient, she gradually became the producer, early taking a pre-eminent stand among the nations of the earth in almost every known branch of manufacture, especially those connected with art. This she has retained to the present day. It is but natural, therefore, that she should have been foremost, at least in the modern world, to originate the idea of Industrial Exhibitions.

The first of which we have any record was that of 1798, born of the Revolution, a reaction as it were from the turbulent spirit of the times, back to the pursuits of peace and industry. The Marquis d'Avèze, shortly after his appointment in 1797 as Commissioner of the Royal Manufactories of the Gobelins, of Sevres, and of the Savonnerie, found the workmen reduced nearly to starvation by the neglect of the previous two years, while the storehouses, in the mean time, had been filled with their choicest productions. The original idea occurred to him to have a display and sale of this large stock of tapestries, china and carpets, and obtaining the consent of the government, he made arrangements for an exhibition at the then uninhabited Chateau of St. Cloud. On the day, however, appointed for the opening, he was compelled by a decree of the Directory, banishing the nobility, to quit France, and the project was a failure. The following year, however, returning to France, he organized another exhibition on a larger scale, collecting a great variety of beautiful objects of art and arranging them in the house and gardens of the Maison d'Orsay for exhibition and sale. The success was so great that the government adopted the idea, and the first official Exposition was established and held on the Champ de Mars, a Temple of Industry being erected, surrounded by sixty porticoes, and filled with the most magnificent collection of objects that France could produce. Here was first inaugurated the system of awards by juries, composed of gentlemen distinguished for their taste in the various departments of art, and prizes were awarded for excellence in design and workmanship.

The government was so satisfied with the good effects resulting from this exhibition, that it resolved to hold them annually; but notwithstanding the circular of the Minister of the Interior to this effect, the disturbed state of the country prevented a repetition until 1801. The First Consul taking the greatest interest in the affair, visited the factories and workshops of the principal towns in France, to convince the manufacturers of the great importance to themselves and their country of favoring the undertaking. A temporary building was erected in the quadrangle of the Louvre, and notwithstanding great difficulties attending the establishment of the exhibition, there were two hundred competitors for prizes; ten gold, twenty silver, and thirty bronze medals being awarded,—one of the last to Jacquard for his now famous loom. Among these prizes, were some for excellence in woollen and cotton fabrics, and improvements in the quality of wool as a raw material.

The third exhibition was in 1802, where there were six hundred prize competitors. These expositions became so popular as to result in the formation of a *Société d'Encouragement*, thus creating a powerful aid to the industrial efforts of the French manufacturers. At the fourth exhibition, in 1806, the printed cottons of Mulhausen and Logelbach, and silk-thread and cotton-lace were first displayed, and prizes were adjudged for the manufacture of iron by means of coke, and of steel by a new process.

Foreign wars prevented further exhibitions until 1819, after which time they became more frequent, being held in 1823, 1827, etc.; the tenth being in 1844, the last, under the reign of Louis Philippe, when three thousand nine hundred and sixty manufacturers exhibited their productions. It was the most splendid and varied display that had ever been held in France. The building, designed by the architect Moreau and erected in the Carré Marigny

Exhibition Building, Paris, 1844.

of the Champs Elysées, was an immense timber shed, constructed and entirely completed in seventy days, at a cost of about thirty cents per square foot of surface covered. We present an elevation showing the royal entrance. It was at this exhibition that the first Nasmyth steam-hammer was shown on the continent, and the display of heavy moving machinery was much greater than had ever taken place before.

In 1849, notwithstanding the political revolution through which France had just passed, she organized another exhibition on a still grander scale than any preceding. The services of the architect Moreau were again called into requisition, and another building, of which we give an engraving, erected in the Champs Elysées, more pretentious in its character than

Exhibition Building, Paris, 1849.

any previous one, covering an area of 220,000 square feet, exclusive of an agricultural annexe, and costing about the same price per square foot as the building of 1844. At this time the number of exhibitors had increased from one thousand four hundred, in 1806, to nearly five thousand, there being no less than three thousand seven hundred and thirty-eight prizes awarded, and the building remained open for sixty days.

Other nations, noticing the beneficial results of the French exhibitions, became active in the matter; the King of Bavaria giving an exhibition at Munich, in 1845, and previous to this time occasional ones had been held in Austria, Spain, Portugal, Russia, Denmark, Sweden, etc.; those of Belgium being numerous and important. In the British Dominions, exhibitions had been held in Dublin as early as 1827, and later at Manchester, Leeds, etc.; but they partook more of the nature of bazaars, or fairs for the sale of the productions of the surrounding country; even that of Manchester, 1849, was of this character.

Each of these previous exhibitions had been strictly national, confined to the products of the special country by which it was held. The idea seems to have been suggested, however, in France, in 1849, of giving an International feature to that exhibition; M. Buffet, the Minister of Agriculture and Commerce, having addressed a circular letter on the subject to various manufacturers, with a view of ascertaining their opinions; but the resulting replies were so unfavorable that the project was abandoned, and France lost the opportunity, which was reserved to England, of the credit of the first really International Industrial Exhibition, in that of London, 1851.

It may truly be said that the great success of this effort was owing to the indefatigable perseverance and indomitable energy of His Royal Highness Prince Albert, who took the greatest interest in the proceedings which gave it birth, from the very commencement, bringing to bear all the influence which attached to his position, his remarkable sagacity in matters of business, and his courageous defiance of all risks of failure. At one of the first meetings held on the subject, on the 29th of June, 1849, at Buckingham Palace, he communicated to those present his views in relation to a proposed exhibition of competition, in 1851, suggesting that the articles exhibited should consist of four great divisions, namely, raw materials, machinery and mechanical inventions, manufactures, and sculpture and the plastic arts; and at a second meeting, on July 14th of the same year, he gave still further suggestions of a plan of operations which he recommended, comprising the formation of a Royal Commission, the definition of the nature of the exhibition and of the best mode of conducting its proceedings, the determination of a method of deciding prizes and the means of raising a prize fund and providing for necessary expenses, etc.; and he also pointed out the site afterwards adopted, stating its advantages, and recommending early application to the government for permission to appropriate it.

After various preliminary proceedings, the Royal Commission was issued, and at the first meeting of the Commissioners, on January 11th, 1850, it was decided to rely entirely upon voluntary contributions for means to carry out the plans proposed.

The appeal made to this effect was answered in a most encouraging manner; a guarantee fund of $1,150,000 was subscribed, one gentleman opening the list with $250,000, and contributions began to come in from all directions.

Upon the security thus provided the Bank of England undertook to furnish the necessary advances. Invitations were issued to architects of all nations to submit designs for a building to cover 700,000 square feet, and although the competitors amounted to two hundred and thirty-three in number, not one design was found entirely suitable for adoption. In this dilemma, the Building Committee prepared a design of their own, and, notwithstanding it was strongly condemned by public opinion as inappropriate and unsuitable in many respects, the committee warmly defended it and advertised for tenders to erect it, requesting at the same time, that competitors would make any suggestions they saw fit, that could in their opinion effect a reduction in the cost.

Messrs. Fox and Henderson availing themselves of this clause, presented a tender for a building of an entirely different character, on a plan proposed by Sir Joseph, then Mr. Paxton, who was at that time engaged in the erection of a large plant-house for the Duke of Devonshire, at Chatsworth. The design fully met the approbation of the Committee and their tender was accepted, on the 16th of May, 1850. Possession of the ground was obtained on the 30th of July, and work commenced forthwith,—the actual erection beginning about the first week in September.

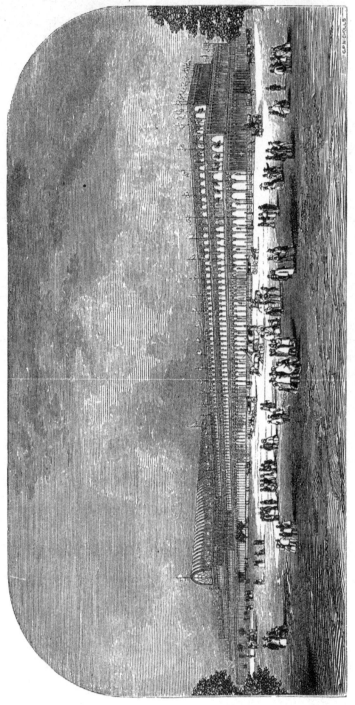

International Exhibition, London, 1851.

Mr. Fox made the working drawings himself, devoting his great experience and skill personally to the work for eighteen hours a day, during seven weeks, and the preparation of the iron work and other material for the construction of the building was taken charge of by Mr. Henderson. As the building progressed, extensive experiments were made to test its strength for the purposes intended, and it was found fully equal to the severest requirements. The contract was not finally consummated until the end of October; but with a courage and enterprise characteristic of this firm, the work was pushed forward for many

Interior of the Transept, from the South Side. Exhibition of 1851.

weeks on faith alone, in order to insure the completion of the building at the time fixed for the opening,—the first of May, 1851. It was opened at the time appointed, by the Queen in person, with great ceremony, although considerable work still remained to be done. A report of the proceedings of the Royal Commissioners was read by Prince Albert as President, which being replied to by the Queen, the blessing of the Almighty was invoked upon the

The Transept of the Exhibition of 1851, from the North Side.

undertaking by the Archbishop of Canterbury, and the ceremonies terminated with the performance of the Hallelujah Chorus by the united choirs of the Chapel Royal, St. Paul's,

Westminster Abbey, and St. George's Chapel, Windsor. The inauguration was one of the most imposing sights that had ever been witnessed in Great Britain. Our engraving gives a view of the building on the south side, extending east and west, and showing the main entrance at the great transept.

In appearance it called to mind one of the old, vast cathedrals, designed, however, in a new style of architecture; not massive, dark, and sombre, but light, graceful, airy, and almost fairy-like in its proportions,—built as if in a night by the touch of a wand,—a true "Crystal Palace," and a noble example of the use of our modern material—iron—for building purposes.

It was obvious that nothing more suitable could have been designed, and that the modern adaptation of one of the oldest architectural ideas—a great rectangular cruciform structure with nave and transepts—was just what was desired, possessing many more of the requirements of a building intended for industrial exhibitions than would appear at first glance. The old cathedral was a place for great ceremonials, for processions, and for exhibitions, in one sense of the word; its walls were covered with pictures and sculpture, and its windows filled with richly stained glass. Extending over a vast area, at the same time it had a grand central point of attraction, visible from all parts, and from which all parts were visible. These advantages were just what was required in an exhibition building, and the fact has been acknowledged over and over again in succeeding exhibitions. It will be seen, further on, that in our exhibition building the same ideas have been carried out, and that the building of 1851 has really been the type for all the most successful buildings erected since.

Fergusson characterized this building as belonging to a new style of architecture, which might be called the "Ferro-Vitreous Style," and states that "no incident in the history of architecture was so felicitous as Sir Joseph Paxton's suggestion." "At a time when men were puzzling themselves over domes to rival the Pantheon, or halls to surpass those of the Baths of Caracalla, it was wonderful that a man could be found to suggest a thing that had no other merit than being the best, and, indeed, the only thing then known which would answer the purpose."

The light appearance of this structure was so strongly marked that many persons, uneducated as to the effect which should be produced on the eye by an iron and glass construction on such a large scale, expressed grave doubts as to its stability. To satisfy these doubts in the public mind, extensive experiments were carried out during the progress of the work, and also after its completion, in the presence of the Queen, Prince Albert, and a number of scientific men, by means of large numbers of workmen, crowding them on the platforms, and moving them back and forth, and also by means of companies of troops, arranging them in close order and marching them on the floors. Frames holding cannon-balls were also constructed and drawn over the floor, and the results of all these experiments were such as to entirely satisfy every one that the building was properly planned and constructed for its purposes.

Passing into the building at the west end, we enter a grand nave 72 feet wide, 1848 feet long, and 64 feet to the roof, crossed by a noble transept of the same width, but crowned by a semi-circular vault, increasing its height to 104 feet at the centre. On each side of the nave and transept a series of aisles spread out the building to a total width of 456 feet, the entire area covered being 772,784 square feet, and, with the addition of the galleries, making a total exhibition space of 989,884 square feet. The quantities of materials in the structure were as follows:—

Cast iron, 3,500 tons;
Wrought iron, 550 tons;
Glass, 896,000 superficial feet, or 400 tons;
Wood, 600,000 cubic feet;

and the total cost was about $850,000; the building remaining the property of the contractors after the exhibition was over. The late Mr. Owen Jones, so well known for his taste in art ornamentation, was entrusted with the decoration of this palace, and the result fully justified the trust reposed in him, and met with very general approval.

United States Department, Exhibition of 1851.

It is said that, in designing the structure, the magnificent transept, with its semi-circular roof, was suggested in consequence of a desire to retain several lofty trees which were on the grounds. Be that as it may, the trees were retained, and we are glad to be able to give an engraving showing the beautiful effect thereby produced. These enclosed trees made a marked feature in the exhibition.

The United States department was quite well represented,—bearing in mind the comparatively small advances which this country had made, at that time, in the higher departments of art manufacture,—and we furnish a view of this department as it appeared. Powers exhibited his celebrated "Greek Slave," shown in the foreground of the picture, of which we believe there are several originals in existence,—one at the Corcoran Gallery in Washington. He also exhibited his "Fisher Boy," a work in every way worthy of the artist, and seen to the right of the "Greek Slave." The piano, on the right, was exhibited by Messrs. Nunn & Clark, of New York. Messrs. Chickering, of Boston, also exhibited a very fine instrument, and even at that time they had obtained a high reputation for power and brilliancy of tone, among European professors. Cornelius & Co., of Philadelphia, exhibited two elegant examples of gas chandeliers, which were very much admired. Some handsome carriages were shown; our celebrated Watson, of Philadelphia, being among the exhibitors. The exhibition of agricultural implements and raw materials was very creditable.

We also present a view of the interior of the transept from the south side, which will aid in giving the reader some idea of the structure and its exhibits. In the centre is seen the curious glass fountain, contributed by the Messrs. Osler, of Birmingham, which attracted so much attention by the novelty of its design, its lightness, and its beauty. Passing on through the building, the visitor came into contact with objects from India, Africa, Asia, the West Indies, and all quarters of the globe; articles of sculpture, textile fabrics, modern and mediæval brass and iron work, animal and mineral products, machinery, works of utility and those of ornament—everything that could furnish delight to man or add to his comfort: a vast collection, exemplifying the great progress which civilization had wrought in the world by the skill of man adapting the materials of nature to his own use.

The exhibition of 1851 was in every way a great success. Upwards of $200,000 had been received from the sale of season tickets alone before the opening. During the six months that it remained open, from May to October inclusive, the average daily number of visitors was 43,536; the total number for the whole time was 6,170,000, and the amount of receipts, $2,625,535; there being a balance of $750,000 in the hands of the Commissioners after all expenses were paid. The exhibitors, coming from all parts of the world, amounted to more than 17,000.

The unique style and acknowledged beauty of this magnificent edifice—the first of its kind—and the delightful recollections connected with its use, combined to preserve it from destruction; and visitors now see the same building, more permanently constructed in a modified and much improved form, at Sydenham, as one of the great pleasure resorts of London. Of those who have been abroad, who does not remember Sydenham?—the beautiful grounds laid out with shrubbery, walks, lakes, and fountains, for the special purpose of making the whole as attractive as possible; the splendid band in constant attendance, the delightful concerts, amusements of all kinds in the most interesting variety, and the vast crowds, wandering about and so thoroughly enjoying themselves. Special excursions are made up, numbering sometimes thousands of people, for a happy day at the Crystal Palace,—a rest from the bustle and turmoil of the city, adding renewed vigor to the tired body to struggle in the battle of life. It is not alone, however, as a pleasure resort, but also as a place of education for the masses, that Sydenham Crystal Palace is worthy of note. Portions of the building are fitted up to represent the styles of architecture of different periods of the world's history, such as the Pompeian Court, the Italian Court, the Renaissance Court, the English Mediæval Court, &c. Another portion contains copies of the works of great sculptors of ancient and modern times, and of paintings of great artists, and down by the lake in the gardens, one finds models, life-size, of the pre-historic animals of the ancient world.

The success attending this exhibition stimulated other countries in efforts to have something of the same kind. Exhibitions, more or less local in character, were projected and held in the large manufacturing towns throughout the British Empire,—at Cork, Dublin, Manchester, &c.

That at Dublin, in 1853, under the auspices of the Royal Dublin Society, which had previously had triennial exhibitions, was the result of a proposition made to the Society by Mr. William Dargan, a well-known contractor, providing a certain fund for the exhibition under certain conditions; and, although international in its features, was not practically as entirely so as the exhibition of 1851. The building consisted of five large, parallel, arched and dome-roofed halls. The great central hall was longer, as lofty, and one-fourth wider than the transept of the Crystal Palace of 1851, being 425 feet in length by 100 feet in breadth and 105 feet in height, with vaulted roof and semi-circular domed ends. We give an elevation of this building, which shows very clearly its general design.

Dublin Exhibition Building, 1853.

The erection commenced August 18th, 1852, and the exhibition was opened by the Irish Viceroy, May 12th 1853, the building occupying in construction about two hundred working days.

The interior effect was spacious and beautiful, and the decoration, notwithstanding the small sum appropriated for it, quite effective,—the prevailing tints being light blues, delicate buffs, and deep ultramarine, with white and red used very sparingly. The columns of the central hall were dark blue, and the skeleton frame of the building was marked out and emphasized by dark and heavy tones of color. The total area covered was 265,000 superficial feet, costing, per square foot, about five-sixths of that of the building of 1851, but the exhibition itself was not a financial success. The collection of art productions was large and particularly fine,—the works coming principally from Great Britain, Germany, Belgium, and France; and the method of lighting the picture gallery was considered very effective, and the best that had been as yet devised.

The most interesting of all the exhibits was the collection of Irish Antiquities, which was very large and arranged with admirable skill, forming something at once valuable and unique. At the close of the exhibition the building was torn down and sold. The materials, however, did not realize more than one-fourth the amount of their valuation; the unwieldy forms of the curved parts being so badly adapted for future use, and the timbers being so injured by nails and the summer heat, and so shattered in taking apart, that very few portions were ever again erected. The result demonstrated two facts: first the expensiveness of temporary buildings for such purposes, and secondly the great increase produced in the cost by the introduction of curvilinear work.

This same year an International Exhibition was also held in the City of New York under the organization of a few influential citizens, as a joint stock company, clothed with sufficient powers by legislation to carry out the objects proposed. This exhibition had in view the comparison of the productions of America with those of other countries, with the object of the promotion of her advancement, it being acknowledged that she had more to gain by such comparison than any other of the great nations of the world. It was liberally assisted by contributions of exhibits from European manufacturers and artists, but misfortune seems to have attended it from the beginning.

It labored under the great disadvantage of professing to be a national undertaking, without receiving support in any way from the government; of exposing itself to the imputation of being a private speculation under the name of a patriotic movement, and was viewed with jealous feelings by many of the great cities of the Union. Great injustice may have been done to the exhibition and its promoters, but still the effect of these adverse influences was perceptible. Although recognized in a semi-official way by the President, and by some of the foreign powers, it cannot be said to have been by any means a success, many exhibitors suffering serious loss. These consequences seem to be inherent at the outset of any great international exhibition that may be held here, from the very nature of our political institutions. Our present exhibition has had its difficulties in this respect. How nobly it has triumphed over all, its record will show. The country is so large, and the interests of the different portions so various, that it requires an anniversary like the Centennial to unite all together in a common celebration.

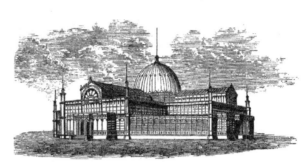

Crystal Palace, New York, 1853.

The opening, although advertised to be early in June, did not take place until the middle of July, in the midst of our hot season; President Pierce formally taking part in the exercises, in the presence of six commissioners of Great Britain, those of many other foreign governments, and all the heads of the various State departments.

The building was erected from designs furnished in competition by Messrs. Carstensen and Gildmeister; the consulting and executive engineers and architects appointed to carry out the plans being Mr. C. E. Detmold, Mr. Horatio Allen, and Mr. Edmund Henry. Although much smaller than the exhibition building of 1851, and possessing considerable originality in architectural effect and constructive detail, it was based upon the same general principles of construction in glass and iron, then so novel, and considered so appropriate for the purpose. Located upon an unfavorable piece of ground, 445 by 455 feet in extent, an octagonal form of building was adopted, changing at the height of twenty-four feet to a Greek Cross with low roofs in the four corners, and crowned by a dome at the centre. The length of each arm of the cross was three hundred and sixty-five feet, five inches, and the width, one hundred and forty-nine feet, five inches. On one side of the building was placed a rectangular, one story annexé, for machinery in motion. The plans which we give, of the ground-floor and galleries, will sufficiently explain the mode of construction. The columns indicated on these plans were placed twenty-seven feet apart each way; there being two principal avenues or naves, forty-one feet, five inches wide, with side aisles and galleries fifty-four feet wide. The dome was one hundred feet in diameter, with a height from the ground-floor to the springing of nearly seventy feet, and to the crown of the arch of one hundred and twenty-three feet, being at that time, the largest dome erected in this country.

The roofs of the building, including the dome, were covered with white pine sheathing boards and tinned, and light was communicated to the interior by the glass construction of the main walls, and by the clerestories of the main avenues and dome. The dome above the clerestory contained thirty-two ornamental windows of stained glass, decorated with the arms of the Union and of the several States, forming quite a conspicuous feature of the interior effect. The exterior walls were constructed of cast-iron framing and panel work, filled in with glazed sash,—the glass used being one-eighth inch thick, enameled, and of American manufacture.

Octagonal turrets were placed at the angles of the building, eight feet in diameter, and seventy-six feet high, containing circular stairways for the private use of the officers and employées of the exhibition. The lower floor was connected with the galleries by twelve stairways, one each side of the four main entrances and four under the dome,—the latter being in reality double stairways with a common half landing.

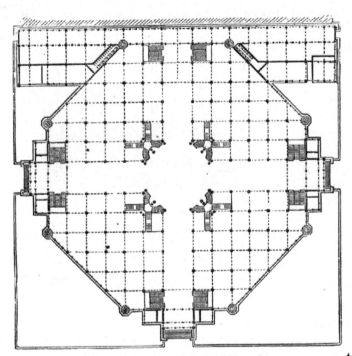

Ground Plan, New York Exhibition, 1853.

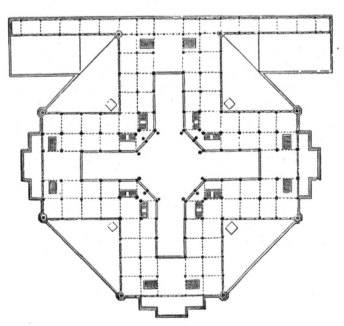

Gallery Plan, New York Exhibition, 1853.

The decoration of the building was considered particularly fine, it having been placed in the hands of Mr. Henry Greenough of Cambridge, near Boston, Massachusetts, excepting only the interior of the dome, which was designed by Signor Monte Lilla. Mr. Greenough started out with the very correct assumption, that the only true method was to ornament the constructive details, following and bringing out the lines therein indicated, without attempting to conceal them by useless and unmeaning decoration.

With the exception of the ceiling of one of the lower corner roofs, and the interior of the dome, which were executed in tempera on canvas, the whole of the exterior and interior work was in white lead in oil, brought to the various tints desired by the admixture of various colors.

Mr. Greenough has given the following rules, to which he states that he mainly adhered in working up the design, and as they were productive of such excellent results, and are so generally applicable, we take the liberty of quoting them:—

"I. Decoration should in all cases be subordinate to construction. It may be employed to heighten or give additional value to architectural beauties, but should never counterfeit them. Being in the nature of an accompaniment, it should keep in modest accordance with the air, and not drown it with impertinent embellishment. Coloring, to be employed with good effect on a building, should resemble the drapery of the antique sculptures, which, displaying between its folds the forms beneath, serves rather to enhance than to conceal their beauty.

"II. All features of main construction should have one prevailing tint, enriched occasionally by the harmonious contrasts of that color. All secondary, or auxiliary construction, may be decorated by the employment of a richer variety of the principal color. This mode of treatment is suggested by the distinction which nature has made between the coloring of the trunk, branches, twigs, and leaves of trees.

Interior View, International Exhibition, New York, 1853.

"III. The prevailing color of the ceilings should be sky-blue, thus borrowing from nature the covering which she has placed over our heads. Monotony may be prevented by the introduction of orange (the natural complement of blue), garnet and vermilion, in such quantities only as may be necessary to recall these colors employed elsewhere.

"IV. Rich and brilliant tints should occur in small quantities, and be employed to attract the eye to the articulations and noble portions of the members, rather than to the members themselves. As in the human figure, variety of color and form is most displayed in the extremities and joints, to which the broader style of the limbs and trunk serve as a foil, so in buildings, the bases and capitals of columns, brackets of arches, and the frame-work of panels, would seem legitimate objects for the reception of rich coloring. Occurring at fixed numerical distances, they are measured out in equal proportions as to space, and afford also a due quantity of brilliant and stimulating tints,—sufficient to enliven the large proportion of mild color, so essential to a general effect of quiet and repose.

"V. All natural beauty of color existing in any material, should, if possible, be brought into play, by using that color itself, instead of covering it with paint of another hue.

"VI. The leading feature of beauty in the Crystal Palace, being that of proportion and geometrical harmony, rather than elaboration of detail, all ornament introduced should be of the same character, mere geometrical outlines and forms, to the exclusion of classical decoration, the characteristic of which is an imitation of the organization of foliage.

"VII. White should be used in large quantities in all cases of simple compositions, not only to give value, by contrast, to the few colors employed, but to reflect light and cheerfulness to the work."

The appearance of the building on its exterior was a light-colored bronze or olive tint, with the purely ornamental features enriched by gilding. The ceilings and dome of the interior had the ground-work of a sky-blue, producing loftiness and airiness, the constructive framing being painted of a rich buff or cream color, harmonizing with the blue and throwing a cheerful tint of sunshine over the whole. These prevailing colors were relieved by the judicious use of the positive colors, red, blue, and yellow, in their several tints of vermilion, garnet, and orange, and in certain parts by gold.

The area covered by the first floor was 157,195 square feet, and by the galleries, 92,496 square feet, making a total floor space of 249,692 square feet, or about 5¾ acres, and the quantities of material used in the structure amounted to 300 tons of wrought iron, 1500 tons of cast, 55,000 square feet of glass, and 750,000 feet, board-measure, of timber.

Marochetti's Statue of General Washington.

We give an exterior and also an interior view of the building, which has now passed away from sight forever, having been entirely destroyed by fire in 1858.

We also present an engraving of the Equestrian Statue of Washington by Baron Marochetti, the largest work shown at this exhibition, and located in a prominent position immediately under the dome. The artist was an Italian sculptor of note, born in Turin, in 1805, long resident in France, and who died in 1867, in London, where he had removed on the outbreak of the French Revolution, in 1848. From the criticisms made on this statue at the time, we should judge that its merit lay only in its size, being two and a half times that of life, and that it was lacking in all the fine attributes of a first-class work of art.

In the Mechanical Department, the exhibits of the United States were, as might have been expected, exceedingly creditable. The high price paid for labor in this country has necessitated the invention of machinery to supersede it, to a much greater extent than in foreign countries, and the result of this is always apparent,—our machines, as a general thing, being more numerous, of better quality, and more varied in their application than those from abroad. The sewing machine was comparatively a new invention at this time, there being in the exhibition of 1851 only three,—one from France, for sewing sacks, one from America, and one from England. At this exhibition of 1853, there were not less than ten varieties by American inventors alone; some using a double and some a single thread, and some adapted to special purposes, as for sewing cloth, leather, etc.

The United States Coast Survey Department made an exhibit of its various instruments, and showed the results of its labors by means of maps, charts, etc., evincing the great progress and honorable position which this country had attained, even more than twenty years ago, in this work.

Gas was supplied to the building primarily for policing purposes only; but it was afterwards arranged to open the building on certain evenings to visitors, and the effect of the interior, when fully illuminated, especially the dome, was exceedingly grand.

France, encouraged by the great success of the London exhibition of 1851,—regretting, perhaps, the opportunity which she lost in 1849, of setting the example of the international feature in exhibitions, and conscious that the exclusive or merely national system which she had previously adopted, would, if continued, be detrimental to the best interests of herself, and contrary to the national pride of her people,—determined to hold an International Exhibition in 1855.

While she had little to fear in the way of competition in those specialties for which she had so long been famous, she also knew that, by bringing before her people those productions of human skill more especially adapted to the *necessities* of mankind, and which heretofore had received so little attention in France, she would benefit her country immensely. The result would be that the French would either improve their own methods of production or make such arrangements by more extensive commercial relations as would insure future supply from those countries best adapted to furnish it.

The Emperor had determined, as early as March, 1852, upon the erection of a large permanent building in the great square of the Champs Elysées, for the purposes of national expositions, and also to be available for great public ceremonies and civil or military fêtes. This building, with temporary additions, it was decided to use for the Universal Exposition of 1855.

The site adopted was authorized by the prefect of the Seine to be given over to the State in July, 1852, and a public company was organized in August of the same year, with M. Ardoin at its head, as "concessionaires" for the erection of the building—the concession to last for thirty-five years, and the receipts from expositions to produce the return for the required outlay of capital.

The buildings for this exposition afford to us an excellent example of the manner in which the French undertook the construction of a permanent building in connection with a great international exposition, and might serve, in some respects, as a precedent for our Memorial Hall.

The first design was prepared by MM. Viel and Desjardins, but it was found to involve great expense in the construction, and an amount of work so immense that it could not possibly be completed by the time fixed upon for the opening. At last, in December 1852, a contract was entered into, by MM. York et Cie. with M. Ardoin et Cie., for the construction of a building—all the work except the decorative painting and sculpture to be completed by a fixed day for a fixed sum,—the contractors to be at liberty to make any alterations in the design they desired, under the conditions that no change was to be made either in "the dimensions, the solidity, or the artistic aspect of the building, considered as a national monument."

The contractors appointed M. Barrault, Chief Engineer to the Palace, and M. Cendrier, Architect to the Lyons Railway, to prepare the modified design, assisted by MM. Bridel and Villain. M. Viel, one of the authors of the original design, was given charge of the masonry.

Main Entrance, International Exhibition, Paris, 1855.

The adopted design was very similar, in general appearance, to the original of MM. Viel and Desjardins.

Although work was commenced immediately, it advanced but slowly—very little being accomplished before February, 1854, and the opening of the Exposition, which was to have been on the 1st of May, 1855, did not take place until the 15th of the same month.

The principal edifice, now known as the "Palais de l'Industrie," and still in use for national exhibition purposes, was a rectangular building, eight hundred and twenty feet long by three hundred and sixty feet wide, exclusive of the central and end projections, containing entrances and stairways, and covered eight acres of ground. It was built of stone, and quite ornamental in appearance,—the main exhibition hall being spanned by a central arched roof of one hundred and fifty-seven feet, with two side arches of seventy-eight and a half feet each, parallel to the centre one; and two of the same span running transversely at the ends, and beyond its gables. At the corners these latter connected by hips and ridges, leaving a clear space underneath. The covering of a large portion of these roofs (about one-third) was roughened glass, which, together with great defects in ventilation, appears to have been a serious mistake in the hot summer climate of Paris—great inconvenience being experienced in consequence, and it being necessary to resort to the expedient of muslin screens. We present an engraving of the front entrance on the Champs Elysées, which will give the reader an idea of the style of architecture adopted.

The structure, as a whole, was framed of iron, designed to stand by itself, without side-walls or anything except the base upon which it rested. The exterior walls were placed around this, being of ashlar masonry, designed in a simple, bold style, encasing and concealing the framed structure within, but having openings for the admission of light.

Our engraving does not give a complete idea of the building, as it comprises only the central entrance of about two hundred and fifty feet, whereas the total length, formed by extensions on each side of this, amounted to over nine hundred feet. The great central roof, although possessing some defects, was, at that period, the noblest specimen of arched roof that had yet been erected, excelling in magnitude, dignity, and true principles of construction. Although Great Britain had then some bold specimens of work, they would not admit of comparison with this.

Fergusson, in giving a criticism on this building, states that the greatest defect in the exhibition building of 1851 was its want of solidity, "and that appearance of permanence and durability indispensable to make it really architectural in the strict meaning of the word." He was of opinion that "the only mode of really overcoming this defect was, probably, by the introduction of a third material. Stone was not quite suitable for this purpose—being too solid and uniform," and "the designers of the Palais d'Industrie seem to have thought so also, as, instead of trying to amalgamate the two elements at their command, they were content to hide their crystal palace in an envelope of masonry, which would have served equally well for a picture gallery, a concert room, or even for a palace." "Nowhere was the internal arrangement of the building expressed or even suggested on the outside, and the consequence was that, however beautiful either of the parts might be separately, the design was a failure as a whole."

The other buildings attached to this exhibition were temporary in character, and were as follows:—a circular building, known as the panorama, in the rear of the permanent building, three hundred and thirty feet in diameter, and covering about two acres; an annexé for machinery, 4,000 feet long by 85 feet wide, covering $7\frac{8}{10}$ acres; and a palace for fine arts, located at a considerable distance from the permanent building and covering 4 acres. The total space covered, including the gallery floors, which we have not considered in giving the several areas, amounted to 29 acres, and the exterior ground devoted to exposition purposes to 6 acres additional, the entire space being greater than used in any previous exhibition. The Panorama, which was a pet of the Prince Napoleon and one of the most attractive spots of the exhibition, containing the exhibits of the products of the French Imperial manufactories, the "Buffet" being also established here, and was a circle of 165 feet diameter; and around this a circular gallery was constructed of timber, in three spans, roofed with sheet zinc and glazed with skylights, increasing the building to the total diameter of 300 feet previously given, and adding some 97,000 square feet to the exhibition in the short space of thirty days from the time that it was first decided upon. A covered passage connected it on the north with the Palais de l'Industrie, and on the south it communicated with the extensive machinery annexé by a covered lattice bridge of three spans, thrown over the Chaussée du Cours la Reine, covered with glass and approached at each end by grand flights of steps. The machinery annexé was built of timber and iron in combination, with masonry foundation,—the end portions of the building being solid blocks of timber, brick and plaster, and presenting quite an imposing appearance. The length of this building was entirely too great, compared with its span, to obtain any good interior effect.

Far greater prominence was given at this exhibition to the Fine Art Department than had ever been previously done. A special building for this purpose was isolated from all the others—as much for greater safety to its valuable contents from fire as by the necessity of the site—and it contained, in addition to a great hall for paintings of 462 by 198 feet, a distinct hall for sculpture of 215 by 72 feet, together with a refreshment department and the necessary store-rooms and offices. It was a timber structure covered with zinc and glass, and lighted from the roof, with an interior ceiling of glass which tempered the light, protected the works of art from leakage, and gave much better opportunities for ventilation than in previous arrangements. The hanging or wall surfaces were very much increased by numerous screens rising from the floor.

The number of exhibitors was nearly 21,000,—France contributing about one-half, and occupying 13½ acres, while Great Britain had 4¼; Germany, 1¾; Austria, 1¼; Belgium, 1; Switzerland, one-half acre; and the United States, one-third acre; the balance of the countries exhibiting decreasing to quite small spaces, and the Republic of Dominica having only two metres. The total cost of the buildings was about $3,373,300.

The Exhibition was closed by the Emperor in person, on the 15th of November, with considerable pomp and ceremonial, and with the distribution of the honors and awards, which were as follows:—for the Industrial Department, 112 grand medals of honor, 252 medals of honor, 2,300 medals of the first class, 3,900 of the second class, and 4,000 honorable mentions; and for the Fine Art Department, 40 decorations of the Legion of Honor, 16 medals of honor voted by the Jury, 67 medals of the first class, 87 of the second class, 77 of the third class, and 222 honorable mentions. The main central nave of the building was fitted up and arranged for the ceremony by removing all exhibits and placing a throne on one side, with a grand central platform, the remaining space being covered with seats rising one above the other, and forming—with the galleries—a vast amphitheatre from which the assembled multitude gazed down upon the gorgeous and exciting spectacle. With such a wonderful advance as shown by this exhibition from the small beginning of 1798, France might well be proud. Here, as before, were found the exquisite tapestries of the Gobelins and of Beauvais, improved and brought to the utmost perfection that art and science combined could make them,—the delicate tints so completely wrought and graded, each into its proper place, with so much mechanical dexterity and artistic skill, that it was difficult to decide whether the original or the copy was most to be admired,—the great softness and perfection of tone and color deciding in favor of the latter in almost every case. Also, here again, were exhibited the porcelain productions of the famous manufactory of Sèvres, excelling all competitors, and fairly astonishing the visitor with the capabilities of the material. The *chef-d'œuvre* was a vase commemorative of the great exhibition of 1851. It was Roman in form, ornamented with antique scrolls in white and gold in low relief, upon an Indian red ground. A collar or fillet supported the body upon a short shaft, which was broken by four masks representing Asia, Africa, Europe, and America; and the body itself was decorated with detached groups of figures proceeding from the back to the front, where Peace was represented as enthroned, with Plenty on one side and Justice on the other. The groups to the left were formed of figures symbolic of England and her colonies, Russia, the United States of America, and China; while those to the right represented France, Belgium, Austria, Prussia, Spain, Portugal, and Turkey. At the back, and dividing the groups, was a figure ingeniously posed in the attitude of sending them on their mission. Olive-leaves in bronze, with gilt fruit, decorated the upper curve of the body and neck, and the words "*Abondance*," "*Concorde*," "*Equité*," were inscribed above the whole.

Savonnière, also, was again represented by her carpets; but, although the work on them was extraordinary and, in one sense, perfection, yet the designs were wanting in adaptation to the true purposes for which carpets are intended,—having too much color, too large forms, and too much relief, or, in other words, not showing an improvement in taste which one would have been led to expect from the advance in other departments.

In the Agricultural Department, under the specialty of Reaping Machines, the United States was in the front rank,—exhibiting a number of very efficient machines. In the trials which were made, that of M'Cormick excelled all others from all countries,—performing the most work in the shortest time, and doing it in the most thorough manner, "evincing much greater perfection in its operations than any of the others whose powers were brought to the test."

In the Machinery Department, the Ribbon Saw—now so extensively used for scroll-sawing—was among the novelties.

The Paris Exhibition of 1855 differed from all previous ones in the "extent of its productions, the variety of its objects, and the facilities afforded for the disposal of the exhibited articles at a fair market-price,—conditions of great value to the exhibitors, in the immense selection submitted to view." It was really "an immense bazaar, from which might be selected every description of manufacture and almost every kind of produce."

"Nothing surprised the observer more forcibly than the beauty and the extent of the articles offered for inspection, and the great skill by which such vast and varied forms of manufacture were produced."

These exhibitions all produced their good results, and in a very marked degree. Fairbairn very truly says of the exhibitions of 1851 and 1855, that "they have shown to the world in every department of industry and of practical science, wherein consists the prosperity of nations, and the happiness of mankind. They have shown how all materials, whether derived from the forest, the field, or the mine, may be turned to purposes of utility; how the labor of man may be multiplied a thousand-fold; how the fruits of the earth may be cultivated and gathered in for man's necessities; and how works of art may be elaborated to increase the happiness and enjoyment of his existence." "All these things were exhibited on a scale commensurate with the greatness of an undertaking so vast in extent, so varied in form, and so characteristic of all the duties and wants of human existence, as to elicit the admiration and praise of astonished multitudes from every country of the civilized world."

In the year 1857, Manchester, England, held an exhibition of Fine Art and Fine Art Manufacture, more particularly confined to the Art Treasures of the United Kingdom,—plans being advertised for in May, 1856, with the conditions that the building must be fire-proof, must cover about 135,000 square feet, or a little over three acres, at a total cost of not more than $125,000, and must be capable of erection within six months.

The design and proposal of Messrs. C. D. Young & Co., of Edinburgh, constructors of corrugated iron buildings, was accepted for the sum of $122,500, the building to be completed by January 1st, 1857, under penalties for delays beyond the 15th of that month. An architect (Mr. Salomons) was appointed to confer with the contractors and modify the design in some respects, so as to improve the architectural effect, if possible, without material increase of cost, and the improved plan, of which we furnish a front elevation, was erected.

The building, in general plan, was a parallelogram, 700 by 200 feet, covered by five roofs running in the direction of the length of the building, the centre and two outer roofs being semicircular. The former was 56 feet span, and the two latter each 45 feet. The intermediate roofs were of the ordinary triangular construction, and each 24 feet span. A transept crossed

Industrial and Fine Art Exhibition, Manchester, 1857.

the building at a distance of 460 feet from the main-entrance end, consisting of a semicircular span of 56 feet, and two side spans of 24 feet each, exactly the same as the centre portion of the main roof, and forming a total width of 104 feet. The structure was supported by cast-iron columns, and the centre arch had a height of 65 feet, the two side arches 48 feet, and the intermediate spans 24 feet. The outside covering was corrugated iron, the sheets being fitted into wave-line recesses in the cast-iron columns without bolts or rivets, and the inner walls were of wood.

The walls and roofs were lined internally with boards, upon which was stretched muslin, and on the latter ornamental paper decoration was placed, the work being under the direction of Mr. Crace, of London. The side-walls of the great halls were a deep maroon, the paneled surface of the roof a warm grayish tint, the whole being relieved by lines and tracery of red and white, and the columns and metal work, bronze with rivet heads, etc., picked out in gold.

The sides of the ribs of the roof were decorated in vermilion on a soft cream-colored ground. The walls of the picture-galleries were of a sage green, with the roof a warm gray, and the border a cream color. The work was considered a remarkable success, combining great repose and beauty.

The facade of the building, up to the springing of the arches, was built of red and white brick, and the ends of the semicircular roofs above were filled in with ornamental work in wood, iron, and glass. Skylights, having an opening of about one-third the span, extended the whole length of each roof, and afforded a most excellent light, especially for the Fine Art Department, but the glass required screening with muslin during the summer months. It seems to be a great desideratum in all large picture-galleries to have the lighting so arranged that, by means of some sort of movable screen or velabrum, it may easily be increased or diminished as necessity requires. The quantity of light at our service varies so much at different periods of the year, and, indeed, at different times of the day, that it is almost impossible to do the lighting to perfection without some such arrangement,—a matter which, as in this case, is too often neglected. The interior effect of the central arched roof—which was constructed entirely open, without any ties or braces to interfere with the line of vision—was exceedingly light and elegant. The total floor space for exhibition purposes was increased by means of galleries, until it amounted to 171,000 square feet.

The Art Treasures included the works of the old masters—commencing with the oldest specimens that could be obtained—and were intended to show the gradual progress in Art from the earliest epoch, on through the periods of Titian, Correggio, and Rubens, up to the modern schools of Art, especially those of England.

Italy—with its principalities freed from the trammels and tyranny of a foreign yoke, and united into one grand nation—resolved upon holding an exhibition at Florence in 1861, for the purpose, perhaps, of inaugurating its new birth, and taking its place among the kingdoms of Europe. Previous exhibitions had been held in various parts of Italy—some at a remote period—but they partook more of the nature of agricultural exhibitions. There had also been one at Naples some years before, but this exhibition, now held, was far superior to any that preceded it, and forming, as it did, an exceedingly attractive display of Italian industrial, fine art, and agricultural products, it seems singular that it did not attract the attention from abroad that its importance deserved.

Exhibition, Florence, 1861.

The classification adopted was based upon that of London and Paris, but more simplified. It was divided into four great departments,—Industrial, Fine Arts, Agricultural, and Horticultural. The main building consisted of a rectangular front portion, built of masonry, as a permanent construction, with a great octagonal building in the rear, covering an interior garden. Into this main building the industrial and a portion of the fine art departments were placed, a detached building containing the balance of the latter. The agricultural department was accommodated in large temporary buildings, and the horticultural display took place in hot-houses and in the gardens surrounding the exhibition.

We present both exterior and interior views of the permanent portion of the main building. The display of the peculiar agricultural products of northern and central Italy was particularly rich, and the fine art collection could not have been otherwise than excellent.

It is to be hoped that Italy—once the centre of the arts and luxuries of the civilized world, and now again rap-

Interior of Exhibition, Florence, 1861.

idly taking her position—will, before long, give another exhibition, showing her progress since she has become united under one head, and—this time—in Rome, where the ancient and modern may be brought face to face, and the faded magnificence of the eternal city seen in contrast with the development and progress of industrial art of the present time.

The advantages which England had experienced from the Exhibition of 1851, had been very great. Before that time, very little had been accomplished in the department of art industry. In fine arts, such men as Reynolds, Gainsborough, Hogarth, Hayman, and Wilson, had achieved great reputation. Gibbons, Wedgewood and others had also been celebrated in their several specialties. These men were, however, all artists, working for themselves, not manufacturers, and their arts died with them. They promulgated no fixed principles and nothing was left to their successors. Art was not imbued into the masses. England was content with styles of art industry that would have shamed a South Sea Islander, and not only was she making no progress, but there was, at one time, an actual deterioration in public taste. She had discovered that, with the mechanical skill and the great producing capabilities which she possessed, she could rapidly accumulate wealth without taking time to attend to points of artistic design, and, in truth it may be said, that the term "industrial art" was really unknown in England before 1832.

Then she awoke to the fact that the artistic ability of France and the Continent was successfully competing with her mechanical superiority in the markets of the world, and she was obliged, in self-defence, to take measures to retain her supremacy. Art schools were accordingly established, and some efforts made to bring the productions of the country up towards the high standard to which the Continental manufacturers had already arrived. But the great difficulty with these schools was the want of practicability in their management. Those employed as teachers were artists, having sufficient influence to give them position, not practical men, and, if an appointment was to be made, it was always a question of blood, not brains. A student was taught not to think and study out a design for himself, but to copy from the designs of others; to reproduce from the French—which was considered the highest standard of taste—and not to originate. The result was an apathy and want of spirit, and, of course, a failure.

On the Continent, practical men were placed to teach practical subjects. Watchmakers were the professors in schools of design for watches. Men of the stamp of Quintin Metsys, who could execute as well as design, were the teachers; and, when in the Exhibition of 1851, England came face to face with the work of such men, the result showed her defects. She became aware that the course she had pursued was not the correct one, and she was even in a worse position, in some respects, than if she had never made any attempts—being obliged not only to commence at the beginning, but also to eradicate the false teaching which her artizans had already received. What was intended as a great display became, in fact, a great teacher, and the improvement in consequence was very marked. The schools of art were reconstructed and improved; a collection of art objects made by purchase from the exhibits of the great exhibition, forming the nucleus of the present Kensington Museum, and a strong progressive movement followed, producing great effects.

Among the direct results were reduced tariffs, increased postal facilities, and a vast increase of industrial prosperity, adding greatly to the commerce of the country. It was but natural, therefore, that England, conscious of the great advantages accruing from this Exhibition of 1851, and seeing also the good results of the French Exposition of 1855, and of her own local exhibitions, should desire, in time, a second great international exhibition; and this desire culminated in the London Exhibition of the Art Works of All Nations of 1862.

On the 14th of March, 1860, a Charter of Incorporation was issued by the Queen to Royal Commissioners for this exhibition, defining their duties and investing them with full powers,—the Prince Consort being made President of this Commission. It was decided, in anticipation, to test the popularity of the undertaking by public subscriptions, and a Guarantee Fund of $1,250,000 was formed with a rapidity beyond all expectation, allowing of the formal execution of the Guarantee Deed to the full amount by the 15th of March, the day after the incorporation of the Commission. This Guarantee Fund was afterwards signed by 1157 persons, in all, to the amount of $2,255,000, and upon this security, the Bank of England advanced $1,250,000 for the expenses of erecting the buildings and making the requisite preparations for the Exhibition.

At South Kensington, within a short distance of the site of the Exhibition of 1851, and at the south end of the new gardens of the Royal Horticultural Society, was a piece of ground belonging to the commissioners of the previous exhibition, and purchased with their surplus funds. This was selected as the location for the buildings of the present exhibition, and arrangements were made for its use. This location was quite favorable in some respects, and unfavorable in others. The new gardens of the Horticultural Society were finely situated, laid out with considerable dignity of style and in excellent taste, and formed a noble and attractive addition to the exhibition. It was imperative, however, in order to provide sufficient space, that the whole of the selected ground should be covered with buildings, and the result was that they were thrown out to the very verge of the street in front,—the street not being of very great width, and already built upon to a considerable extent on the other side,—so that no matter what the elevation of a building necessarily so long, it could never be seen to advantage, and no opinion could be formed of its proportions, whether good or bad. The approaches were also restricted, few, difficult and dangerous for the great multitudes which such an exhibition would draw together. In fact, the arguments opposed to the selection of this site seemed to preponderate very much over those in favor of it. Having determined upon the site, the Commissioners decided not to allow open competition for designs, and during a consideration of the propriety of permitting a limited competition, a plan was presented to their notice, designed for the site in question by Captain Fowke, of the Royal Engineers, an officer of great skill and experience, who had been in the British Department of the Paris Exhibition, and who had prepared this plan so as to meet many practical defects, found, in his opinion, to exist in the buildings of the exhibitions of 1851 and 1855. This plan met with so much favor that it was immediately adopted, and Captain Fowke appointed sole and responsible architect for the Exhibition buildings; the Commission thus passing over the whole engineering and architectural professions of the country, including those who had

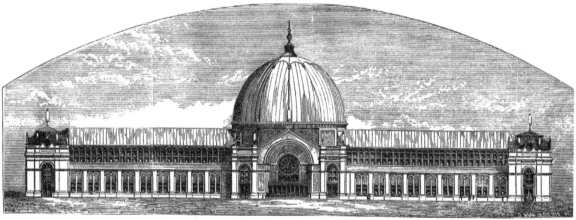

Front Elevation, International Exhibition, London, 1862.

been so honorably connected with the previous exhibition, and creating much jealous feeling and disappointment. The plans were somewhat modified, in order to keep the cost within certain figures fixed upon by the Commissioners: bids were received, and the work was let on the 23d of February, 1861. We present an exterior view of the building from the Albert Road, which will give the reader a very fair idea of its appearance. The design was severely criticised at the time; the frontage on the Cromwell road,* showing to the right on our picture, especially being condemned as featureless and ugly; and the Art Journal characterized the building as "the wretched *shed* that was the Fowke version of the Paxton Crystal Palace." But it must be remembered that the site was determined upon, and that the question of cost was fixed, precluding any expense beyond a certain amount. There seems also to have been

---

* Vide page xliii.

an intention of making a certain portion of the buildings so permanent that it could be finished up after the close of the exhibition as a national gallery of Fine Art. Any architect, under these conditions, would have worked to great disadvantage. And in reference to the front on the Cromwell road, it may be said that there would have been very little use in finishing it up expensively and artistically, as no one could see the building on this side, except in small portions at a time.

Designating the south face of the building on the Cromwell road as the main front, we may describe the building as follows:—

This main front occupied an extreme length of 1150 feet 9 inches, and a depth of 50 feet, and was constructed in brick, with a grand central entrance consisting of three arched openings. The wings on either side were built in two stories, the upper being used for picture-galleries; and the face walls were pierced with arched window openings, filled in on the lower story with glass, and on the upper with blank panels, so as to allow an uninterrupted wall-space in the interior for pictures. At the ends of these wings, as will be seen by the perspective view, were double corner towers. Passing into the central entrance, grand stairways led to the upper floor, where in the centre was a sculpture gallery, 150 feet in length, with entrances leading to the picture-galleries on either side.

These galleries possessed noble proportions and were effective and useful for their purposes. On the east and west sides, on Prince Albert and Exhibition roads, brick fronts extended north from the corner towers, each having a face of about 700 feet and a large central arched entrance, and really presenting a better appearance than any other portion of the building. The wings on the sides of the central arch were only 25 feet wide, and were built in two floors, the upper forming auxiliary picture-galleries, and the lower being used for offices, retiring rooms, etc.

The picture-galleries, all together, produced about 4600 feet lineal, or two acres superficial, of hanging space. A grand nave extended through between the central entrances on the east and west sides, 800 feet in length, 85 feet in width, and 100 feet in height from the floor to the ridge of the roof. At either end of this were large octagonal spaces 135 feet in diameter across the faces of the octagon, crowned by great duodecagonal glass domes 150 feet in diameter. We give a view of the interior of the great nave, looking west.

Two transepts crossed the nave at the domes, extending north and south, having the same width, height and manner of construction as the nave, and nearly 600 feet in length, right through. The nave and transepts had arched timber roofs, supported by double columns of iron. The domes rose to a height of 200 feet, with gilded finials 55 feet higher, and were constructed of wrought-iron framing, covered with glass. They presented a very light appearance, and were quite transparent when viewed from a near point of sight, showing the skeleton of the framing through the glass. The best view of them was that from some point a mile or two distant. Between the nave and the south front, and also on the north side of the nave up to the gardens of the Horticultural Society, the whole area was roofed over with glass and traversed with galleries.

Annexes, 200 feet in width, extended north for a distance of about 900 feet, on each side of the gardens, being prolongations, as it were, of the east and west fronts. That on the west front was devoted to machinery, and the one on the east to agriculture,—the latter having an open court in the centre. These annexes were of timber framing, very lightly constructed, the outside walls being of plaster on lathing, and the roof consisting of a series of four consecutive arches of 50 feet span each, boarded and covered with tarred and sanded felt. Each arch had a continuous glazed skylight for its whole length. A range of refreshment rooms was placed at the north end of the Horticultural Gardens, constructed over the arcades of the entrance, and connecting the ends of the two annexes. The view from here over the gardens was the most beautiful that the whole ground afforded.

The decoration of the building was placed in the hands of Mr. J. G. Crace, a gentleman of considerable reputation in his special art; the same who had decorated the Manchester

Interior View of the International Exhibition, London, 1862.

Exhibition Building, and who also had been specially selected by Sir Charles Barry to carry out the decorations of the Houses of Parliament. The work was completed in three months and gave, with one or two exceptional points, very general satisfaction. A light gray was adopted in the main portion of the building for the interior roof surface, and the timber framing marked out in colors more or less decided, each piece forming the polygonal rib, being painted in red or blue alternately, so arranged that in consecutive ribs, like sides of the polygon, were of different colors, and red showed against blue, or *vice versa*. It was intended, in taking a view of the roof, that these colors should mix and balance each other and produce a soft effect. The result was not as expected, and it would have been better to have painted the ribs of one uniform color. The sashes, and much of the wood-work on the sides below the roof, were of vellum color; the cast-iron work of columns and girders light bronze green; and the capitals of columns picked out red, blue and gold. The portions of the building below the arches were made quiet in color, so as not to interfere with the brilliancy and richness of the exhibits, while the vividness of coloring in the roof was intended to carry up, in some degree, the gaiety of the scene below.

The walls of the vestibule, stairways, etc., intended for sculpture, were colored in tints of maroon and quiet reds, with some green. Those of the picture-galleries were nearly all a subdued sage green, relieved along the cornices and string-mouldings by stenciled ornaments in a sort of cream or vellum color. Under the domes, the large supporting iron columns—nearly 100 feet in height—were a dark maroon, with the capitals gilt; and the panels between the arches and frieze were in shades of red, relieved by colored lines, the names of the four quarters of the globe being inserted in four of the compartments, with the initials of Victoria and Albert below. On the eight spandrils of the four main arches, medallions were placed, emblematic of manufactures, commerce and the various arts and sciences. The moulding of cornice and facia was of vellum color, relieved by gilding; the trusses gold-color, with the facia between them red, and the broad facia below, blue, and inscribed with scriptural sentences in gold letters. In the domes proper, the main ribs were painted bright red, with spaced black and white at the edges, and a fine gold line in the centre, spreading at intervals into lozenges and circles containing gilt stars on a blue ground. At the ring-plate above, the red was carried round, the points of intersection being painted black and white, and above that the eight main ribs were painted deep blue, relieved with red, gold and black, until they met in the centre pendant, which was gilt, bordered with red. The covering above was light blue with gilt rays diverging from the centre.

The domes of this building were by far the most costly part of its construction, and were thought by many to be quite a useless and unnecessary expense. The roof covering adopted was found much better than the glass covering of previous exhibitions, resulting in a much more equable and pleasant temperature in the interior.

The total area roofed in was 988,000 square feet—larger than that of any previous exhibition; but the total area of space, covered and uncovered, and available for exhibition purposes, was not as great as that of Paris, 1855; the proportions standing 1,023,000 in the present case, to 1,500,000 in the other. The total cost was not less than $2,150,000, equal to about $2.18 per square foot. Including the expenses of the exhibition, during the time it was open, the total amounted to $2,298,155, and the entire amount received by the Royal Commissioners amounted to precisely the same sum, making no loss or no gain,—the exhibition just self-sustaining and no more.

By great exertions, the exhibition was opened upon the day appointed,—the 1st of May, 1862. One great loss was felt in the death of Prince Albert, to whom so much was due for the favor and encouragement he had given to international exhibitions, and to whom they really owed their origin.

The contrast between the administration of the Exhibition of 1851, under his charge, and that of 1862, after he had been called away, was very marked; and of the great throngs who crowded into and around the building on that day of opening, not one but felt his absence. The Queen, of course, was not there, and although the ceremonies were very stately and imposing, a gloom was cast over the whole which nothing could entirely dispel. Apart from his royalty, Prince Albert was a very popular man,—endearing himself to the people by the active part he took in all industrial and art matters,—and hence the loss to the nation was felt all the more keenly.

The Queen was represented by the Duke of Cambridge, who received and replied to the address of the Commissioners, and to whom was handed the master-key, which opened all the different locks on the various doors of the exhibition building. After this, the grand orchestra, consisting of 400 instruments and 2000 voices, opened with a grand overture by Meyerbeer, followed by a chorale, composed by Sterndale Bennett, to the words of an ode written for the occasion by Tennyson, and then by Auber's "Grand March." After a prayer by the Bishop of London, Handel's choral hymns —the "Hallelujah" and "Amen," from the *Messiah*—followed, and the National Anthem was again sung in conclusion. The Duke of Cambridge then rose and proclaimed the exhibition open; a prolonged *fanfare* from the trumpets of the Life-Guards saluted the announcement, and the ceremony ended.

International Exhibition Building, London, 1862:—Showing Cromwell Road.

The display from the United States at this exhibition was very small—owing to the troubles at home—but what was exhibited, was very creditable, and—as in the Paris Exhibition—agricultural machines took a conspicuous position. McCormick's Reaper, with its self-raking attachment, was exhibited, and published as one of forty thousand made and sold from one establishment; and Russell's Screw-power Reaping Machine also attracted considerable attention. A very novel and ingenious invention—and one that received much notice—was the "Improved Cow-Milker," of Messrs. Kershaw & Colvin, of Philadelphia. Two machines for Boot and Shoe Stitching, invented by Mr. L. R. Blake, were remarkable for their simplicity of construction and efficiency and rapidity of production. Sewing-machines—which were novelties in 1851—had improved and increased in variety to a very great extent, and a large number of United States manufacture were exhibited.

Hoe & Company, of New York, exhibited their famous Printing Machines, by a model provided with ten impression-cylinders, as then used by the London *Times* and *Telegraph;* and the Composing and Distributing Machines of Mitchell were wonderful specimens of American ingenuity.

In the Machinery Department, Mr. Ramsbottom, of England, exhibited his admirable invention for supplying locomotive tenders with water while at full speed, now adopted in this country, and used with so much success for express trains on the Pennsylvania Railroad. It consists of a dip-pipe, or scoop, attached to the bottom of the tender, its upper end running into the upper part of the water-tank, and the lower end curved forward and dipping into water contained in a shallow, open trough lying longitudinally between the rails. The Giffard Injector—now in such universal use—was also among the new inventions at this exhibition.

A very efficient apparatus was a Folding, Pressing and Stitching Machine, from Switzerland, registering and folding sheets of paper with far greater precision than the most experienced hand-labor could do, at the rate of 1400 to 1500 sheets per hour, and at the same time pressing and stitching them.

Among the notable exhibits was Babbage's Calculating Machine, which could work quadrations and calculate logarithms up to seven places of figures, and, with the improvements of Schentz, of Stockholm, print its results. The Calculating Machine of M. Thomas —the Babbage of France—was also shown, dividing 16 figures by 8 figures in half a minute, or giving the square root of 13 figures in one minute, although not larger than a musical snuff-box.

The exhibits in reference to Electric Telegraphs, and electrical apparatus, showed a great advance in this department of science.

The steel exhibits were remarkably fine; Bessemer Steel, now so extensively employed for railway bars, then just coming into use; and the greatest progress was shown from the time of the previous exhibition.

The display of Chemicals was the finest that had ever been made,—far exceeding that of 1851. The Pharmaceutical Society, of London, exhibited a splendid collection of drugs.

The coal-tar dyes, then newly discovered, were among the most important of the exhibits. Aniline, but a few years previously so rare as to be known among chemists almost only by name, had now become an article of commerce, and a circular block about 20 inches high and 9 inches in diameter, was shown, which was the whole product of no less than 2000 tons of coal, and was sufficient to dye 300 miles of silk fabric. Those beautiful blue and purple dyes which are obtained from lichens were also exhibited.

The number, variety and beauty of the articles in Pottery was very great, although in the English department the designs of the ornamentation still showed a predominance of French ideas. The Majolica and Tile exhibits of Messrs. Maw & Co., and Messrs. Minton, were exceedingly fine. The majolica fountain of the latter—under the eastern dome—the largest exhibit of its class, and executed from designs of the sculptor Thomas, although a work of great expense, elegant, symmetrical and bold, and, so far as workmanship went, of great merit, was not considered a success, and fully exemplified the non-adaptability of the material to the purpose for which it was used, giving a lesson of warning what to avoid rather than what to copy. The Sèvres Porcelain exhibit maintained its standard of excellence, the leading feature of this display being the sea-green ware, or *céladon changeant*, which first appeared in the Paris Exhibition of 1855; a gray, dull sea-green as a body-color, more like what one might expect to find in old oriental ware—more easily recognized than described—on which is penciled with a similar but white paste, designs of leaves and flowers, standing out in slight relief, as white upon a céladon ground. The céladon changeant is a variety which possesses the singular capability of reflecting local color.

England made a superb exhibit of Glassware, being first in quality of material and artistic development, and far outstripping Austria and France, which, in 1851, held the supremacy.

In Furniture, the advance made by England since 1851 was very marked, the designs departing from the French, or rococo renaissance, which had been the order of the day, and partaking of the Italian school, being much purer in tone, simplicity and taste, and showing greater progress than by any other nation.

In Metal Work, the progress had also been rapid, the British outstripping all competitors, and developing an inherent strength, artistically, as well as mechanically. M. Ducel, of Paris, exhibited some remarkable figure castings in iron. Works in the precious metals showed great advance, and in this department the French were far ahead of the English.

Among the Sculpture exhibits, we may mention Fuller's bronze statue of "The Castaway," representing a shipwrecked man—faint, bruised and exhausted—floating on a piece of wreck, raising himself up and holding his hand aloft as he makes a last desperate effort to attract assistance. It was a work of great merit, gaining for its author a high reputation.

The "Reading Girl," by Pietro Magin, of Milan,—which the writer had the pleasure of seeing at Milan, several years ago,—was another one of the gems of the exhibition. A girl of no decidedly idealized type, loosely draped, as if partly prepared to retire for the night, is seated on a common rush-bottomed chair, sideways, and reading a book, supported on its back. The position is so entirely free from affectation, and the attitude and expression so natural, that it appeals to the heart at once, and no one could fail to notice and appreciate it. Gibson exhibited a colored "Venus," a work of elaborate and exquisite execution, and exceeding beauty and refinement,—the coloring, by many, however, was considered a failure. It was not merely a tone given to the marble, but polychromatic, and too weak,—not approaching nature sufficiently to give human expression, and yet sufficiently tinted to take away the divine purity of the simple marble. Miss Hosmer exhibited her "Puck," and "Zenobia Captive;" and Powers, his "California."

The exhibition closed on November 1st, a day of fog and drizzling rain. There was a very large number of persons present, among them Prince Napoleon, the Duke of Cambridge, and many others of distinguished rank, but no special ceremonial took place, in the usual acceptance of the term. As an exhibition, its success was not equal to that of 1851, either in fitness of edifice, novelty of articles exhibited, or in financial results.

Whatever may be said of the Emperor Napoleon III., all will admit that he systematically labored to advance the interests and promote the happiness of the people under him, continually engaging in projects for the development of the great natural resources of his empire; originating and giving an impulse to national industries, before unknown, and taking every opportunity of pleasing the inherent tastes of his people, and gratifying their pride by improving and adorning Paris, until it grew to be called the most beautiful city of the modern world—the very Heaven of the pleasure-seeker. In strict accordance with his expressed views, and with the characteristic features of his reign, he decided upon holding a great International Exhibition in Paris, in 1867, and on the 22d of June, 1863, an imperial decree was issued to this effect; the "Universal Exposition," as it was called, being intended to comprise typical examples of works of art, and of the industrial products of all countries, and to include every branch of human labor or skill. The invitation was extended to artists, manufacturers and workers of all nations, to take part in the Exposition, and it was expressly stated that the decree had been issued so early in order to afford all desiring to enter the Exposition ample time for mature consideration and reflection, and for arranging and carrying out the necessary preparations. This was followed by a second

International Exhibition and Grounds, Paris, 1867.

decree in February, 1865, confirming the previous one, explaining in full such details as had become at that time necessary, and defining the leading features of the proposed exhibition. An Imperial Commission was appointed, a Guarantee Fund provided, Commissions and

Committees formed—at home and abroad—and a comprehensive system of co-operation organized and brought into service. The Presidency of the Commission was confided to Prince Napoleon, the Emperor by this selection bearing high testimony to the importance which he attached to the success of the Exposition. Formal invitations were issued to Foreign Governments; and in reference to these, it was required as an absolute condition for the admission of any exhibitors from any country, that the government of such country should first accept the invitation extended to it, and assume the responsibility of forming the exhibition of its section.

In arranging the plan of the exhibition, two fundamental points were determined upon by the Commissioners: first, that a two-fold classification should be adopted, allowing the contributions from each country to be kept separately in one mass, while, at the same time, all the productions of a class from the various countries should be grouped together; and secondly, that the building should be so constructed, and of such ample dimensions that the whole display could be made upon the main floor, without the use of the galleries.

The site selected for the exhibition was the "Champs de Mars"—the same spot upon which was located the first French Exposition of 1798—a rectangle of 119 acres, to which was attached, also, the Island of Billancourt, affording an additional area of 52 acres, or 171 in all. The main building was located upon the former, and the latter was used for the Agricultural Department. An elliptical form of building was adopted, or, in reality, a rectangle with rounded ends; the length of the straight portion between the curved ends being 360 feet, the total length 1,608 feet, and the width, 1,247 feet. The total area within the outer limits of the building was $37\frac{8}{10}$ acres, and an open garden of $1\frac{1}{2}$ acres occupied the centre, reducing the amount under roof to $36\frac{3}{10}$ acres. The building was composed of a series of vast concentric oval compartments, each one story in height, the inner one encircling the centre garden as an open colonnade. The whole list of objects exhibited was divided into ten groups; of these, seven were provided for in the main building, a compartment being appropriated to each special group. There were, therefore, seven principal compartments; and the arrangement of area under roof was as follows, proceeding from the centre outwards:—

Promenade around centre garden	17	feet wide.
Gallery de l'Histoire du Travail	28	"
1. Gallery of Fine Arts	49	"
2. Corridor for the Liberal Arts	20	"
Passage-way	16	"
3. Corridor for Furniture	76	"
Passage-way	16	"
4. Corridor for Textile Fabrics	76	"
Passage-way	16	"
5. Corridor for Raw Materials	76	"
6. Gallery for Machines	115	"
Gallery for Restaurants	33	"

The spaces devoted to the different countries were arranged in a wedge-like form, radially from the centre of the building to the outer edge, and the visitor, by proceeding around one of the concentric oval departments, passed through the different countries exhibiting, one after the other, always keeping in the same group of subjects; but if he walked from the centre of the building outwards, radially, he traversed the different groups of the same country. The arrangement of double classification required was, therefore, by this plan, completely accomplished, and afforded great convenience and facility for study and comparison.

The area encircling the Industrial Palace—amounting to 81 acres—was divided into the Park and the Reserve Garden, and in the former, numerous structures, constructed by the different nationalities, grew up, in all varieties of style,—from the hut of the Esquimaux to the palace of a Sultan—the workmen or attendants at each being almost universally peculiar to the special country, and imparting additional interest to them. The Champs de Mars, in a short space of time, changed like magic from a dry and arid plain—useful only as a place for manœuvres of troops—to a charming Park, containing a city in the midst of groves and green lawns; a place such as the author of the "Thousand-and-one Nights" alone could have imagined—groups of buildings so violent in their contrasts as to produce harmony only by reason of their oddity, and leading the visitor to imagine that he had been transported to dream-land. Turkish and Egyptian palaces; mosques and temples of the Pharaohs; Roman, Norwegian and Danish dwellings by the side of Tyrolese chalets; here, a specimen of the Catacombs of Rome—there, a group of English cottages; workmen and farmers' dwellings; light-houses, theatres, a succession of hundreds of constructions, as unlike each other as possible; restaurants and cafés everywhere, for all classes of people; noises of all kinds filling the air; concerts, orchestras, the ringing of bells and the blowing off of steam-boilers; such was the Park of the Champs de Mars during the Exposition Universelle.

The Reserve Garden contained the botanical, horticultural and piscicultural collections. Nothing so charmed or rested the eye as the green lawns spread out so extensively before the visitor; nothing so picturesque as the chance glimpses of ground beyond, that intercepted the horizon; as the shrubbery, the grottoes, the cascades, the conservatories, some so grand, and others so *petite* and pretty. No one who saw the Exposition could forget all the beauties of this spot; the aquariums, the diorama, the pavilion de l'Impératrice, or, above all, the aristocratic restaurant of the Jardin réservé.

An iron coliseum grew up in the midst of all this, far exceeding in magnitude the ancient Coliseum of Rome itself, gathering beneath its roof nearly 50,000 exhibitors from all parts of the world.

Flowers, statuary and fountains adorned the open garden in the centre, and a central pavilion contained an exhibition of the weights, measures and moneys of all countries. The outer compartment of the building was the highest and broadest of all, having a width of 115 feet, and a height to top of roof of 81 feet. The roof was of corrugated iron, supported by iron columns; and along the centre of the whole length of the compartment was an elevated platform, carried upon iron pillars, and forming a promenade, at once safe and convenient, from which to view the machinery below.

The vast supply of water necessary for the use of the exhibition, for the display of the fountains, etc., was obtained from the Seine, and raised by means of powerful steam-pumps to a reservoir on high ground on the opposite bank of the river.

The Government surrendered the site to the Commissioners on the 28th of September, 1865; the first iron pillar was raised April 3d, 1866; and, although the building was not entirely completed by the time fixed—the 1st of April, 1867—the opening ceremonies, nevertheless, took place, as per appointment, with considerable pageantry.

The Emperor and Empress arrived at two o'clock in the afternoon, accompanied by the Ministers of State, the Prefect of the Seine and the Imperial Commission. Entering the Palace by the *Porte d'honneur*, facing the Bridge of Jéna, they traversed the grand Gallery of Machinery, commencing at the French Department and terminating at the English. They then passed through all the galleries, and having received the artists and authors of distinction in the Salon des Beaux-Arts, they visited the Imperial Pavilion, and resting a

INTERNATIONAL EXHIBITIONS.

while, then entered their carriage and departed, amidst vociferous acclamations from the assembled multitude, and the Exhibition was open to the world.

The day was perfect, and everything combined to make the opening a success; the bright sun, the deep Italian blue of the sky, the varied and rich costumes of the multitudes, the gorgeous decorations, the oriflammes waving in the breeze, and the music from the orchestras floating through the air —all united to produce that elated, happy, contented feeling which one experiences at times—a true enjoyment — the struggles and toils of this world forgotten almost entirely in one real day of pleasure.

In two weeks' time everything was in order, and the exhibition had developed from its unfinished state into perfection, —an object of beauty and instruction to all who passed within its boundaries.

In passing through the exhibition, the first portion that attracted attention — after leaving the central garden—was the Gallery *de l'Histoire du travail*. This department was intended to exhibit the various phases through which each country had passed before arriving at the present era of civilization, and was a grand idea as a preface to the Exposition. It was exceedingly interesting, although not as complete as it might have been, and not carrying the connecting links quite up to the present date. The French Department was the most perfect, being divided up into a series of halls, or apartments, to represent the different periods. The first hall represented the Stone Age, and here one found the collections from the lake-dwellings of Switzerland, the bone-caverns and the peat-bogs. Next came the relics of the Bronze

Period—objects of ornament and utility, bracelets, agricultural implements, etc., extending down to the Gallo-Roman. Following, were the relics of the Celtic and Gallic races; the works of the Middle Ages, seals, caskets, croziers and illuminated missals; and after that came the Renaissance Period, embracing curious locks, spherical watches and a handsome exhibit of the enamels of Limoges, from the collection of Baron Rothschild. In the sixth hall were productions of the seventeenth and eighteenth centuries. In the contributions from other countries were some very curious articles—the cradle of Charles XII, of Sweden, fine collections of ancient arms and armor, etc.

The Department of Fine Arts—which occupied the next gallery—was one of great interest both to artists and amateurs, the different nations having almost universally furnished the best productions of their most eminent artists in both painting and sculpture. Some countries were very much crowded in the space assigned to them, and erected special buildings—outside the main building—for their exhibits. The statuary, from all countries, was very much scattered through different parts of the building, and over all parts of the Champ de Mars.

In Paintings, the French were well represented by Gérome, Meissonier, Corot, Cabanel, Hamon, Yvon, in his "Taking of the Malakoff," Rosa Bonheur, Fromentin and others. Among the *genre* subjects, Plassan, Fichel, Toulmouche and Welter were represented by some exquisite pictures. The Belgian exhibit—a very fine collection—was outside, and consisted of contributions by Leys, Stevens, Willems, Verlat, Clay and others. The government of Holland—also outside—exhibited 170 pictures, the artist Israels standing foremost in rank among the contributors, and distinguished by his delicacy of sentiment and simplicity of expression. The Belgian and Holland schools showed strong inclination towards the French, neglecting the styles of their ancestors, with the exception of Leys, who was the pre-Raphaelite prophet of the Netherlands. Switzerland and Bavaria also had their own buildings in the Park, and showed large exhibits.

It was a little singular that the exhibit from Italy—the cradle of art—consisting of fifty-one oil paintings, should have been scarcely above mediocrity. The collection from the United States was a very creditable one, the foundry scene of Weir being the best work of its kind in the Exposition. Bierstadt, Church, Kensett, Broughton, Huntingdon, Hart, Healy and others were well represented.

The influence of the French school was very apparent in all the Continental collections. The English and American pictures were quite different, showing much more character and individuality, the difference in system of study throwing the artist entirely on his own resources, and thereby bringing out his peculiar style, which, under the Continental method of teaching, might never develop.

The Mosaic Work, contributed by Russia from the atelier of Michael Chmielevski, of St. Petersburg, was the finest, by far, in the exhibition.

The exhibition of Sculpture showed the influence of the realistic school over the classical, the best artists availing themselves of the good points of both schools without binding themselves to either. The gem of the classical school was of American origin, "The Sleeping Fawn," by Miss Hosmer. One of the most striking statues of the realistic school was "The Last Days of Napoleon I," contributed by an Italian, who received a gold prize for his work.

Passing on to the corridor for the Liberal Arts, one came into contact with books and printing, paper and stationery, lithography, photography, musical instruments of all kinds, medical and surgical apparatus, appliances for teaching science, mathematical instruments, maps and geographical and cosmographical apparatus.

Among the Photographic exhibits was a fine series of views of the Yosemite Valley, by E.

Watkins, of San Francisco; also, Rutherford's photographs of the moon and the solar spectrum, attracting great interest from the savans, and receiving a silver medal.

Among the Musical Instruments, Steinway & Sons, of New York, and Chickering & Sons, of Boston, were considered as having the best pianos in the Exposition, and although the Jury of Awards had only four gold medals to award to this class, they each received a gold medal, and the fact of two going to America, under the circumstances, was a great honor. Mason & Hamlin's cabinet organs were objects of great interest on account of their superior workmanship and singularly pure tone, and received a silver medal.

The exhibition of Surgical Instruments made by the Surgeon-General of the United States was very complete and interesting, consisting of ambulances, medicine wagons,

field-hospitals, artificial limbs, and every species of apparatus which had been invented or improved by the exigences of our late war. A very ingenious orrery was exhibited from the United States, showing the planetary system in a very exact manner, not only giving the rotation of the earth round the sun, but at the same time that of the moon around the earth.

In passing through the gallery devoted to Furniture, one could not fail to notice the great degree of perfection to which the industries here represented had arrived.

The French glass works of Baccarat and the Compagnie des Cristalleries de St. Louis, and those of England and Venice; the Italian *faience*, the art bronzes of Paris, the productions of Sèvres, of Beauvais, and the Gobelins, the pottery, goldsmith-ware, cutlery, perfumery and other celebrated articles of Paris, and numerous other specialties of acknowledged merit, were here all displayed in profusion. The exhibit of English white crystal Glass was far finer than at any previous exhibition, showing a remarkable advance since the Exhibition of 1862, and distinguished for its purity and brilliancy of color. The French displayed an immense variety of colored, gilded and painted glass; but the white glass, when compared with the English, had a clouded and gray appearance, owing to a far less quantity of lead being used in its composition. Baccarat exhibited some effects in decoration, produced by giving to crystal glass a deep-colored surface, and then etching on this a design to different depths, producing different shades of color down to the clear, white glass itself. The effect was excellent, and the process evinced great capabilities. The most remarkable exhibit of Austrian glass was that of Lobmeyr, of

INTERNATIONAL EXHIBITIONS.

Vienna, the designs being in perfect taste, and the material first-class. The Bohemian glass was superb; the decorations in gold, especially those in raised gold, without an equal in either execution or artistic effect; and gilding and coloring were applied in such a way as not to be at variance with either the material or the purpose for which the article was intended to be used. Dr. Salviati, of Venice, showed some wonderful specimens of modern glass manufacture, inaugurating a revival of the glories of the old Venetian glass, and imitating the peculiarities of that production, such as gold metallic particles floating in the material, thread work, dainty touches of color, etc., in such perfection as to attract the attention of all lovers of art work.

Pottery stands among the earliest of art manufactures, and in none has there been less change; the finest designs of the present day being of the same forms as in use two thousand years ago. Taking a material possessing primarily less value than almost any other used in the arts, the manufacturer, by the exercise of labor, skill and taste, produces forms ministering greatly to the necessities of man, and often of untold value, ranging from objects of everyday use to the porcelain of Sèvres. We engrave on page xlix a vase produced from the Imperial manufactory of Sèvres, a beautiful work of art and an excellent specimen of the gems which are created in that school of pottery so creditable to the government which has established it.

In the display of Textile Fabrics, carpets and tapestries occupied a prominent place. Carpets from Persia were more like shawls in their exceeding beauty of texture and the style and color of their designs; and in the French Department those of Savonnerie and the Gobelins still held their own against all competitors. The Imperial manufactories of the Gobelins and of Beauvais had on exhibition exquisite specimens of tapestry, and those of the different manufacturers of Aubusson were of the highest merit. Among the varied collection of table-covers were those of Philip Haas & Sons, of Vienna, the most eminent and extensive manufacturers of Tapestries, Carpets and Curtains in Austria, and we engrave on pages xlix and li some specimens of their work, which were of great elegance, and so much admired that one exhibit was almost hidden from view by the vast number of cards attached, on which were written orders for similar pieces of work.

Mr. Harry Emanuel, of London, exhibited in *répousée* silver, Tazze of Night and Morning, designed by the eminent artist Pairpoint, of which the engravings we give on page lii convey an excellent idea.

An exquisite dessert service in turquoise and gold was exhibited by Messrs. Goode of London, and manufactured for the Duchess of Hamilton—also Princess of Baden—and we show engravings on page li of parts of this service, on one of which will be seen represented the arms of the Duchess.

What has been designated by many as the best work of its kind in the exhibition was the famous Milton Shield of Messrs. Elkington, London, from a design by Morel Ladeuil, one of the grandest works of its class that had ever been produced, admirable in conception, and perfect in execution. We understand that this shield will form part of Messrs. Elkington's exhibit this year, and give an engraving of it on the following page.

In Bronzes, France—especially Paris—had at this time achieved the highest reputation, which was fully sustained in this exhibition, the French Bronze Court surpassing anything of the kind ever before seen, either in extent or variety. The admirable collection of M. Barbedienne stood unrivaled, being fine art work in every sense of the term, the use of various tints of bronze, and gilding and silvering where required, displaying great decorative and artistic taste. M. G. Servant, of Paris, also exhibited excellent specimens of bronze work, and the Boudoir Mirror we engrave on page lv, was one of his productions.

The display of Furniture proper was very extensive, and remarkable for great variety of

style, excellence of workmanship and rich diversity of material, coming from all quarters of the globe, and representing all peculiarities of taste. The English showed simplicity of treatment and improvement in design. The French was very lavish in ornamentation, the use of caryatides and uncouth human figures, and although perhaps pleasing the popular eye, was unquestionably degenerate in taste. The German was solid and heavy, and the Belgian bold and effective, but too naturalistic and unartistic in the ornamental work.

An ebony Cabinet, of great beauty, and a production of the very highest order of art manufacture, was exhibited by Herr Türpe, of Dresden, and is engraved on the next page. The bas-reliefs were of pear-wood, and the sculptured figures were the handiwork of a true artist.

Some charming works in Carved Wood were shown by Mr. G. A. Rogers, whose father, W. G. Rogers, had achieved a great reputation in this specialty. The design and carving of the specimen we show on page lvii were both by Mr. Rogers and exhibit the same pure feeling for which his father was so celebrated. Switzerland has attained great reputation for wood carving and none of her

contributors have a wider renown than MM. Wirth Bros., of Brienz, whose manufactures are true art productions, no two of them being ever exactly alike, and always the work of artists. On page lvii will be found several specimens of their work.

Mr. Charles J. Phillip, of Birmingham, one of the leading British manufacturers of ornamental gas-fixtures, exhibited fine specimens of his work, one of which we engrave on page lviii.

Passing on to the next gallery, we enter the Department of Textile Fabrics, comprising articles for clothing; goods in cotton, wool, silk, flax, hemp, etc.; materials and tissues collected together, from the most marvellous silks of Lyons to the cheapest cottonades; from the cashmere of the Indies — worked in gold — down to the merino scarf; from the robe of Alençon lace, or the point d'Angleterre, to the tulle which may be purchased for a few cents per yard. Here jewelry flashes in the

light, gleaming diamonds, emeralds, pearls and coral; there are displayed French artificial flowers so perfect as to excite even the jealousy of nature. In one portion of this department were life-size figures dressed to display the peculiar costumes of the various nations, those of Sweden and Norway being distinguished for their perfect execution.

The display of Lace and Embroidery was very profuse and beautiful. From the time of Marie dé Medici to the present day, nothing has been found to take the place of this costly fabric, lace; and nothing else can give to a lady's toilette the same finish and elegance. Its manufacture has attained great perfection in France, Belgium and England, and it is also made to a small extent in other parts of Europe, but not of so fine a quality. In Italy, the manufacture once so extensive, has degenerated, and the point lace of Venice and Genoa, so celebrated in the sixteenth and seventeenth centuries, has disappeared.

In France the principal varieties manufactured are the Point d'Alençon, the black lace of Normandy and the laces of Auvergne, of which Le Puy is the centre, and those of Lorraine at Mire-

court, with the light fabrics of Lille and Arras. The Normandy lace is made in the most perfection at Bayeux. MM. Lefébure, the eminent lace manufacturers of this place, exhibited some beautiful specimens of their work, of which we engrave part of a curtain on page lix, in the style of the the old Venetian point, of scroll pattern, with birds and flowers introduced.

In Belgium, which may be termed the "classic land of lace," the manufactories are

at Brussels, Mechlin, Valenciennes and Grammont. The especial lace of England is Honiton. Embroidery comes from Nancy, Switzerland and Saxony, and an important branch of industry in Switzerland is the fabric of net and muslin curtains, embroidered in crochet.

The display of Cashmere Shawls, both of Indian and French manufacture, was magnificent, showing great elegance of pattern and beauty of execution.

In Goldsmiths' Work and Jewelry, Froment-Meurice—whose father was styled the Cellini of France—exhibited beautiful specimens of work, and we engrave on page lx three examples of his ordinary every-day productions, which are always characterized by beauty, richness and great artistic taste. Some excellent and solidly-manufactured work was shown by Messrs. Tiffany, of New York.

The next corridor, adjoining, was that for Raw and Manufactured Materials, obtained directly from nature; products of the soil and mine; of the forests, and industries pertaining to the same; of the chase and fisheries; uncultivated products; agricultural products not used as food; chemical and pharmaceutical products—specimens of chemical processes for bleaching, dyeing, printing and dressing of textile fabrics; leather and skins. Here one found collections and specimens of minerals and metals of all kinds, from all countries; coal

and fuel of all sorts; rock-salt, sulphur, sponges, metal manufactures, stearine, soap, paints, wool, cotton, silk in the raw state, furs, tobacco, seeds, various varieties of wood, etc.

The Prussian salt-mines of Strassfurt were represented by a quantity of the salt cut into large blocks and built up into the form of a half-dome. Spain exhibited blocks of cinnabar from the famous mine of Almaden; and Russia displayed large vases and candelabras made from malachite, jasper and rhodonite; great varieties of rough and polished precious stones, models of meteorites, etc. Alibert exhibited remarkable specimens of graphite from his mines in Siberia, now in such extensive use for the celebrated Siberian pencils; and a mass of malachite weighing over two tons was shown from the mine of Prince Demidoff.

There was a large and creditable mineral exhibit from the United States; coal, iron, lead, copper from Lake Superior, quicksilver, silver and gold from Idaho and California, and emery from Massachusetts. The exhibit of wrought-iron, in all forms of manufacture, was very great; enormous plates, bars and girders; cast-steel from the Krupp Works of Essen, Prussia; ornamental castings, etc. The ornamental cast-iron productions of Durenne, of Paris, were particularly noticeable for beauty of design and excellence of work. We reproduce on page lxi a specimen of railing exhibited by him. None had greater renown in iron castings at that time than Barbezat & Co., of Paris, and many fine designs were exhibited by them, of which we engrave one, a street-lamp, on page lxii. Some specimens from the establishment of Count Dimeidel, in Prussian Silesia—the famous foundry of Lauchaumer—were art castings of a high order of merit, exquisite in design, and remarkably sharp and brilliant in finish. One of them, a stove, which excited universal admiration, we engrave on page lxiii.

The exhibit of Furs was very extensive and in great variety, ranging from the rarest kinds of sable down to the ordinary, cheap, glossy rabbit skins. France and Russia had fine assortments; and Messrs. Gunther, of New York, displayed some excellent specimens of North American furs.

The next gallery was that for Machines or Apparatus and Processes used in the Common Arts. This was the highest and largest gallery of the Exposition, and on entering it for the first time, the *coup d'œil* was certainly striking. Gigantic masses of manufactured metal articles, arranged in the form of

trophies, rose up on all sides, and a multitude of machines in motion, gave forth a thousand noises of all kinds, bewildering the mind and perplexing the ear. The elevated platform—passing around in the centre of the gallery—was a favorite promenade, and gave an excellent general view of the exhibits.

It would be impossible, in our limited space, to enter into any detail concerning the immense number of machines which had been brought together under one roof from all quarters of the world, and testifying to the inexhaustible inventive genius of man in its endeavors to supply the increasing wants of the age. We will, however, mention a few of the most important. Among the machines for drilling of rocks, the Diamond-pointed Drill occupied a prominent position, and now forms the basis of the most important machines of this kind in use. Traction engines were conspicuous in the English Department, and reapers and mowers in the American, the latter carrying off the prize in two trials made at the Emperor's farms at Vincennes and Touilleuse.

Under the head of Machine Tools, the principal exhibitors were France, England, Prussia and America, the novelty of form and excellence of workmanship of America being admitted to be equal to that of any other nation. The planing-machines, exhibited by Messrs. William Sellers & Co., of Philadelphia, were unsurpassed by any in the Exposition, and were remarkable for many novelties. Their screw-cutting machine was also of an entirely new character and an excellent tool. The display of Messrs. Bement & Dougherty in machine tools was first-class and showed many points of excellence. The lathes of Harris and the American Tool Company possessed several very interesting peculiarities.

The principal improvements which this Exhibition showed to have taken place in machine tools during the preceding twelve years may be mentioned as follows:—greater simplicity, perfection and solidity of construction, and more frequent adoption of automatic motions; better adaptation of form to the materials employed; increasing tendency to completion of products by mechanical means alone; adaptation of machines to more universal

use, allowing several operations to be performed on the same piece of material without dismounting it; construction of portable machinery; increase in rapidity of motion of the tools; and a general improvement in the execution of small tools; and greater simplicity in the means of transmitting motion.

Apparatus was shown for processes in carding, spinning, weaving and the preparation of textile fabrics generally. Sewing-machines, machines for shoemaking and for making of felt hats, were especially noticeable—an entire revolution in the machinery for the latter industry having been made within a few years. Machinery for furniture manufacture showed great improvement, and printing-machines of all varieties—for our daily morning paper, for lithographic work, for stamping of textile fabrics, for various kinds of printing and decoration on paper, etc.—were displayed in profusion.

There was a very interesting exhibit of Railway Apparatus, and thirty-two locomotives were exhibited; the Grant locomotive, from Paterson, New Jersey, attracting much attention from the general observer, owing to its exceedingly handsome appearance, being covered with polished brass and German silver, with ivory handles to the different cocks, and various other details of fine workmanship, which, by the more practical men, were considered out of place and not particularly adapted to actual service. American ingenuity and invention again occupied a prominent place in the exhibition of telegraphic apparatus and processes.

In Civil Engineering, Public Works and Architecture, the display of France was simply superb. There were handsome models—complete in every detail—of bridges, viaducts, reservoirs, docks, etc. In the Italian Department were plans and sections of the Mount Cenis Tunnel—then not completed;—and in the American Department plans were exhibited showing the method adopted, and now in use, for supplying the city of Chicago with water from Lake Michigan—a bold and most successful scheme of engineering. The Suez Canal exhibit was full of detail and of great interest.

The seventh, and outer gallery of the buiding, was devoted to food, either fresh or preserved; and in almost every instance, a restaurant was connected with each country, where the various foods could be practically tested. Visitors were waited upon by young girls in the costumes of the different nationalities, and one met here the blondes of Bavaria, the gay Austrian, the pretty Russian, crowned with a tinsel diadem, the Mulatto offering cocoa and guava, Greeks, Swiss, Neapolitans, Italians, Indians, and even the Chinese women, with their little tea shop. All languages mingled strangely together on this promenade, and all nationalities elbowed each other, from the elegant Parisian to the Bedouin in his burnous; and the animated aspect of the surroundings of the Exposition will always be remembered by those who were fortunate enough to see it.

Down on the banks of the Seine were displayed

models of all kinds of naval artillery, from enormous steel cannon for iron-clads, to little bronze pivot-guns for gunboats, and every specialty in reference to maritime affairs, pleasure and life-boats, yachts that were *chefs d'œuvre* of great beauty and elegance, gondolas, Egyptian caïques, painted, and gilded and manned by their Oriental crews, steamers, monitors, etc. A complete history of naval constructions was exhibited in a temporary building, by means of models in relief.

We have already spoken of the Agricultural annexé on the island of Billancourt, and this deparment was on a much more extensive scale than ever given before at any international exhibition, in fact, forming an exhibition of itself, presenting exhibits of all kinds of agricultural implements, and the finest breeds of live-stock—horses, cattle, sheep and other domestic animals—the exhibits being changed every fortnight, and making a succession of fourteen competitive exhibitions.

The distribution of prizes took place at the Palais of the Champs Elysées,—the permanent building which remained after the Exhibition of 1855,—on the 1st of July, and was accompanied by all the pomp and ceremony characteristic of the Empire. The building had been decorated for this occasion with great magnificence. The stage was hung with velvet, covered with gold bees, and surmounted by a gigantic imperial crown.

Down the centre of the nave were placed ten trophies, formed of the principal products in each department of the industries to which prizes were awarded. The glass roof was covered with white vellum striped with green and starred with gold, and from it hung ten banners bearing the colors corresponding to the ten groups into which the exhibits were divided. The columns of the gallery were decorated with the flags of the various nations represented at the exhibition. On the imperial platform were seated the Emperor and Empress and the Prince Imperial, accompanied by the grand dignitaries of the crown. Around their Majesties were the Sultan and three young princes of his family, the Prince Napoleon and the Princess Clotilde, the other members of the Imperial family, the Prince of Wales and Prince Arthur, of England, the Prince Royal of Prussia and various others of the royal visitors, including a brother of the Tycoon of Japan. The audience was composed of representatives from all nations, and numbered about seventeen thousand persons. At the moment of the entry of their Majesties, the orchestra executed the "Hymn to the Emperor," a work composed expressly for the occasion by Rossini. M. Rouher, Minister of State, then presented his report on the Exposition, and after an address by the Emperor, the names of the persons, the establishments and the localities to which were decreed the new order of awards for "Social Harmony," were read. This order of awards had been instituted by the Emperor in favor of persons, establishments, or localities where, by special institutions, good harmony had been promoted among those who carry on the same labors, and the material, moral and intellectual well-being had been thus secured among the operatives. These awards were ten prizes of one hundred thousand francs each and twenty honorable mentions. Following, were read the names of the exhibitors who had obtained the grand prizes for the groups of Beaux-Arts, Agriculture and Industry.

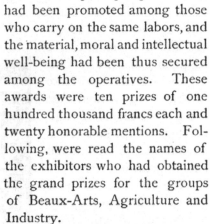

The awards granted by the juries of the Exposition were, sixty-four grand prizes, eight hundred and eighty-three gold medals, three thousand six hundred and fifty-three silver medals, five thousand five hundred and sixty five bronze medals, and five thousand eight hundred and one honorable mentions. The number of these awards is not surprising when it is recollected that the exhibitors numbered forty-five thousand, and that they were comprised of the élite of the artists and industrial workers of the entire world.

There were at this exhibition over twelve millions of entrance tickets recorded, representing at least four millions of different visitors. The total cost of the main exhibition building was $2,356,605, or $1.43 per square foot of surface covered. The total expenses of every kind from the commencement of the construction of the buildings—February 1st, 1865—to, and including the restoration of the Champs de Mars after the close of the exhibition, were $4,688,705, and the total receipts, including the subsidies from the government and from the city of Paris of $1,200,000 each, were $5,251,361, leaving a net profit of $562,654, of which dividends were declared of $553,200, and the balance of $9,456 was held for unforseen events and finally used for the public good.

During an interval of several years after the Paris Exposition of 1867, a number of minor local and general exhibitions were held in various places, among which we may mention that of the Central Union of the Fine Arts applied to Industry, in Paris, in the old Palais de l'Exposition, in 1869; an exhibition in Dublin, and also one at Leeds, the latter a purely fine art and loan exhibition, similar to the one held at Manchester in 1857. Exhibitions were also held at Copenhagen and Moscow, in 1872, and one of Domestic Economy in Paris, the same year. These exhibitions were all more or less of a local character, that at Copenhagen being confined to the products of Sweden, Norway and Denmark. The Moscow Exhibition, which was on a considerable scale, was held under the auspices of the Moscow Polytechnic Society, with the favor and protection of the government. It was too far distant to receive much attention from this country.

In England, a series of annual international exhibitions were organized in 1871, and held regularly afterwards, in a permanent building erected for the purpose at South Kensington, flanking the Royal Horticultural Gardens. These exhibitions were only moderate in size, but of special interest, great care being taken in the selection of exhibits, and the trade interests always set aside in favor of the encouragement of progress.

Awards have been given at these annual exhibitions with great judgment and discretion, very much enhancing their value, and the exhibitions have resulted in considerable benefit to England.

Austria, anxious to keep pace with the other great powers of Europe, had early had her attention drawn to the consideration of the subject of International Exhibitions, even previous to the time of the Paris Exhibition of 1867. Various causes, however, had combined to prevent any special action in the matter for several years, until the subject again came up in 1870. The city of Vienna within the last decade had changed from an old time town to a modern metropolis. The ancient fortifications had been taken away and replaced by the magnificent Ringstrasse. Inducements of every kind had been offered to those who would improve and embellish the city, and splendid buildings had grown up in all parts, especially along the Ringstrasse and its tributaries;

a noble opera-house had been built; a New Vienna had arisen and a time had arrived to display its glories to the world by devising an exhibition which it was proposed should outrival all previous efforts in this direction.

Active measures for an international exhibition to be held in Vienna in 1873, were first taken by the Trades' Union of the city, an organization of great opulence and influence, having Baron Wertheimer—a wealthy manufacturer—at its head. According to the original arrangement, a guarantee fund was formed of $1,500,000, and subscriptions to this amount were obtained—chiefly among members of the Society—it being supposed that the receipts from the exhibition would nearly, if not quite, meet the expenditures, and that this fund would cover all possible deficiencies. At this stage of the proceedings, the government was induced to give its patronage and support to the undertaking, and a decree was issued by the Emperor—May 24th, 1870,—announcing that "under the august patronage of His Imperial and Royal Majesty, the Emperor, an International Exhibition would be held at Vienna in the year 1873, having for its aim to represent the present state of modern civilization and the entire sphere of national economy, and to promote its further development and progress."

An Imperial Commission was formed with Archduke Charles Louis as Protector, Archduke Régnier, President, and Baron William von Schwarz-Senborn as Director-General; the total number of members being one hundred and seventy-five, and selected from the chief officers of the departments of the government, and from the leading men of science, art and industry in the empire. Money was appropriated by the government to the amount of $3,000,000 towards an exhibition fund, to which was added the guarantee fund previously obtained by private subscription, and all income from the exhibition itself. One-half of the amount furnished by the government was considered a regular appropriation, and the other half an advance made, without interest, and it was provided that if the total receipts from the exhibition and the government appropriation were not sufficient to cover the total expenses, the government would call in the guarantee fund. As the work progressed, it was found that the cost was greatly underestimated, and a supplemental grant of $3,000,000 additional was made by the government, although given under strong protest.

At no previous exhibition had so much interest been evinced by foreign governments, and their commissioners were chosen from their most talented and eminent men.

The site selected for the buildings was the Imperial Park—called the Prater—situated just outside of the city; as convenient a location as could possibly have been obtained, possessing within itself many attractions, and a favorite resort of all classes of citizens. On the north side flowed the Danube River, spreading out into numerous arms, some so

The Rotunda, Vienna Exhibition, 1873.

shallow as to be entirely unnavigable, others so full as to flood the flat country for miles around upon the least rise in the water. To the south lay the Donau Canal, a natural arm of the river, improved by art to a uniform width of one hundred and fifty feet, and the only available channel for navigation. Great improvements were in progress at this time, consisting in straightening and forming a new bed for the river nearly a thousand feet broad and one-half mile nearer the city, reclaiming land from floods and properly protecting the same by embankments, constructing docks, quays, warehouses, etc., and increasing the facilities for navigation and commerce in a marked degree. The work performed for the exhibition was expected to be of permanent value to the Danube improvement, and it was this, more than anything else, which induced the government to lend its aid to the enterprise. The Machinery Hall was intended to be used eventually as a freight or grain depot for the Great Northern Railway, and the grand rotunda of the main building was considered

Main Entrance, International Exhibition, Vienna, 1873.

the future corn market of the city. The total area of ground for exhibition purposes comprised within the surrounding fence was about 280 acres.

In arranging a method for grouping the exhibits, the double classification—as used in Paris in 1867—was not considered entirely satisfactory, and it was finally decided to adopt a purely geographical arrangement—each nation to be kept to itself—and no systematic classification to be recognized except such as might be obtained by providing separate buildings for specific purposes, and exemplified in the Machinery Hall, the Art Gallery, etc.

The principal buildings for the exhibition were the Palace of Industry, or main exhibition building, for miscellaneous manufactures, the Gallery of Fine Arts, the Machinery Hall and the Agricultural Building. In addition to these were various other buildings for minor purposes, similar to those distributed around the Main Exposition Building of Paris in the Champs de Mars. These were of unprecedented variety and importance, representing on a scale of great splendor and completeness the habits, manners, customs and methods of construction of various nations. At the Paris Exposition of 1867 this idea was first worked out as an international feature; here, it was on a still grander scale, and the rivalry of the nations of the Orient resulted in producing especial magnificence. The Palace of the Viceroy of Egypt was one of the most noticeable of these buildings. Designed by an Austrian architect long resident in the East, and constructed by native Egyptian workmen with great skill and truthfulness, it presented an appearance at once interesting and instructive. One saw here a sumptuous mosque, decorated in the richest manner, an ordinary dwelling-house, and then a regular farm and stable department stocked with dromedaries and other domestic animals of Egypt. Then there were also on the grounds specimens of the national habitations of Turkey, Persia, Morocco, Japan, Sweden, etc. Farmers' or peasants' homes from all countries, restaurants and refreshment saloons, the Imperial Pavilion, the Jury Pavilion, and special exhibits of all sorts, amounting in the aggregate to more than two thousand buildings, each one presenting something novel and pleasing.

The Palace of Industry was designed in the style of the Italian Renaissance, elaborately ornamented and finished on the exterior with that plaster-work which in Vienna has attained such perfection. It had for a main central feature a grand rotunda, covered by an immense

conical wrought-iron dome or roof of 354 feet in diameter, a *chef d'œuvre* of its designer, Mr. Scott Russell, of England, and the largest by far that had ever been constructed before, that of St. Peters, at Rome, being only 156 feet in diameter, and those of the London Exhibition of 1862 only 160 feet. It was supported upon 32 wrought-iron rectangular columns resting upon base-plates and founded upon concrete, and it was crowned by a central lantern of 101 feet in diameter and 30 feet high, provided with side-lights and a similar conical roof to the main dome. On top of this was another lantern 25 feet in diameter and 30 feet high, which was surmounted in turn by a gigantic copy of the crown of Austria, formed of wrought-iron plate, gilded, and decorated with glass imitations of the crown jewels.

Extending east and west from this central rotunda was a nave of 82 feet 10 inches in width and 22 feet 6 inches in height, with a total length from east to west through the rotunda of 2953 feet. A circular corridor or passage, half the width of the nave, ran all around the rotunda, connecting with the nave on both sides, and the columns carrying the dome, standing between this passage and the rotunda, were finished in ornamental plaster on wooden framing, with arches from one to the other, producing an exceedingly handsome effect. The floor of the rotunda was lower than that of the rest of the building, and in the centre was a highly ornamental fountain, adding very much to the general appearance. The interior of the conical roof was covered with canvas, stretched as a velarium over the whole of its under surface, divided into panels and decorated with colors in oil, each panel having painted on it in the centre an angel twenty-one feet long, and the whole of the interior work being elaborately picked out in gold and neutral colors.

There were cross-transepts, thirty-two in number, at intervals throughout the whole length of the nave, extending through both on the north and south sides, and having a length from face to face of 246 feet 3 inches. At the east and west ends of the nave the pairs of transepts adjoining were connected together next to their outer faces, and treated architecturally as one, producing an effective exterior appearance. The four transepts next to the circular passage around the rotunda, were also joined together by courts parallel to the nave, forming with these transepts a square of 676 feet exterior to the rotunda. The main entrance of which we give a view on page lxvi, was in the middle of the south side of this central square. It was designed like a grand triumphal arch, having a central arched opening, flanked on the sides by pairs of pilasters decorated between with niches, figure-subjects and medallions of the Emperor and Empress, and the whole crowned by a group of emblematic figures in plaster. The wings on the sides were arranged as arcades, and at the ends or corners of the square were small pavilions designed in the same general style although on a smaller scale, as the central entrance.

Concrete foundations were used under the permanent portions of the building, consisting

of the central dome and its surrounding courts, but the balance of the building was founded upon timber piles. The framing of the side walls of the nave and transepts consisted of vertical wrought-iron lattice columns of the lightest possible construction, standing on cast-iron foundation-plates, which rested upon the piles below. Upon these columns were fixed the trusses of the roof, consisting of segmental arches of the same lattice construction, connected by timber purlins covered with sheathing braids and zinc roofing-metal. The spaces on the sides of the building, between the vertical columns, were filled in with brickwork, plastered on both sides, the outer flanges of the columns being encased in the brick. The weight of the brick caused the outer foundations to settle more than the inner, consequently bulging the inner flanges of the columns out of position, which was remedied by fixing solid pieces of circular timber to them to stiffen them. These were finished with light wooden pedestals and mouldings, and plaster capitals painted to resemble bronze, the smooth portion of the columns being covered with crimson canvas ornamented with spiral lines of gold. Each transept was lighted by twenty-six windows of 11 x 14 feet each, and in the nave were five windows of 15 x 16 feet in each wall-space between the transepts, no skylights being used in any part of the building.

The iron-work of the interior was painted an olive-green, the wooden cornices a creamy gray color picked out with gold, and the under side of boarding of roof was calsomined.

The lower portion of the side walls under the windows was painted in panels of a light neutral green, and the parts between the windows covered with canvas in its natural color, on which was printed an arabesque pattern in dark blue and orange. The interior decorations were largely executed with colored canvas, the architect availing himself of an invention of an Italian—M. Bossi, of Milan—who discovered how to print patterns on canvas with great rapidity, producing, when put in position, all the effects of fresco at a very reasonable cost. This style of decoration was exceedingly gorgeous in appearance and accorded well with the tastes of the Vienna people.

The exterior effect of the temporary part of the edifice was not very striking. The plaster work was moulded and laid off in blocks to represent stone, and the general appearance was that of a long low line of gray buildings, broken at intervals by the transepts, the whole covered by the monotonous, arched zinc roof. The transepts were of much smaller dimensions than the nave, the crown of the roof coming just under the eaves of the roof of the nave, and in the end of each transept was a doorway surmounted by the coat-of-arms of the particular country exhibiting within. The grand central rotunda was a necessity, not only as a great hall for the opening and other ceremonies, but as the one redeeming feature in the architectural effect to relieve the tameness that would otherwise have been produced. After the construction of the building, many of the garden courts, between adjacent transepts, were covered over to provide additional room for the vast influx of exhibits.

In reference to the arrangement of the articles exhibited, the southern half of the central courts and a portion of the nave and eight transepts east of the centre were occupied by Austria. The other countries were arranged according to their geographical positions, east or west of Austria. Thus, Germany took the central courts north and west of the rotunda. Then, going west, came Holland, Belgium, France, etc., to the United States, which occupied the extreme west end; and on the east—next to Austria—were Hungary, Russia, etc., to Japan which occupied the extreme east end. Any one possessing a knowledge of geography could thus easily find the exhibits of any country he desired. The effect was to make a little exhibit in itself of the display from each nation, the whole being a continuation of a series of small exhibitions. The system adopted, however, made it extremely difficult to make comparisons of similar products from different countries, especially those at a distance geographically from each other.

To the east of the Palace of Industry was situated the Gallery of Fine Arts, entirely disconnected from it except by two galleries of communication. It was a building of brick, covered with cement and plaster so as to produce an ornamental appearance, and about 650 feet long by 115 feet wide. It proving too small to contain all the exhibits, two annexes were built, connected to it by covered passages, these passages containing works of sculpture. Western Europe, Austria, Hungary, Germany, America and Greece were accommodated in the Main Art Gallery; Italy and Northern Europe in the annexes. The arrangements for lighting were very successful and a great credit to the architect.

To the north of the Main Building and lying parallel to it, was the Machinery Hall, a building 2615 feet long and 164 feet wide, consisting of a nave about 92 feet wide and two side aisles of 28 feet width in the clear each, the balance of the total width being taken up by the walls, which were very heavy. The nave was used for machinery in motion and the side aisles for machinery at rest.

The Agricultural Department was divided into two separate buildings, occupying together

about 426,500 square feet. They were built of timber, upon pile foundations and answered their purposes very well.

Although the exhibition was far from being ready, yet it was opened at the time specified, at twelve, noon, May 1st, with great splendor, notwithstanding an unfavorable state of the weather. At the dawn of day immense crowds of people wended their way to the grounds, every street and alley leading to the Prater being thronged. By nine o'clock an uninterrupted string of carriages blocked the avenues, and many a man who desired to be present, and had spent the whole morning on the road, was obliged—notwithstanding the rain and wind—to leave his carriage and go on foot in order to reach the site in time for the opening. Thousands of people filled the enormous space under the dome, and precisely at the given hour, the coming of the Emperor was announced, and amid hymns from the United Musical Societies of Vienna and the acclamations of the people, he passed into the splendidly adorned entrance, escorted by the Director-General—Baron Swartz-Senborn—and accompanied by the Crown Princess of the German Empire. Following in the train came the Crown Prince of the German Empire and the Empress of Austria, then the Prince of Wales, the Crown Prince of Denmark,

the Duke of Flanders, and numerous other royal personages. The Grand Duke, Carl Ludwig, as Protector, then addressed the Emperor and handed in his report of the undertaking. The Emperor replied, followed by music. The President-Minister and the Mayor of Vienna then addressed the Emperor, thanking him in the name of the people of Austria for the foundation of the Exhibition and the assistance extended by the government to the great work. A chant composed by Joseph Weiler and set to the "Song of Victory," in Handel's *Judas Maccabeus*, was then executed by the United Societies and the Exhibition was declared open.

In making a cursory review of the articles exhibited at Vienna, we may state that the display was the most extensive that had ever previously been made in any part of the world, and

the admirable way in which the exhibition had been carried out gave to it additional interest.

An examination of the departments of all the nations gave evidence of the rapid extension of the knowledge of practical art and science to all parts of the world, equalizing civilization, increasing the energy and creative power of mankind in general, and tending to ameliorate the condition of the human race.

In reference to the machinery exhibits, great improvements had been made since the exhibition of 1867. Germany came out in great force, and the American display, although much smaller than that of many other countries, was full of original ideas and devices. Messrs. Sharp, Stewart & Co., of the Atlas Works, England, and Messrs. William Sellers & Co., of Philadelphia. stood as the typical machine manufacturers of their respective nations, and made

most admirable displays. The American productions, generally, were noted for originality, the novelties being all improvements leading towards precision of work and saving of labor. In drills, America still took the lead, and the Sellers' drill-sharpening machine was a work of especial merit.

France made great displays through Deny and Arbey, of Paris, and the finest pair of marine engines, perhaps, ever produced by any country were those exhibited by Schneider & Co., of Paris. Switzerland exhibited a most remarkable lace-making machine capable of working a hundred needles, and an object of great attraction both to experts and the general public. Probably the finest and most beautiful heavy lathe was that of F. Zimmermann, of Buda-Pesth, in Hungary.

In those special and peculiar tools required in the manufacture of sewing-machines, revolvers, firearms of every variety and fine instruments of all kinds, two firms, those of Pratt & Whitney, of Hartford, and Brown & Sharpe, of Providence made unexcelled displays.

Messrs. Jones & Laughlin, of Pittsburgh, exhibited specimens of cold-rolled shafting that attracted universal attention. A tub or bucket-making machine, by Baxter D. Whitney, was one of the most interesting American exhibits, manufacturing a bucket complete in the short time of five minutes. In reference to the exhibit of Stationary Engines, one of the most noticeable facts was the great favor which the principle of the American Corliss Engine

seemed to have obtained in Europe, and the numerous imitations and modifications of it displayed.

Never before had there been so fine an exhibit of Agricultural Machinery made as at Vienna, and the display from Great Britain was very superior. The American Department consisted more particularly of reapers and mowers.

In Pottery and Porcelain Ware the display was remarkable, and no branch of art had shown so much improvement and the beneficial effects of international exhibitions as this did.

We illustrate on page lxvii two elaborate specimens of plates by the Messrs. Mintons, whose ceramic display was immense and in the highest style of art. A curious and interesting collection of Moorish pottery was exhibited by Dr. Maximilian Schmidl, Austro-Hungarian Consul in that country, showing the soft, friable pottery manufactured there in every different

style of decoration, from the refined moresque to the bizarre mixtures of green, yellow and blue enamels. Hans Macht, of Vienna, exhibited a beautiful little box in Limousine enamel, of which we engrave a side view on page lxviii. Some beautiful water-jars and mugs were exhibited by F. W. Merkelbach, which are shown on page lxix, the designs of which were considered remarkably fine. An enameled vase by Christofle & Co., of Paris, engraved on page lxx,—very graceful in form and beautifully ornamented with birds and flowers, was admired by all who saw it.

In reference to Ornamental Terra-Cotta for building and decorative purposes, the establishment of Herr Paul March, of Charlottenburg-by-Berlin, had no superior. His principal exhibit was a raised garden-alcove seat, the floor laid in encaustic tiles of the most har-

monious colors and tasteful designs, the seat and its back in glazed *faïence*, arranged in a semicircle and decorated with fruit and leaves in majolica, and on a low wall, terra-cotta columns of the most exquisite design, upon which was placed a wooden trellis for climbing plants.

Among the most remarkable of the Porcelain specialties from France were the decorative plaques shown by M. Léon Parvillée, a celebrated architect of Paris. The designs were made after the very best period of Moorish art, and M. Parvillée's reputation is so great in this respect that even Turkey itself has made use of his skill. The peculiarity in the enamels he uses is such that they will not run, however highly they are fired, and the result is that the outlines of the designs are preserved in all their beauty, producing almost the effect of cloisonné enamels. Japan made one of the most creditable, interesting and instructive displays of porcelain and pottery exhibited by any nation, and obtained many medals of award.

In the Department of Glassware, no previous exhibition ever made a display equal to this. Situated as Vienna is, with Hungary, Bohemia, Venice and Bavaria in proximity, it was but natural that all should strive to attain great excellence, and anticipation in this matter was not disappointed. France and Great Britain, perhaps owing to their greater distance from the scene of action, did not make the display that might have been expected of them, although what Great Britain did send was good. The exhibits of Mr. James Green and Messrs. Pellatt & Co., of London were unsurpassed. A superb chandelier, by the former, and a large ewer and wash-hand-basin, by the latter,—probably the largest piece of cut flint-glass ever manufactured in England—were among the specimens. Many of the designs exhibited gave evidence of the high position which Japanese art has gained within the last few years in the tastes of the European world; and some of the specimens designed in this style were exceedingly charming and artistic.

M. Constant-Valés, of Paris, exhibited imitation pearls so perfect as to deceive the eye completely, and for which he obtained a progress medal. MM. Regat & Sons, of Paris, also received a medal for their exquisite imitation gems.

In the Italian Section, the Venetian glass of Dr. Salviati was one of the greatest attractions of the Exhibition. By using the works of the old masters as models, studying by every means in his power to equal them, Salviati has, year by year, approached nearer and nearer to perfection.

Nothing approaches the Venetian glass in its creative fancy. Professor Archer, in his official report on Glass to the British Government, says: "The glass-blower of Venice, like a child blowing bubbles, throws them off with ease and rapidity, producing with every touch of his fingers new forms of beauty, which gladden his own eyes as much as the ever-differing rainbow hues of the child's soap-bubble. In everything appertaining to the blown, pinched and moulded glass of the Venetian artist there is an exuberance of fancy, and he conjures up forms always new, and always graceful and beautiful."

The greatest specialty of the Salviati Company was their mosaics, of which they exhibited some magnificent pieces. Tomassi e Gelsomini, of Venice, also displayed some beautiful glass cloths of spun glass and beads, resembling embroideries.

Among the German exhibits of glass, which, as a general thing, were not specially remarkable, we may mention those of H. Wentzel & Son, of Breslau, of which we engrave specimens on page lxxi.

The display of Austrian, Hungarian and Bohemian glass was immense; but the Bohemian glass, although very superior, was not equal to the Venetian, lacking the beautiful transparency of the material, and the artistic forms which may be produced.

We present from the Furniture Department an engraving—shown on page lxxii—of an exceedingly ornate grand-piano in ebony and gilt, after a design by Storcks, executed by Boesendorfer, in Vienna. Some chairs in stamped-leather work by B. Ludwig, of Vienna,—which we engrave on page lxxiii—were among the handsome exhibits. We also give an engraving on page lxxiv of a cabinet or case for hunting apparatus, of excellent design, executed in stained oak by H. Irmler, of Vienna, from drawings by C. Graff.

The display of Carpets was very great and varied. We engrave on page lxxv a design exhibited by Shuetz & Juet, of Wurzen, which shows great taste.

A beautiful flower-vase in gilt-bronze, executed by Hollenbach from a design by Claus, of

lxxviii   *HISTORICAL INTRODUCTION.*

Vienna, was among the exhibits, and we are glad to be able to give a picture of it, which is represented on page lxxvi.

We close the very few engravings of the exhibits which our limited space has allowed us

to present, by a representation—seen above—of an Album-cover, in enamel painting, in possession of the Grand Duke Rainer, the design for which was made by J. Storch and F. Laufberger, of Vienna, and fully explains itself.

There were five different medals awarded at Vienna:—

1. Medal for Fine Arts.
2. Medals for Good Taste.
3. Medals for Progress.
4. Medals for Co-operators.
5. Medals for Merit.

These medals were all of the same size and of bronze, bearing on the obverse the portrait of His Majesty, the Emperor, with the inscription, "Franz Joseph I, Kaiser von Oesterreich, Kœnig von Boehmen, etc. Apost. Kœnig von Ungarm;" and on the reverse side artistic emblems, varying with the different medals.

The announcement of awards was made August 18th, with very little ceremony. There were in all two thousand six hundred and two awards, as follows:—

421 Diplomas of Honor
3,024 Medals for Progress.
10,465 Diplomas of Honorable Mention.
8,800 Medals for Merit.
326 Medals for Good Taste.
978 Medals for Fine Arts.
1,988 Medals awarded to Workmen, etc.

The Society of Arts and Manufactures in Vienna also distributed, on the 27th of September, in the beautiful hall of "Gewerbevrein," in the presence of Arch-Duke Charles Louis, Arch-Duke Rainer, several Ministers of State, Baron Schwarz-Senborn and others, a number of silver medals to deserving foremen of all the countries represented at the Exhibition. There were one hundred and thirty-four silver medals, with diplomas, awarded, exercising a most excellent moral effect.

The Exhibition closed on November 2d, the total number of exhibitors being about seven thousand.

The total cost of buildings and accessories amounted to $7,850,000, and the total receipts for visitors, from the opening until the close, amounted to $1,283,648.78. There were considerable additions to the revenue from other sources—rents for space, concessions for various purposes and the sale of the buildings—but far from enough to cover the total cost and expenses, and a heavy deficit had to be met by the government. The Main Building itself, from its peculiar form and mode of construction, was unnecessarily expensive, and was not a success either in interior or exterior effect.

While the indirect benefits to Vienna and the rest of Austria may have been great, the direct result was a positive loss and a considerable disappointment.

In the United States, local exhibitions had been a common event for many years. The Franklin Institute, of Philadelphia—founded in 1824—early initiated a system of exhibitions for the purpose of promoting the Mechanic Arts, awarding medals and premiums to inventors, manufacturers and mechanics. Its first exhibition was held in Carpenters' Hall, in the autumn of 1824, attracting large crowds of people, and was attended with most fortunate results.

These exhibitions were continued, at intervals, for many years, increasing in public favor and usefulness. The last was held on the fiftieth anniversary of the Institute in 1874, in a building covering an area of two acres available space on the ground-floor, with a large cellar for storage, and a four-story corner building for offices. It was the largest exhibition ever held in Philadelphia, the profits added greatly to the revenues of the Institute, and in every respect it was a complete success.

The American Institute, of New York, has for many years held similar exhibitions with the most satisfactory results; and, of late years, both Cincinnati and Chicago have held annual Expositions of Industrial Art in large, permanent buildings erected for the purpose, resulting in great success, both financially and in regard to the advantages derived from them by the exhibitors.

THE INTERNATIONAL EXHIBITION, 1876.

RATHER more than two hundred and fifty years ago, a veteran navigator from the old world, in voyaging along the coasts of the then newly-discovered Western Hemisphere, drifted into a magnificent and hitherto unknown inlet, the exit of a noble river. The navigator was Henry Hudson—the inlet was Delaware Bay.

A few years later, the Dutch Government—at that time the great commercial nation of the age—perceiving the great advantages that might accrue by the ownership of this location, acquired the right to it by purchase, and incorporated a company for trading purposes, taking possession of the ground and erecting a stockade called Fort Nassau, at a place now known as Gloucester, on the east shore of the river, some three miles below the site of the present city of Philadelphia.

The banks of the river and bay were rapidly colonized, principally by Swedes and Dutch, each party claiming for its own government the land upon which it settled, and contentions continually took place between the two nationalities, until, finally, the whole west bank of the river passed under the control of the Dutch, who held possession of it until 1664, when it came under the jurisdiction of the English government, on articles of capitulation to Sir Robert Carr for his Royal Highness, the Duke of York, afterwards King James the Second. In 1672, by the fortunes of war, it again fell into possession of the Dutch, but only for a few months, when, by the terms of a treaty of peace between England and the States General, the country came back once more under British rule.

In the early part of the seventeenth century, a religious sect had arisen in England under the guidance of one George Fox, whose adherents were remarkable for their simplicity of manner and dress, great mildness and forbearance, fine moral nature, mutual charity, the love of God, and a deep attention to the inward motions and secret operations of the spirit. They were characterized by great disposition to peace and opposition to violence and warfare, and were in every way a veritable "Society of Friends." Suffering persecution in their own country, they desired rest and happiness on a foreign shore.

In the year 1680, a distinguished member of this fraternity, William Penn, whose father had been an admiral in the British Navy, petitioned King Charles the Second, in consideration of large public debts due his father by the Crown, to grant him from his possessions in the New World that tract of land now known as Pennsylvania, and bounded on the east by the Delaware River, including, therefore, the possessions of the Duke of York on the west shore, and already settled by Swedes and Dutch. Here he hoped to establish a settlement where the members of his society could obtain that peace which they were unable to procure at home.

The King granted the desired letters-patent in 1681, and the considerations under which

Jury-Pavilion.

the grant was given were "the commendable desire of William Penn to enlarge the British Empire and promote useful commodities; to reduce the savage natives by just and gentle manners to the love of civil society and Christian religion," together with "a regard to the memory and merits of his late father."

Penn—having obtained a release from the Duke of York of his previous claim upon the province—immediately despatched a small number of emigrants to take possession of the country, and the following year sailed himself, landing at New Castle, in Delaware, on the 24th of October, 1682. The original settlers—of which there were quite a number at various points along the coast, the Swedes predominating—received him with every manifestation of welcome, "judging that all conflicting pretensions to the soil would now cease," promising to "love, serve and obey him," and adding "that it was the best day they had ever seen." On the 4th of December he called an assembly at Upland (now Chester), and passed all the laws which had been agreed upon previously, and also others, the law concerning "Liberty of Conscience" being placed at the head of the list.

Philadelphia, the city of "Brotherly Love," was immediately laid out, and as the site selected was already in possession of the Swedes, an exchange was proposed and accepted

Court of Finance Building.

by them for other land in the vicinity. The plan, covering a space of twelve and a-half square miles, was afterwards considered on entirely too extensive a scale, and it underwent considerable modifications in 1701, reducing the area to two square miles and limiting the boundaries to the Delaware on the east, the Schuykill on the west, Vine Street on the north, and Cedar (now South) Street on the south. Beyond Cedar Street were the Swedish settlements, and some of their old landmarks remain to the present day, notably the old Swedes' Church, consecrated on the 2d day of July, 1700.

Time has proved that Penn was wiser than those who came after him, since, in less than two hundred years, the city has stretched out far beyond the limits imposed upon it in 1701, and now the thickly-inhabited portion alone occupies more than four times the space originally determined upon for its area by Penn.

The city grew and prospered under its friendly and liberal rule, and although it received accessions to its inhabitants from all countries and of all sects, yet the Quaker influence predominated, and gave that solid, steady tone to society and aversion to mere outward display for which Philadelphia was so famous, traces of which may be found to the present day. When the troubles arose with the mother-country—nearly a century after its foundation—the city took an active part in colonial affairs. It had at this time increased to a population of 28,000, contained nearly 5,500 dwellings, had an extensive commerce, and ranked as first among the cities of the Colonies. The first Continental Congress assembled here in 1774, holding its meetings in Carpenters' Hall, a building situated south of Chestnut Street, between Third and Fourth,—still standing and kept in excellent preservation by the Carpenters' Company, to whom it belongs.

During the Revolution, Congress continued to hold its meetings in Philadelphia with but few exceptions, and the Declaration of Independence was adopted here July 4th, 1776, and first read publicly from a stand in the State-House yard by John Nixon, July 8th, following. The old Independence Bell, cracked and out of use, is still preserved in the hall of the State-House, as a memento of the times when it "proclaimed liberty throughout the land, and to all the inhabitants thereof." In this place the present Constitution of the United States was adopted by the Convention which met for the purpose in May, 1787; the first President of the United States resided here; and on this spot Congress assembled for some ten years after the adoption of the Constitution, until the removal of the seat of government to Washington.

When, therefore, the Centennial Anniversary of this great Republic approached, and the success of its form of government had become no longer an experiment, even in the eyes of the old monarchies of Europe, but an established fact, it seemed expedient that some effort should be made to properly celebrate this great event,—this birthday of freedom. A hundred years ago this young nation had struggled for existence; now she has established her position as one of the great powers of the world. What more fitting, then, than that she should commemorate this centennial of her life by an International Exhibition of Arts, Manufactures and Products of the Soil and Mine.

Inviting all the other principalities of the globe to unite with her in a competitive display, she could show for herself the greatest progress that had ever been made in the world's history in the same length of time,—an advancement without a parallel, fully entitling her to a foremost position among the nations of the earth. And what locality more eminently suitable for this celebration than Philadelphia? the birth-place of the nation, and the hallowed site of so many passages in her early history.

As the anniversary approached, the project was discussed in an informal way by many, and it only needed a move to start it into action. This initiatory move was taken by the Franklin Institute, the subject having been first brought forward at a regular meeting of the Board of Managers, held August 11th, 1869, and the discussion which followed led to the appointment of a special committee for the purpose of considering the question, and the advisability of memorializing Congress in regard to such an exhibition, to be held in Philadelphia in 1876 under the auspices of the Institute. At the next regular meeting of the Board, the month following, this committee reported, and stated that it did not consider it expedient for the Franklin Institute to place itself in the prominent position of patron to this enterprise, although at the same time it was of opinion that the Institute should use its utmost efforts to secure the proposed National Exhibition in the city of Philadelphia, and the committee also stated that "if such a celebration were combined with an exhibition of those arts and manufactures for which this country is so justly celebrated, and to which she owes so much of her material prosperity and greatness, there would be an additional reason for adopting this site, as no other city possessed such advantages as are afforded by the vast industrial works of Philadelphia." The action taken by the Board resulted in the appointment of a new committee to take the subject in charge, and this committee was instructed, on December 8th, to prepare a letter to the Select and Common Councils of the city of Philadelphia, explaining the action of the Institute and the reasons therefor, and requesting Councils to memorialize Congress on the subject. This letter, which was duly presented to each chamber of Councils through the mayor of the city, so clearly enunciated, even at this early date of the enterprise, the objects which are now being carried out, that we consider it worthy of reproduction in full, as follows:—

"The Franklin Institute of the State of Pennsylvania (the first founded of institutions of its kind in this country), being mindful of what may conduce to the credit and prosperity of the city of its location, has resolved through its Board of Managers that it will be expedient to celebrate the Centennial Anniversary of our national existence by an International Exhibition of Arts, Manufactures and Products of Soil and Mines, to be held upon grounds which, it is hoped, may be obtained within Fairmount Park for this purpose.

"It would seem eminently proper that such an Exhibition should be the form of celebration selected, and that this city should be the spot chosen by the nation for a national celebration at that time. There, was written and given to the world that Declaration which called our nation into existence; there, the laws which guided its infancy first took place; there, it began its march to benefit the human race. Under the laws there established, and in the nation there created, all arts and sciences have progressed in an unparalleled degree, and it is believed that the form of celebration indicated would be emblematic of their progress. The historical relations alone of our city should entitle it to selection for such a celebration; but apart from its claim as the birthplace of our Government, its geographical position, its railroads and navigation facilities, and its abundant means of accommodation for large numbers of strangers, all add to its claim and fitness to be selected for such a purpose.

"In consequence of these conditions the subscribers have been appointed a committee to bring the subject to your notice, and to request that your honorable bodies will memorialize Congress upon the subject for the purpose of obtaining that aid which will make such an Exhibition truly international in its character.

(Signed), WILLIAM SELLERS,
FREDERICK FRALEY,
ENOCH LEWIS,
COLEMAN SELLERS,
B. H. MOORE."

Entrance, Main Building.

The communication was received with favor by Councils and warmly supported, a committee of nine being appointed from each chamber to take charge of the matter, and to arrange for laying it before Congress. The question was also brought up before the State Legislature at Harrisburg, and similar action was taken, a committee being delegated from each House to act in conjunction with the committee of City Councils. These committees and also a special committee from the Franklin Institute, acting jointly, visited Washington and had an interview with the House Committee of Congress on Manufactures, presenting a memorial prepared for the occasion, which was favorably received, and

INTERNATIONAL EXHIBITION, 1876.

View of the Main Building from the Jury Pavilion.

a draft of an Act prepared and presented to Congress through the committee, resulting in the passage of the following Act of Congress, approved March 3d, 1871:—

"An Act to provide for celebrating the One Hundredth Anniversary of American Independence by holding an International Exhibition of Arts, Manufactures and Products of the Soil and Mine, in the City of Philadelphia and State of Pennsylvania, in the year eighteen hundred and seventy-six.

"Whereas, the Declaration of Independence of the United States of America was prepared, signed and promulgated in the year seventeen hundred and seventy-six in the city of Philadelphia; and whereas, it behooves the people of the United States to celebrate, by appropriate ceremonies, the Centennial Anniversary of this memorable and decisive event, which constituted the fourth day of July, Anno Domini seventeen hundred and seventy-six, the birthday of the nation; and whereas, it is deemed fitting that the completion of the first century of our national existence shall be commemorated by an exhibition of the national resources of the country and their development, and of its progress in those arts which benefit mankind in comparison with those of older nations; and whereas, no place is so appropriate for such an exhibition as the city in which occurred the event it is designed to commemorate; and whereas, as the exhibition should be a national celebration, in which the people of the whole country should participate, it should have the sanction of the Congress of the United States; therefore—

"SECTION 1. Be it enacted by the Senate and House of Representatives of the United States of America in Congress assembled, That an exhibition of American and foreign arts, products and manufactures shall be held under the auspices of the Government of the United States, in the city of Philadelphia, in the year eighteen hundred and seventy-six.

"SECTION 2. That a Commission, to consist of not more than one delegate from each State, and from each territory of the United States, whose functions shall continue until the close of the Exhibition, shall be constituted, whose duty it shall be to prepare and superintend the execution of a plan for holding the Exhibition; and, after conference with the authorities of the city of Philadelphia, to fix upon a suitable site within the corporate limits of the said city, where the Exhibition shall be held.

"SECTION 3. That the said Commissioners shall be appointed within one year from the passage of this Act by the President of the United States, on the nomination of the governors of the States and territories respectively.

"SECTION 4. That in the same manner there shall be appointed one Commissioner from each State and territory of the United States, who shall assume the place and perform the duties of such Commissioner and Commissioners as may be unable to attend the meetings of the Commission.

"SECTION 5. That the Commission shall hold its meetings in the city of Philadelphia, and that a majority of its members shall have full power to make all needful rules for its government.

"SECTION 6. That the Commission shall report to Congress, at the first session after its appointment, a suitable date for opening and for closing the Exhibition, a schedule of appropriate ceremonies for opening or dedicating the same, a plan or plans of the buildings, a complete plan for the reception and classification of articles intended for exhibition, the requisite custom-house regulations for the introduction into this country of the articles from foreign countries intended for exhibition, and such other matters as in their judgment may be important.

"SECTION 7. That no compensation for services shall be paid to the Commissioners or other officers provided by this Act, from the treasury of the United States; and the United States shall not be liable for any expenses attending such Exhibition, or by reason of the same.

"SECTION 8. That whenever the President shall be informed by the Governor of the State of Pennsylvania that provision has been made for the erection of suitable buildings for the purpose, and for the exclusive control by the Commission herein provided for, of the proposed Exhibition, the President shall, through the Department of State, make proclamation of the same, setting forth the time at which the Exhibition will open, and the place at which it will be held; and he shall communicate to the diplomatic representatives of all nations copies of the same, together with such regulations as may be adopted by the Commissioners, for publication in their respective countries."

The enterprise had now been placed upon a foundation, and the first really progressive step had been made in the work. The various Commissioners were in due time appointed, but no provision had been made to call them together until the city of Philadelphia, in October, 1871, issued an invitation for them to meet in March, 1872, and made an appropriation to cover their expenses. This invitation was accepted, and the first meeting of the Commission was held March 4th, 1872, at the Continental Hotel in Philadelphia. It continued in session until March 11th, a thorough organization being effected, the Hon. Joseph R. Hawley, of Connecticut, elected President, and the necessary special committees appointed and assigned their respective duties. The location for the proposed Exhibition was also fixed at Fairmount Park, and a Committee on Plans and Architecture instructed to report at the next session, and furnish sketches of plans for a building adapted to a double classification, similar to that of Paris, 1867, and to cover fifty acres of floor-space, estimates of cost to be furnished at the same time.

The second session commenced on May 22d following, and continued until May 29th. It was discovered that very little material progress could be made without pecuniary means, and that the first requisite was to take some measures for obtaining the funds required. This, as the special work of the Executive Committee, the Hon. D. J. Morrell being chairman, received most attentive consideration, and on the recommendation of this committee it was decided not to ask for National or State aid, but to rely upon the people, trusting that their patriotism, ability and will could be depended upon, under a proper and systematic business organization, to provide the money needed for the enterprise, and also to furnish that general support and coöperation so essential to secure nationality and success to the Centennial celebration.

To this end "It was concluded to apply to Congress for the charter of a corporation to be called the 'Centennial Board of Finance,' which should have power, under the direction of the Centennial Commission, to raise money upon the sale of stock, and to attend

Horticultural Hall.

to all duties necessary to bring the work of the Exhibition to a successful issue." The Act creating this corporation passed Congress and was approved June 1st, 1872. Its distinct purpose was declared to be that of raising funds for the preparation and conduct

INTERNATIONAL EXHIBITION, 1876.

Horticultural Hall.

of the Exhibition, and it was empowered to secure subscriptions of capital stock not exceeding ten millions of dollars, to be divided into shares of ten dollars each, the proceeds from the sale of this stock and from all other sources to be used for the erection of suitable buildings and fixtures, and for all other expenses required to carry out the Exhibition as designed. The Centennial Board of Finance to prepare the grounds and erect the buildings, all plans, however, to be previously adopted by the Centennial Commission, and the Commission also to fix and establish all rules or regulations governing rates for "entrance" and "admission" fees, or otherwise affecting the rights, privileges or interests of the exhibitors or the public. No grant conferring rights or privileges of any description connected with the grounds or buildings, or relating to the Exhibition, to be made without the consent of the Commission, which would have the power to control, change or revoke all such grants, and appoint all judges and examiners, and award all premiums. It was also provided that the Centennial Board of Finance should, as soon as practicable after the close of the Exhibition, convert its property into cash, and, after the payment of all liabilities, divide its remaining assets among the stockholders pro-rata, in full satisfaction and discharge of its capital stock.

At the close of the third session of the Commission, which took place in December, 1872, very little real progress had been made in the organization of the Centennial Board of Finance. The Executive Committee had been occupied in this work, and in publishing and issuing circulars and addresses to the people, informing them what had been done, and calling their attention to the mode of making stock subscriptions. A design for a seal was at this time adopted by the Commission, it being circular in shape, about two inches in diameter, with the official title, "The United States Centennial Commission," between inner and outer concentric circles, and in the centre a vignette view of Independence Hall as it appeared in 1776, and beneath the vignette the prophetic sentence, "Proclaim liberty throughout the land unto all the inhabitants thereof," which was cast on the Statehouse bell that rang out the first announcement of the adoption of the Declaration of Independence.

Nothing had been accomplished by the Committee on Plans and Architecture, as no funds were at its disposal, and nothing could be done without them. The Committee received instructions, however, to advertise for plans whenever the necessary funds could be obtained, expending a sum as they deemed best, not exceeding twenty thousand dollars.

To cover the incidental expenses of the Commission, Philadelphia appropriated the sum of fifty thousand dollars, and permanent offices were secured and books opened for subscriptions to stock in accordance with the provisions of the Act creating the Centennial Board of Finance. At the end of the year 1872, the prospects of the Exhibition looked very discouraging, many of those best qualified to give an opinion declaring its success exceedingly problematical, and some going so far as to say it was impossible. Fortunately, however, the Commission possessed an Executive Committee of great ability and tenacity of purpose, who were determined to give the matter a vigorous trial.

The Citizens' Centennial Finance Committee, which had been previously organized, and through whom had been obtained the above-mentioned appropriation of fifty thousand dollars from the city of Philadelphia, was given charge of the work, and under this Committee were placed sub-committees of the citizens of every trade, occupation, profession, and interest in Philadelphia, whose object was to obtain subscriptions to the Centennial stock. Every means was used to awaken the interest of the city, and through it, that of the country at large, and within sixty days all doubts were dispelled and success assured. The city of Philadelphia promptly subscribed one-half million of dollars, and the State of Pennsylvania one million, conditioned upon the subscription of the city, the whole to be appropriated for the erection of a permanent building in Fairmount Park, to remain perpetually as the property of the people of the State for their improvement and enjoyment, a depository of articles valuable either on account of association with important national events, or as illustrating the progress of civilization and the arts in this new country, and a worthy memorial of an event of which any nation might be proud.

On the 22d of February, 1873, an imposing mass-meeting was held with the most beneficial results, and before the time of the fourth session of the Commission in May, more than three millions of dollars had been subscribed, including the State and city donations; public interest had been aroused everywhere; information had been scattered by the press in all directions, and inquiries as to what was proposed to be done came pouring in from all quarters, and even from foreign countries. The question was taken up by the various States, a number of them strongly commending the project, promising their hearty coöperation, and issuing instructions to their members of Congress to support all measures requisite for making the Exhibition a success worthy of the nation and of the great men and events it was intended to commemorate.

It seemed especially desirable that full information should be obtained by the Commission in reference to the organization and working of the Vienna Exhibition then in progress, and the Executive Committee for this purpose sent abroad early in March one of their own members, Prof. W. P. Blake, a gentleman who had been principally in charge of the work of classification, and who was thoroughly conversant personally with all details of the Paris Exhibition of 1867. It was important also in connection with this that complete plans should be obtained of the buildings of the Vienna Exposition, and thorough data as to their mode of construction, adaptability to their purposes, &c., and that similar information should be procured concerning all previous great exhibitions. For this object Mr. Henry Pettit, an accomplished civil engineer, highly recommended, was appointed and sent abroad about the same time as a special agent. It may be mentioned in this connection that Mr. Pettit generously gave his services gratuitously to the Commission, with an allowance of only actual expenses.

In the winter of 1873 one of the most effective helps that the cause ever had was organized under the auspices of the Citizens' Centennial Finance Committee, in the shape of the "Women's Centennial Committee of Pennsylvania." Thirteen patriotic women, residents of Philadelphia, were appointed an executive committee and officially recognized on February 24th. Mrs. E. D. Gillespie, a lady of wonderful talent and administrative ability, a descendant of Franklin, was elected president, and continued to occupy that position, throughout the entire time of its organization, with marked skill and surprising success—the great work accomplished by Mrs. Gillespie and her zealous aids being one of the most prominent features in the history of the Centennial.

The organization of the Centennial Board of Finance was fully completed in April, 1873, a board of twenty-five directors being elected by the stockholders—John Welsh appointed president; William Sellers, first vice-president; and Thomas Cochran, temporary secretary; and the Exhibition work was fairly started upon a sound business footing, with a considerable capital already subscribed, a corps of officers of remarkable efficiency and ability and the highest standing, and every prospect of success. Mr. Frederick Fraley, of Philadelphia, a gentleman distinguished for his abilities and integrity, was afterwards regularly appointed secretary and treasurer, and continued to hold that position permanently.

Exhibition of English Rhododendrons.

Funds being now provided, invitations were issued on the first day of April of this year for preliminary designs for the Main Exhibition Building and Art Gallery. In order to induce any one who had an idea on the subject to bring it forward, so that the Commission could, if it wished, avail itself of every suggestion that might be offered, whether by a professional man or not, this invitation was made as broad as possible, and architects and others were requested to submit sketches and plans under an unlimited public competition. A detailed specification of what was desired was issued to competing parties, and it was requested that the designs be handed in before noon on the 15th day of July following.

It was during the fourth session of the Commission in May that the position of director-general was created, the Hon. Alfred T. Goshorn of Ohio being chosen at the annual election to fill the place.

The eventful public ceremony of this year was the formal transfer by the Park Commission, on the 4th of July, of the grounds which had been selected for the use of the Exhibition, at Fairmount Park, to the Centennial Commission. The ceremonial was performed in the presence of the various official dignities of the Government, the State

The Elevated Railway across the Ravine.

and the city, the members of the Centennial and Park Commissions, and numerous invited guests. After assembling at Independence Hall, and being formally presented to the mayor of the city, they were driven out to the Park, where a handsomely decorated stand had been erected on the site intended for Memorial Hall, and in front of which was a flagstaff, with a flag furled at the top and ready to be thrown to the breeze at the proper moment. Beyond lay the Lansdowne plateau, scattered over with crowds of people and troops.

After the ceremony had been opened with prayer by Bishop Simpson, the Hon. Morton McMichael, President of the Park Commission, delivered an eloquent address and made a formal transfer of the grounds to the United States Centennial Commission. President Hawley in accepting the transfer, replied by an able speech, closing as follows: "In token that the United States Centennial Commission now takes possession of these grounds for the purpose we have described, let the flag be unfurled and duly saluted." As the last words fell from the speaker's lips, the flag of the nation was thrown to the breeze and saluted by thirteen guns. Announcement was then made by Governor Hartranft, of the State of Pennsylvania, to the effect that in accordance with the conditions of the Act of Congress in relation to the Centennial Celebration, as sufficient provision had been made for the erection of suitable buildings for the purpose of the International Exhibition, he felt it his duty to certify the same to the President of the United States, and had forwarded him a certificate to that effect duly signed. The Hon. George M. Robeson, Secretary of the Navy, and delegated representative of the President of the United States, who was absent on account of the death of his father, then presented, in the President's name, a "Proclamation," announcing the holding of an International Exhibition in the city of Philadelphia in 1876, and commending the same to the people of the United States and to all nations who might be pleased to take part therein. The ceremonies were concluded by Secretary Robeson, who stated that "in making this proclamation the President desired to express his deep personal interest in the objects of the great enterprise, his sympathy with the patriotic endeavors being made, and his appreciation of the fitness of the place and the occasion designated, his earnest desire that 'all nations' would take part in this triumph of human industry and skill, on the great memorial occasion of a people whose energies are drawn from every land, and his hope and confidence that in its spirit and its success the 'Exhibition and Celebration' would remain a lasting illustration of peace and civilization, of domestic and international friendship and intercourse, and of the vitality of those great principles which lie at the foundation of human progress, and upon which depend our national strength, development and safety." The proclamation and a copy of the general regulations of the Commission were forwarded officially to each foreign Government and also to each minister of the United States accredited to a foreign Government.

In response to the invitation issued for plans, it was announced on July 16th that forty-three plans had been submitted. Of these, ten were selected as admitted to a second competition and worthy of the award of $1000 to each. The names of the successful competitors were made known on August 8th, and the conditions, requirements and awards of the second competition on August 11th, not differing materially from those of the first competition, which were still in force.

The second competition designs were put in on September 30th, and the awards upon them were decided about the end of October, as follows:—

Collins and Autenrieth,	1st award,	$4000
Samuel Sloan,	2d award,	$3000
John McArthur, Jr., & Joseph M. Wilson,	3d award,	$2000
H. A. & J. P. Sims,	4th award,	$1000

The Committee reported that all of these designs showed great care, skill and labor on the part of the several engineers and architects in carrying out the requirements of the specifications, each possessing so many points of excellence that the Committee was very much embarrassed in its efforts to arrive at a practical conclusion in the matter. It stated that many additional points of great importance had presented themselves in regard to the buildings, after the issue of the specifications for the second competition, which would necessitate more or less modification in any design adopted. In making the awards, however, the relative merits of the different designs were decided upon, solely with regard to their meeting the requirements stated in the specifications. This action was, of course, the only just one to the competing parties, but resulted in giving the awards to some designs which were radically different from what the Committee at the time of the award deemed it advisable to erect. No one of the designs, "in its judgment, could be considered as representing in an entirely satisfactory manner what was required for the Centennial buildings;" and the Committee, in examining the designs and considering the subject in all its bearings and requirements, came to the following conclusions: That it was not feasible to erect an Art Building and Memorial Hall as two distinct structures, but that the Memorial Hall should be built separate from the Main Exhibition Building, and used during the Exhibition for the purposes of an Art Gallery, a building covering one and a half acres of ground being ample for the requirements (the original specification required five acres of space in the Memorial Hall); that the Main Exhibition Building should be a temporary construction, covering at least thirty acres of ground, and capable of extension if required, rectangular in plan and without galleries, the interior arrangement to allow of vistas and attractive promenades, and in the construction the reduplication of parts to be an essential feature, iron and brick being largely used to secure against risk of fire, and the material to be worked up in such details of construction that it could be sold for fair prices after the Exhibition closed; vertical side-light to have preference to overhead-light; domes, towers and central massive features to be ignored as too ambitious and expensive, and the building to trust for its impressiveness to its great size and proper treatment of its elevations, and to its interior vistas and arrangements, and not to any central feature erected at great expense for only a few months. They also decided that there should be a separate building for a Machinery Hall, covering ten acres; one for the Agricultural Hall, covering five acres; and a Conservatory.

The Committee had a modified plan prepared for the Main Exhibition Building and presented for adoption, being an adaptation of a plan submitted by Messrs. Calvert Vaux and George Kent Radford, of New York, for the first competition, and to which no award was given in the second competition, owing to the requirements of the specifications not being complied with. This adaptation also embodied the principal idea presented in the design of Messrs. H. A. & J. P. Sims.

In reference to the Memorial Hall, the Committee stated that they "now entertained grave doubts as to whether the Centennial Commission had, or were even intended to have, any supervision over the plans or construction of the 'Permanent Centennial Exhibition Building,' or any interest in the manner of the expenditure of the appropriation made by the State and city." They considered this a matter for the State Centennial

Supervisors, and recommended that the plans for the Memorial Hall be transferred to them, with the suggestion that if they approved of a plan it should as nearly as possible conform to the requirements indicated by the Committee; and if they determined not to proceed with the construction of a "Permanent Centennial Exhibition Building," as provided, then the Committee would at once prepare and submit a design for an Art Gallery.

The plan for the Main Exhibition Building, as submitted November 6th, 1873, to the Executive Committee, was accepted and approved. At the same time the Board of State Centennial Supervisors communicated its desire that the plan for the Memorial Building should be prepared under the direction of the Commission, and upon this request the Director-General procured a design from Messrs. Collins and Autenrieth, of Philadelphia, which was submitted to the Executive Committee, December 17th. The Committee

Interior Agricultural Hall: Department of Brazil.

approved of the plan in its general features, but the estimated cost was in excess of the appropriation, and it recommended that it be erected only upon the condition of its cost being within the appropriated sum, and requested the Director-General to transmit the design to the Board of the State Centennial Supervisors, where it remained without farther action until the spring of 1874.

As soon as the Executive Committee had approved of the modified plans submitted for the Main Exhibition Building, they were placed in the hands of Messrs. Vaux and Radford, who were selected as the architects, for further elaboration and estimates, the results of which were given to the Committee. It was claimed by these results that Messrs. Vaux and Radford's system of construction throughout would be preferable, and not more expensive, than if combined with that of Messrs. Sims. This arrangement was

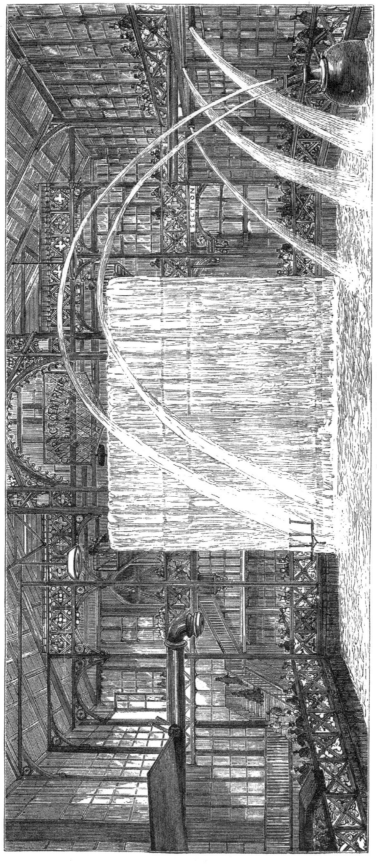

Cataract, Machinery Hall.

approved of, and the architects were instructed to obtain propositions from various iron firms for the furnishing and erection of the building in iron material, and also for the purchase and removal of the building after the close of the Exhibition.

About this time, the Secretary of State of the Government saw fit to give such construction to that portion of the Act incorporating the Centennial Commission which related to the participation of foreign nations at the Exhibition as would necessarily cause serious embarrassment, and probably entirely defeat the international features connected with it. His interpretation of the Act was, that while it stated "that an Exhibition of American and *foreign* arts, products, etc., shall be held," and instructed the President "to make proclamation, through the Department of State," and to communicate the proclamation and regulations of the Commissions "to the diplomatic representatives of all nations," yet it did not really authorize the government to invite anybody from abroad to attend; and he considered it necessary to issue special instructions to this effect, directing the diplomatic officers that they must confine themselves carefully to *commending* the celebration to all nations who might be pleased to take part therein, without *inviting* them to do so. That "with the exception that Congress created the Commission into a body corporate, and that the Commissioners were confirmed by the President, and that Congress authorized the proclamation made by the President and sympathized with the people in the success of the Exhibition, the national government had no connection with the Commission, no control over it, and was in no way responsible either for its management or its results." This interpretation was at entire variance with the understanding of the Commission on the subject and called for immediate action. A bill that would cover the whole question clearly and without doubt was at once prepared and introduced into Congress, passing the House almost unanimously, but meeting with delay and postponement in the Senate, until June 5th, 1874, when it was finally passed and approved, and the proper invitations were extended to the foreign governments. They met with a prompt response, and the international features were fully and firmly secured. In the autumn of 1873, that great financial panic, of which the effects are still seen, swept over the country, embarrassing all business operations and very seriously interfering with the procurement of subscriptions to Centennial stock. It was deemed, therefore, by the Executive Committee, of the utmost importance that pecuniary aid should be obtained from the Government. Every effort was made in this direction, a bill for the purpose being introduced into Congress on April 16th, 1874, but it failed to pass, and the Commission was obliged to place its dependence only upon voluntary subscriptions, which, up to May 1st, 1874, had amounted to $1,805,200, and the appropriations, which had been made by the State and municipal corporations, which were as follows:—

State of Pennsylvania, for permanent building,	$1,000,000
City of Philadelphia, " "	500,000
City of Philadelphia, for a conservatory,	200,000
City of Philadelphia, for a machinery hall,	800,000
State of New Jersey, conditional upon a sufficient sum being obtained from other sources to carry out the Exhibition,	100,000
Total municipal and State appropriations,	$2,600,000

It was, therefore, absolutely essential that the cost of the Main Building should be kept to a minimum. Acting on this, bids were received for the work in both wood and iron construction, and the excess in cost of iron precluded its use. Another plan was then prepared by the architects for wood protected partially by galvanized iron, the cost of which was found to be about $103,000 per acre. This plan was approved and handed over to the Board of Finance for execution, but the Building Committee refused to erect it on account of its combustible nature, and referred it back to the Executive Committee, who instructed Messrs. Vaux & Radford to re-design the structure in wrought iron, and by a reduction of the spans endeavor to keep within a more reasonable cost. The architects were unable, however, to get the cost below about $182,000 per acre, and attention was accordingly directed to the consideration of some more simple form of building than had as yet been presented. Two prominent manufacturers of iron constructions combining together then came forward and laid before the Committee plans and proposals upon two separate designs, one of which would cost $182,000 per acre—the same amount as for the architects' last plan—and the other $128,000 per acre. The latter was a simple shed construction of too monotonous and ordinary appearance to be acceptable, and the former was not considered so desirable as the architects' plan, although costing the same sum. The Executive Committee, therefore, approved of Messrs. Vaux & Radford's last design and transmitted the drawings to the Board of Finance, requesting that the work be placed under contract by May 15th, 1874, if possible. The Board, however, anxious to decrease the cost still further, obtained yet another plan from the architects, the cost of which was now reduced to $124,000 per acre. As successive efforts had resulted in successive reductions of cost, it seemed feasible to do still more in this direction, and it was decided that the building should in no event exceed in cost $100,000 per acre. Messrs. Vaux & Radford were then instructed to prepare new plans on this basis, and while these were being furnished Mr. Henry Pettit, the consulting engineer of the Commission, rcommended to the architects and advised the adoption of yet another modification of design for the building, embodying pavilions in the centre with wings of shed construction, allowing of any extension that the future wants of the Exhibition might make desirable. Messrs. Vaux & Radford not working up this idea satisfactorily, Mr. Pettit was requested to prepare plans and procure estimates at the same time as the architects. This he refused to do, as the Board of Finance already had a contract with Messrs. Vaux & Radford to prepare any plans they required, but he willingly offered to co-operate with these gentlemen in every way possible to further the work. The designs of the architects were three in number, as follows:—

No. 1. Pavilion plan throughout, with groined arch ribs in iron.

No. 2. A design consisting of three parallel galleries, each 150 feet span, with intermediate aisles, the roof of the 150 feet spans being flat arches with parallel extrados and intrados filled in with diagonal bracing, and the main tie-rod curved and supported from the arch by radial rods.

No. 3. Same as No. 2, except that straight, triangular roof trusses were used, the design being represented by a single tracing, and intended to embody the suggestions of Mr. Pettit.

Music Pavilion, Central Transept, Main Building.

INTERNATIONAL EXHIBITION, 1876.

Machinery Hall from the Jury Pavilion.

The Building Committee, after a full examination of these plans, again requested Mr. Pettit to work up a design according to his suggestions, and, under the circumstances, he could do nothing else than acquiesce. He accordingly furnished sketches and specifications which were designated as Design No. 4.

These four designs were presented to the public for bids, from June 17th to 25th, 1874. In comparing the amounts given by the lowest bidder for the several designs, there appeared to be a difference of only $2,824 between Nos. 2 and 4—in favor of the former—but the Committee decided that No. 4 possessed advantages over No. 2, which made it preferable even at the same price. The cost of Plan No. 1 was in excess of No. 4 by $520,733, and although the Committee was of opinion that the interior effect of No. 1 would be superior to that of No. 4, still, it felt that the great difference in cost would outweigh any advantages in this respect, and it therefore adopted Mr. Pettit's plan on June 30th, the Director-General giving his approval on July 4th, by order of the Executive Committee. This was the first design upon which the Board of Finance and Executive Committee both agreed, and was the final result of the successive efforts of many talented in their profession, developing step by step from the grand ideas of the original requirements to a practical basis which could be met by the resources at hand. All those who contributed towards the attainment of this end—be it more or less—are entitled to due credit for it.

The contract was awarded to the lowest bidder, Mr. Richard J. Dobbins, of Philadelphia for $1,076,000—exclusive of drainage, plumbing, decoration and painting—the area to be covered being eighteen acres; and Messrs. Vaux & Radford were authorized to proceed with the execution of the design. A professional issue arising, Messrs. Vaux & Radford declined to execute the work, and their contract with the Board of Finance was closed. Arrangements were then made with Mr. Henry Pettit and Mr. Joseph M. Wilson to act as joint engineers and architects to the Centennial Board of Finance, for the Main Exhibition Building and for the Machinery Hall.

Actual work commenced immediately, prospects became encouraging from this day forward, and it was soon evident that the space allowed for the Main Building was too little. It was therefore increased to twenty acres, and, at the same time, the central portion of the building was raised and towers added for exterior effect, the cost being increased to $1,420,000.

According to the agreement made with the contractor, it was provided that one wing of the building should be erected by September 1st, the other by October 1st, and the central portion by November 1st, the whole building to be completed by January 1st, 1876.

In reference to the Memorial Building, the designs as so far prepared by the selected architects did not appear satisfactory, considerably exceeding in cost the appropriations at command, and a plan presented by Mr. H. J. Schwarzmann, one of the engineers of Fairmount Park, was finally adopted, a contract being effected with Mr. Richard J. Dobbins on July 4, 1874, for the execution of the same, at a cost of $1,199,273, the sum being covered by the appropriations of the State of Pennsylvania and the city of Philadelphia.

Messrs. Pettit and Wilson proceeded at once under instructions to prepare a design for the Machinery Hall, which, being completed and adopted, was submitted to bidders, and the contract awarded to Mr. Philip Quigley, of Wilmington, Delaware, January 27th, 1875, for the sum of $542,300, including drainage, water-pipe, plumbing, etc., but exclusive of outside painting, the building to be finished by October 1st of the same year.

A design had already been prepared by Mr. Schwarzmann for a Conservatory Building, and bids being received, the contract fell to Mr. John Rice, of Philadelphia, for $253,937, exclusive of heating-apparatus, the papers being signed January 1st, 1875, and the work to be completed by September 15th.

Mr. James H. Windrim was selected as architect for the Agricultural Building, and his design being approved, the contract was awarded to Mr. Philip Quigley on June 16th, 1875, for the sum of $250,000, the work to be completed by January 1st, 1876.

The area covered by these buildings was as follows:—

Main Building,	21.47 acres.
Art Building,	1.50 "
Horticultural Building,	1.50 "
Machinery Building,	14.00 "
Agricultural Building,	10.15 "
Total,	48.62 acres.

Thus, by indefatigable perseverance on the part of the Board of Finance, the five principal buildings for the great Exhibition were at last fairly under way, and a most important step taken in advance towards a successful issue. The work proceeded rapidly, fully realizing all expectations, and with far greater speed than many even well versed in such matters deemed possible. Additional buildings soon began to spring up; the United States Government commenced the erection of a building, under Mr. Windrim as architect, for the collective exhibits from the different Government departments; offices were projected and started for the Executive departments of the Centennial Commission and the Board of Finance; State pavilions; buildings for special exhibits, etc., etc., began to dot the enclosure at point after point, increasing rapidly in number as the time for the opening of the Exhibition approached, and rivaling those of all previous Exhibitions, at least in multitude if not in architectural variety and national characteristics. A fence-line of some sixteen thousand lineal feet was constructed around the grounds, enclosing two hundred and thirty-six acres, this area being exclusive of the stock-grounds for the display of horses, cattle, sheep, etc., and located at another site. Walks and roads were laid out within the enclosure, comprising a total length of over seven miles; an artificial lake of water formed, covering an extent of three acres; fountains, statuary and vases erected, and shrubbery planted; a complete system of drainage designed and constructed for buildings and grounds, and the whole area so transformed, changed and beautified far beyond the already natural loveliness of the location as to be hardly recognizable even by those most familiar with it.

The necessity of including the Lansdowne and Belmont ravines within the Exhibition grounds required the construction of two bridges for the use of the public park roads, which were designed by and constructed under the direction of of Messrs. Pettit & Wilson. That over the Lansdowne ravine was of considerable engineering pretensions, and afforded an opportunity for quite an artistic construction. In order to secure an abundant supply of water entirely independent of the city department, temporary pumping-works were erected on the west bank of the Schuylkill River, a large and commodious brick building being constructed

Buildings of the British Commissioners.

and furnished with a Worthington steam-pump of a capacity of six million gallons of water per day, and an auxiliary pump of one million gallons additional. The necessary stand-pipe and a circulating system of pipes, amounting to about eight miles in total length, were provided, the designing and erection of the whole being under the care of Mr. Frederick Graff as Chief Engineer. Gas mains were laid out to the principal buildings from the city system, so as to afford the full supply desired.

As to transportation facilities, no previous Exhibition ever had so perfect arrangements. About three and a half miles of tracks were laid within the grounds to the several buildings, and there connected, by means of the Pennsylvania Railroad Company's lines, directly with the wharves on the Delaware and Schuylkill rivers, and with all the railroads entering the city, rendering no transhipments necessary except from vessels to cars.

Interior of British Commissioners' Building.

In the meantime the progress made by the Commission and the Board of Finance in their labors during the year 1875 was most satisfactory. The general classification as arranged was—

I. MINING, III. EDUCATION AND SCIENCE, V. MACHINERY,
II. MANUFACTURES, IV. ART, VI. AGRICULTURE,
VII. HORTICULTURE,

and the adaptation of this classification to the principal buildings placed the first, second and third departments in the Main Building, the fourth in the Art Gallery, the fifth in the

Machinery Building, the sixth in the Agricultural Building, and the seventh in the Horticultural Building. The public sentiment developed in favor of the Exhibition was such as to warrant the most liberal provision for its success, and the increased number of co-operative agencies established throughout the world tended greatly to overcome all difficulties.

The usual annual report required from the Commission by Congress was made to the President on January 20th, 1875, "setting forth the progress of the preparations for the Exhibition, and respectfully presenting the claims of the Commission for financial aid to properly execute their trust." Appropriations were asked for specific purposes, the expenses of which it was thought should rightly be borne by the Government, as follows:—

> For expenses of the United States Centennial Commission, . . . $400,000
> Awards and expenses incident thereto, ... 500,000
> Protection (police, etc.), ... 600,000
> $1,500,000

But Congress did nothing. It did make an appropriation, however, of $505,000 for the use of the Board representing the United States Executive Departments in preparing a collective exhibition, and the Board, having this appropriation, proceeded to the erection of a suitable Government building, previously mentioned, to contain the exhibits.

The Women's Centennial Executive Committee, under the able direction of Mrs. E. D. Gillespie, greatly enlarged its influence and usefulness, forming one of the most important volunteer organizations which had come to the aid of the Commission. It rendered exceedingly important service not only in procuring stock subscriptions, but in obtaining money by other means, and in awakening popular interest, performing a large share of the labor towards insuring the success of the undertaking. In addition to the large sums collected and handed over to the Board of Finance, this Committee raised by voluntary contributions of the American women the separate sum of $35,000, which it appropriated to the construction of a special building for the exclusive display of women's work, erecting a structure creditable to the enterprise of the ladies, and a useful and ornamental addition to the list of Exhibition buildings. We hope to give more particulars concerning this Committee hereafter.

It was soon found necessary to organize the various administrative bureaus which would be required to properly attend to the direct duties of the Exhibition under the supervision of the Director-General. The bureaus formed with their respective functions and chiefs were as follows:—

> FOREIGN—Direction of the foreign representation, ... The Director-General.
> INSTALLATION.—Classification of applications for space-allotment of space in Main Building—Supervision of special structures, ... Henry Pettit.
> TRANSPORTATION.—Foreign transportation for goods and visitors—Transportation for goods and visitors in the United States—Local transportation—Warehousing and customs regulations, ... Adolphus Torrey.
> MACHINERY.—Superintendence of the Machinery Department and building, including allotment of space to exhibitors, ... John S. Albert.

AGRICULTURE.—Superintendence of the Agricultural Department, building and grounds, including allotment of space to exhibitors, ... Burnet Landreth.

HORTICULTURE.—Superintendence of Horticultural Department, conservatory and grounds, including allotment of space to exhibitors, ... Charles H. Miller.

FINE ARTS.—Superintendence of the Fine Arts Department and building, including allotment of space to exhibitors, ... John Sartain.

The subject of awards received very careful attention from the Executive Committee, the experience of those connected with previous Exhibitions being solicited and given due consideration. A system was finally decided upon, widely different from any ever used before; and instead of having several grades of awards, causing disputes among the recipients as to their comparative importance, a single uniform medal was adopted, which was in each case to be accompanied by a report and diploma stating the nature of the merit for which it was awarded. It was determined to have only a small body of judges, one-half of whom should be foreign and one-half from the United States, and to insure the presence and attention of men practically conversant with the subjects on which they were to report, it was decided to provide an allowance to each, designed to cover actual expenses.

A final effort was made at the Congressional session of 1875-6 towards obtaining the appropriation asked for at the previous session, and after considerable opposition it was successful, the sum of $1,500,000 being granted on condition of its being paid back to the Government out of the proceeds of the Exhibition in advance of any dividends from profits being given to any other claimants. This gave immediate relief from all chance of pecuniary embarrassment, avoiding the necessity of perhaps mortgaging the buildings or receipts in advance, which might have been required otherwise.

The Centennial year began to draw near; the buildings towards which so many eyes were turned grew up and approached completion; events crowded one on the other until it was impossible for the coolest head to avoid being stirred up with enthusiasm. The 1st of January was ushered in with illuminations and rejoicings such as had never before been known. Foreign representatives, of which there had been a few for some time, now began to arrive in numbers, and exhibits commenced to appear on the grounds. The writer well remembers the interest occasioned by a lot of Japanese goods which were among the first to come, and were unpacked in Machinery Hall. They came by way of San Francisco, and were parts of the building afterwards erected by that Government for the use of its officers, so curiously put together by native workmen, who appeared to do everything exactly the opposite way from which it was done in this country, possibly from living in a reversed position on the other side of the globe.

The winter of 1876 was fortunately very mild. Planting was possible almost continuously, and the erection of the numerous buildings proceeded without interruption. By the time of the opening-day, everything was in readiness with the exception of a few of the exhibits which had suffered detention. The buildings had all been completed and ready for the reception of goods at the dates designated, occasioning no delays on their part, a fact never before accomplished at any previous Exhibition. Patriotism had been fully aroused, and for weeks before the 10th of May the people were busy decorating with flags and draping with bunting, until Philadelphia wore a gala look such as she had never done before and may never do again. Chesnut Street was one mass of color—red, white and blue—as far as the eye could reach. It was a pageant, a raree-show, such as few see twice in a life-time. The poorest little shanty in the town had its penny flag hung out, and even now the thought of those days stirs up one's feelings and bears evidence of that depth of love of country which always shows itself when the occasion arises.

The 9th of May was dark and cheerless, but all were busy placing the last flag and

Kansas Building.

giving the last touch until far into the night. The 10th opened at early dawn still cloudy and uncertain. Nevertheless all were stirring, for was not this the opening-day of our great celebration, where we were to show to the world the progress that a free country under self-government could make in a century of life? The rain held off; the crowds began to gather. The whole area in front of the Memorial Hall facing the Main Building had been arranged with seats on platforms, and apportioned off into sections, and here were grouped the President of the United States and Cabinet, the Senate and House of Representatives, the Supreme Court, the Diplomatic Corps, the Governors and other officers of States, the Centennial Commission and Board of Finance, the Foreign Commissioners, the Women's Centennial Committee, the Board of Judges and Awards, other Boards and Bureaus of the Exhibition, the Army and Navy, the various city officers, etc., etc.—

forming a brilliant assemblage such as only an occasion like this can draw together. In the centre of the front was the platform for the President and those distinguished officers and guests who were to take active part in the ceremonies. At the entrance to the Main Building, opposite and facing Memorial Hall, was the platform for the immense orchestra of one hundred and fifty pieces, under the leadership of Theodore Thomas, and around this was grouped the grand Centennial chorus of one thousand voices, one of the great results of the good work of the Women's Centennial Committee. In the rear, in the interior of the Main Building, but with the large arched windows of the façade open, was

Mississippi Building.

the noble Roosevelt Organ, the first instrument of its kind in the history of International Exhibitions to take part in the opening ceremonies in combination with the grand orchestra, and mingle with it its glorious tones in one melodious whole.

The Main Building, Memorial Hall and Machinery Hall were reserved for officials, invited guests and exhibitors until the conclusion of the ceremonies. Invited guests entered through the Main Building, and other gates to the grounds were opened to the public at nine o'clock A. M., at the established rate of admission, fifty cents. The avenue between the centre exit of the Main Building, on the north side, and the Memorial Hall was kept open, and guests passed by this to their places, which were to be occupied by quarter past ten o'clock.

Let us take our stand of observation in the outside balcony of the Main Building, in the rear of the orchestra, where we can see and be above everything. As the hour approaches the excitement increases. The clouds lighten up, and the grounds become gradually covered with a dense mass of good-humored people, who crowd up towards the platforms until they threaten to entirely close the passage between the two buildings, necessitating the utmost efforts of the police to keep them back, taking the pushing and shoving, however, with that remarkable good nature for which the American citizen is so noted. As far as the eye can reach, the people are seen pouring forward. A perfect sea of heads meets the view on every side. Every one looks pleased, and expectation rises to the highest pitch.

From below, the buzz and hum of the crowd floats up to the ear; the balmy air and freshness of the spring morning delights the senses, and one feels perfectly happy. The seats on the opposite side are gradually filled; distinguished visitors arrive one after the other, and are received with acclamations. There goes His Excellency Dom Pedro, of Brazil, that man who is every inch a true emperor, with the Empress—the only crowned heads who grace our opening. We remove our hats in compliment to these our royal guests. There comes the British Commission in full uniform, and following are the representatives from other countries, all decked in their most gorgeous official dresses, and decorated with their medals and honors; the Japanese Embassy, the French, the Austrian, the Swedish, the German, and all the nations of the earth, to join with us in this our triumphal day. The Emperor and Empress take seats on the central platform reserved for the President of the United States and distinguished visitors. The hour of opening has arrived, and the grand orchestra strikes up the national airs of all nations. The moment we have dreamed of for the past three years of labor and toil has come, and our work is consummated. One feels in his heart, O happy day! that I have lived to see it and had it come in my time! Music is heard in the distance; it draws nearer. It is the President, who comes escorted by Governor Hartranft, of Pennsylvania, with troops. They enter by the rear of Memorial Hall, and passing through to the front, the escort forms in two lines down the passage between the buildings, while the President joins the Emperor and Empress. Acclamations rend the air, and at this moment the clouds break away, and a burst of sunshine illuminates the animated scene—a happy omen for the success of the great undertaking.

The orchestra begins Wagner's Centennial Inauguration March, of which so much was expected, another gift from our noble women. To one who is an enthusiastic admirer of Wagner, it must be confessed that it is somewhat disappointing. Still, it *is* Wagner. None can dispute that. The grand clashes, the sounds from the brass instruments, the volumes of tone swelling up and up until they almost overtop the heavens themselves. Then all is hushed, and Bishop Simpson asks God's blessing on our work, gives thanks for all our past successes, and beseeches his kind guidance in the future. Whittier's hymn follows, with the grand chorus, the orchestra and the organ. The place, the day, the tumultuous feelings within one combine to produce an effect never to be forgotten, as a thousand voices swell up on the bright morning air—

INTERNATIONAL EXHIBITION, 1876.

Our fathers' God! from out whose hand
The centuries fall like grains of sand,
We meet to-day, united, free,
And loyal to our land and Thee,
To thank Thee for the era done,
And trust Thee for the opening one.

Here, where of old, by Thy design,
The fathers spake that word of Thine,
Whose echo is the glad refrain
Of rended bolt and falling chain,
To grace our festal time, from all
The zones of earth our guests we call.

Be with us while the new world greets
The old world thronging all its streets,
Unveiling all the triumphs won
By art or toil beneath the sun;
And unto common good ordain
This rivalship of hand and brain.

Thou, who hast here in concord furled
The war-flags of a gathered world,
Beneath our Western skies fulfill
The Orient's mission of good will,
And, freighted with love's Golden Fleece,
Send back the Argonauts of peace.

For art and labor met in truce,
For beauty made the bride of use
We thank Thee, while, withal, we crave
The austere virtues strong to save,
The honor proof to place or gold,
The manhood never bought nor sold!

O! make Thou us, through centuries long,
In peace secure, in justice strong;
Around our gift of freedom draw
The safeguards of Thy righteous law;
And, cast in some diviner mould,
Let the new cycle shame the old!

The buildings are then presented by the Centennial Board of Finance, through its President, Mr. John Welsh, to the Centennial Commission, and the presentation is followed by Sidney Lanier's Cantata—

From this hundred-terraced height
Sight more large with nobler light
Ranges down yon towering years:
Humbler smiles and lordlier tears
  Shine and fall, shine and fall,
While old voices rise and call
Yonder where the to-and-fro
Weltering of my Long-Ago
Moves about the moveless base
Far below my resting-place.

Mayflower, Mayflower, slowly hither flying,
Trembling Westward o'er yon balking sea,
Hearts within *Farewell dear England* sighing,
Winds without *But dear in vain* replying,
Gray-lipp'd waves about thee shouted, crying
  No! It shall not be!

Jamestown, out of thee—
Plymouth, thee—thee, Albany—
Winter cries, *Ye freeze: away!*
Fever cries, *Ye burn: away!*
Hunger cries, *Ye starve: away!*
Vengeance cries, *Your graves shall stay!*

Then old Shapes and Masks of Things,
Framed like Faiths or clothed like Kings—
Ghosts of Goods once fleshed and fair,

Grown foul Bads in alien air—
War, and his most noisy lords,
Tongued with lithe and poisoned swords—

  Error, Terror, Rage, and Crime,
  All in a windy night of time
  Cried to me from land and sea,
  No! Thou shalt not be!

  Hark!
Huguenots whispering *yea* in the dark!
Puritans answering *yea* in the dark!
Yea, like an arrow shot true to his mark,
Darts through the tyrannous heart of Denial.
Patience and Labor and solemn-souled Trial,
  Foiled, still beginning,
  Soiled, but not sinning,
Toil through the stertorous death of the Night,
Toil, when wild brother-wars new-dark the Light,
Toil, and forgive, and kiss o'er, and replight.

Now Praise to God's oft-granted grace,
Now Praise to Man's undaunted face,
Despite the land, despite the sea,
I was: I am: and I shall be—
How long, Good Angel, O how long?
Sing me from Heaven a man's own song!

Agricultural Hall.

"Long as thine Art shall love true love,
  Long as thy Science truth shall know,
Long as thine Eagle harms no Dove,
  Long as thy Law by law shall grow,
Long as thy God is God above,
  Thy brother every man below,

So long, dear Land of all my love,
  Thy name shall shine, thy fame shall glow!"

O Music, from this height of time my Word unfold:
In thy large signals all men's hearts Man's Heart behold!
Mid-heaven unroll thy chords as friendly flags unfurled,
And wave the world's best lover's welcome to the world!

The basso solo is sung by Myron W. Whitney, of Boston, whose powerful and superb voice floats out clearly and distinctly over the space, even to the most distant parts of the platforms, and such bravos are raised as to require a repetition to render satisfaction. After this the Centennial Commission by its President, General Joseph R. Hawley, presents the Exhibition to the President of the United States, who replies in a brief address, and declares it open to the world. The flag unfurled, and the sublime Hallelujah Chorus bursts forth with orchestra and organ, and the simultaneous salute of one hundred guns and ringing of the chimes.

A procession is formed, and the President of the United States, conducted by the Director-General of the Exhibition, and followed by the guests of the day, passes into and through the Main Building. The various foreign commissioners, having gone in advance, join the procession at the sections of their respective countries, and the whole body passes on to the Machinery Hall, through the military escort which forms in two lines between the buildings, to the great engine which the President and the Emperor of Brazil, assisted by Mr. George H. Corliss, set in motion, starting all the machinery connected with it, and completing the ceremonies. The restless, happy crowds separate and wander over the buildings and grounds; the restaurants are filled to overflowing and taxed far beyond their capacity, the number of visitors exceeding all calculations, and the day closes with a sudden shower of rain, dispersing all to their homes. So ended the first day.

For six months thus the Exhibition continued open—a time long to be remembered by those who passed through it—those pleasant days of May and June, when one strolled through the aisles of the Main Building and listened to the strains from Gilmore's band, or heard the tones from the grand organ swelling up and dying away in the distance. No matter where one went, good music delighted the ear at all times, greatly enhancing the enjoyment of the visitor. The ever-varying crowds, sometimes more and sometimes less, all classes, so interesting as studies, all happy and enjoying themselves; or, when passing into Machinery Hall and standing by the famous Krupp guns, one saw the surprise and astonishment in the faces of the people at these tremendous messengers of death, or observed the curiosity and interest displayed by those around the weaving-machines; or, if present among the number who crowded about the great Corliss engine after the mid-day rest, one noticed the desire manifested to see it started to work by the movement of a hand, so quietly and steadily—so much power, all so completely under the control of one human being. One could not but feel the immense pleasure and benefit given to the masses of our people by this method of celebrating our great Centennial, and acknowledge the wisdom of those who so strongly defended and labored for it.

Then the hot days of July, with the grand torchlight procession on the night of the

The International Exhibition of 1876, Ground Plan.

INTERNATIONAL EXHIBITION, 1876.

South-east Section.

1. Main Exhibition Building.
2. Memorial Hall (Art Gallery).
3. Annex to Art Gallery.
4. Photographic Gallery.
5. Carriage Building.
6. Centennial National Bank.
7. Public Comfort (clothes room).
8. Swedish School-house.
9. Penna. Educational Departm't.
10. Singer's Sewing Mach. Build'g.
11. Lafayette Restaurant.
12. Hunters' Camp.
13. Milk Dairy Association.
13A. Extension to Milk Dairy.
14. Bible Society.
15. Public Comfort.
16. Phila. Municipal Headquarters.
17. Soda Water.
18. Moorish Villa.
19. German Government Building.
20. Brazilian Government Build'g.
21. Kittredge & Co.
22. Soda Water.
23. Philadelphia "Times" Build'g.
24. Glass Factory.
25. Cigar Stand.
26. American Fusee Company.
27. Centennial Photographic Association.
28. Penna. R. R. Ticket Office.
29. Centennial Medical Departm't.
30. Judges' Hall.
31. Department of Public Comfort.
32. Japanese Government.
33. Kindergarten.
34. Soda Water.
35. Public Comfort Station.
36. Cigar Stand.
37. Stand Pipe.
38. French Government Building.
39. Stained Glass.
40. Vienna Bakery.
41. Bankers' Exhibit.
42. Empire Transportation Co.
43. Centennial Fire Patrol No. 2.
44. Portuguese Governm't Build'g.
45. Pavilion of French Art.
46. Burial Casket Building.
47. Public Comfort (clothes room).
48. Police Station.
49. Police Station.
49A. Music Stand.
49B. French Ceramic Pavilion.

South-west Section.

50. Machinery Hall.
51. Shoe and Leather Building.
52. British Boiler House.
53. Boiler House.
54. Corliss Boiler House.
55. Weimer's Furnace.
56. Boiler House.
57. Stokes & Parrish Machine Shop.
58. Boiler House.
59. Nevada Quartz Mill.
60. Gas Machine.
61. Yale Lock Company.
62. Brick Working Machine.
63. Storehouse.
64. Artesian Well.
65. Rock Drilling Machinery.
66. Jesse Starr & Son.
67. Gunpowder Pile Driver.
68. Automatic Railway.
69. Tiffany's Gas Machine.
70. Pennsylvania Railroad.
71. Engine House.
72. Emil Ross Saw Mill.
73. Gillinder & Son Glass Factory.
74. Annex (Saw Mill).
75. Saw Mill Boiler House.
76. Campbell Printing House.
77. Fuller, Warren & Co.
78. Liberty Stove Works.
79. Boston "Herald" and "Advertiser."
80. Catholic Total Abstinence Fountain.
81. Kiosque.
82. Turkish Cafe.
83. Pennsylvania State Building.
84. Pop Corn.
85. Rowell's Newspaper Build'g.
86. Lienard's Relief Plans.
87. Public Comfort Station.
88. Soda Water.
89. New York "Tribune."
90. French Restaurant.
91. Sons of Temperance Fount'n.
92. Colossal Arm of Liberty.
93. World's Ticket Office.
94. Catalogue Office.
95. Loiseau's Prepared Fuel Co.
96. Office Board of Finance.
97. Office U. S. Centennial Com.
98. Bartholdi's Fountain.
99. Jerusalem Bazaar.
99A. Vermont State Building.
99B. Chilian Machine Building.
99C. Police Station.
99D. Statue of Elias Howe.
99E. Columbus Monument.
99F. Averill Paint Company.

North-west Section.

100. U. S. Government Building.
101. United States Hospital.
102. United States Laboratory.
103. Cigar Stand.
104. Tent.
105. U. S. Signal Service.
106. Bishop Allen's Monument.
107. Soda Water.
108. Cigar Stand.
109. Canada Log House.
110. Arkansas State Building.
111. Spanish Building.
112. West Virginia State Building.
113. Spanish Government Build'g.
114. Spanish Government Build'g.
115. Japanese Building.
116. Mississippi State Building.
117. George's Hill Restaurant.
118. California State Building.
119. New York State Building.
120. }
121. } British Government Build'gs.
122. }
123. Public Comfort Station.
124. Tunisian Camp.
125. Centennial Fire Patrol No. 1.
126. Ohio State Building.
127. Indiana State Building.
128. Illinois State Building.
129. Wisconsin State Building.
130. Michigan State Building.
131. New Hampshire State Build'g.
132. Connecticut State Building.
133. Massachusetts State Building.
134. Delaware State Building.
135. Maryland State Building.
136. Tennessee State Building.
137. Iowa State Building.
138. Missouri State Building.
139. Block House.
140. Fire Patrol.
141. Rhode Island

North-east Section.

150. Agricultural Building.
151. Agricul'al Annex (Wagons).
152. " " (Pomology).
153. Brewers' Building.
154. Butter and Cheese Factory.
155. Tea and Coffee Press Build'g.
156. American Restaurant.
157. Kansas State Building.
158. Southern Restaurant.
159. New Jersey State Building.
160. Horticultural Hall.
161. Women's Pavilion.
162. Gliddon Guano Building.
163. New England Log House.
164. Pop-Corn.
165. Cigar Stand.
166. Cigar Stand.
167. Soda Water.
168. Bee Hives.
169. School House.
170. German Restaurant.
171. Virginia Building.
172. Boiler House.
173 to 183. Wind Mills.
184. Police Station.
185. Hay Packing.
186. Practical Farmers' Office.
187. Public Comfort Station.
188. Centennial Guards.
189. Public Com'ort (cl. room).

KEY TO ABOVE GROUND PLAN.

3d, and the military parade and special celebration in Independence Square on the 4th; and when the latter days of August came, and the throng of visitors began to swell and steadily increase, insuring the financial success of the undertaking, fresh interest was aroused, culminating in that red-letter Thursday in September, the great Pennsylvania Day, with its two hundred and sixty-seven thousand visitors, a result far beyond what had been done at any previous Exhibition.

The flowers and plants in the Horticultural grounds grew and flourished, waxing strong and beautiful, fully equaling those exquisite displays made in the royal pleasure-grounds of Europe, until the frosts of October cut them down and gave the first intimation that there was to be an end to all this fairy-like spectacle.

Let us now select a fine day and observe the Exhibition more in detail, passing from building to building, and noting the particular characteristics of each. To enable the reader to follow us more intelligently, we give an engraving on page cxviii showing a general plan of the grounds, on which all the principal features are marked. We will enter from Belmont Avenue. On the right and left are two similar buildings, used as offices for the Centennial Commission and the Centennial Board of Finance. They are of one story, constructed of frame and plaster, entirely surrounded by porticos, having arched openings filled in with sawed scroll-work. Open courts are arranged in the centre of each building, planted out with flowers and trailing vines, and one of our engravings shows the lovely effect produced by them. The buildings are very pretty, and remind one of Vienna.

Within the grounds we enter a large open space about five hundred feet square, flanked on the east by the Main Building, and on the west by Machinery Hall, laid out geometrically with walks, and having in the centre the great Bartholdi cast-iron fountain, consisting of female figures standing on dolphins, and supporting a large basin, with gas-lamps grouped in among the water-jets, and intended to produce a brilliant and novel effect, which does not appear to be completely realized. A few short months before the opening, no one would have supposed that this then barren space could have been in so little time so thoroughly transformed. We move towards the Main Building as the primary object of attraction, and we notice that its outlines are characterized by extreme simplicity. In the construction of this building the necessities of the case have required the omission of everything which would entail extra expense over what was strictly essential to satisfy the demands of the Exhibition. The problem given to the engineers and architects was to cover a rectangular piece of ground of a certain area with a building, constructed of certain materials, for the lowest possible cost, and the requirements have been strictly carried out. There are no projections, no recesses—all such accessories to architectural effect being rigidly excluded on the score of economy, and every foot covered by the building has been made fully available for the purposes of the Exhibition. The roofs are made no higher than practical use requires, and of the simplest and cheapest forms of trusses; high arched roofs, so effective architecturally, being of course prohibited. It is believed that the general effect of the building is quite satisfactory, taking all of these restrictions into account. Nothing like monumental grandeur or solidity was feasible. The amounts of material used had to be kept strictly to the requirements for proper strength and no more.

The monotony has been very much broken by the manner adopted of working up the façades with central features extending to a considerable height, having arcades upon the ground-floor and large arched openings above, towers being placed on the corners, connected with

Art-Gallery Annex, Spanish Court.

the central part by low-roofed wings. The arrangement of the roofs in varying heights for different spans, and the raising of the central part of the building and introduction of four high towers at the corners of this central portion as a crowning feature to the whole building, have aided very considerably in the production of a pleasing and satisfactory structure.

The building is located in the original position selected as its site, with its greatest dimension parallel to Elm Avenue, and distant one hundred and seventy feet from it, and the ground plan comprises a rectangle of eighteen hundred and eighty feet in length by four hundred and sixty-four feet in width, measured to centres of exterior columns. In general arrangement it consists of a central longitudinal nave of one hundred and twenty feet span, with two side avenues of one hundred feet each, separated from the central nave by intermediate spans of forty-eight feet, and having, on the exterior side of each, one span of twenty-four feet; the total width of four hundred and sixty-four feet being thus made up of two spans of twenty-four feet, two of one hundred feet, two of forty-eight feet, and one of one hundred and twenty feet.

In order to break the great length of the roof-lines, a cross transept of one hundred and twenty feet span intersects all of the longitudinal avenues at a distance of nine hundred and seventy-six feet from the east end of the building, having on each side of it cross avenues of one hundred feet span, separated from it by spaces of forty-eight feet. The governing dimension or unit of span of the building is twenty-four feet, nearly all measurements conforming to this unit, the exceptions being in the case of the spans of one hundred feet, the spacing of some of the trusses—in the central portion of the building, and the arrangement of columns at the main entrances.

The spans of one hundred and twenty feet in nave and transept have a height to square at top of columns of forty-five feet, and to ridge of roof of seventy feet, and at their intersection at centre of building they produce a space of one hundred and twenty feet square, which is raised to a height on square of seventy-two feet, or to ridge of ninety-eight feet six inches, and is spanned by diagonal trusses, the roofs of nave and transept on each side of this square for a distance out of thirty-two feet being raised to the same height.

The spans of forty-eight feet on each side of nave have a height to square of twenty-seven feet six inches, and the intersections that would be made by these spans, if extended, with the forty-eight feet spaces on each side of the transept, produce four interior courts of forty-eight feet square at the four corners of the central diagonal roof, on which are constructed square towers rising to a height of one hundred and twenty feet. This forms the central feature before mentioned, adding so much to the effect as a whole.

The spans of one hundred feet have a height to square of roof of forty-five feet, and at their intersections with each other and also with the spans of one hundred and twenty feet, they produce eight open spaces, four of them one hundred feet square, and four one hundred feet by one hundred and twenty feet, entirely free from columns. The small spans of twenty-four feet have a height to square of twenty-two and one-half feet. The main central avenue or nave has an extreme length of eighteen hundred and thirty-two feet, exclusive of the portions at the ends occupied by entrances, and is the longest avenue of that width ever introduced into an exhibition building to this time.

There are four main entrances in the centres of the four façades, the east entrance forming the principal approach for carriages, while the south entrance is the principal approach for those arriving by street-cars. The north and west entrances are within the grounds, the former connecting directly with the Art Gallery, and the latter with Machinery Hall.

Upper floors devoted to offices and galleries of observation have been constructed at the main entrances and also in the towers; but with the exception of some area assigned in the west gallery to the American Society of Civil Engineers, and in the eastern one to the Massachusetts Educational Department, the entire Exhibition space is upon the ground floor. The four central towers have stairways, and one of them an elevator, all extending to the top, and bridges over the roofs of the building connect the towers together, forming a favorite resort for visitors to enjoy the cool breezes and the fine view of the grounds.

The areas covered by the flooring are as follows:—

Ground floor,	872,320 square feet	= 20.02 acres.
Upper floors in projections,	37,344 "	= 0.85 "
Upper floors in towers,	26,344 "	= 0.60 "
Total,	936,008 "	= 21.47 "

The foundations of the building consist of piers of solid rubble masonry, well laid, each pier under a column, being finished off with a granite block one foot thick, neatly dressed on upper and lower faces. Between the piers on the exterior lines of the building, foundation-walls are laid, finished off with a base course of dressed stone, on which brick-work about six feet in height is constructed, laid in ornamental patterns of red and black, and pointed on both faces with colored mortar. The outer columns of the building extend down through this brick-work to the foundation masonry. The four main entrances have piers and side jambs of ornamental pressed brick and tile, with string courses of stone and ornate galvanized iron capitals, there being stone sills between the piers. The wrought-iron columns of the building extend through these piers to the foundations. The entire frame-work, or skeleton of the building, including towers, is of wrought iron, there being wrought-iron vertical columns resting upon the foundation masonry, and connected and braced to each other by wrought-iron wind-trusses and beams; wrought-iron roof-trusses with wrought-iron ridge connecting members, and all necessary lateral wrought-iron bracing and ties; the whole structure properly joined together, and stiffened to resist the heaviest winds.

The two outer rows of columns are connected to the foundation masonry by anchor-bolts extending nearly to the bottom of the masonry, and all other columns have cast-iron bases with lugs let into the cap-stones on piers. The roof-trusses are made entirely of wrought iron except the heel-blocks connecting with the columns, which are of cast-iron, and these trusses are computed for a load of thirty-five pounds per horizontal square foot of surface covered, exclusive of the weight of structure. The roofs of spans of one hundred and of one hundred and twenty feet are constructed upon the French triangular system with straight rafters; the spans of forty-eight feet are straight, double-intersection triangular girders, and those of twenty-four feet are sloping triangular trusses. The four main entrances have intermediate ornamental cast-iron columns, with brackets, lamps and wrought-iron gates, the openings being finished above with arches. These arches and the ornamental face-work above the brick-work up to the foot of the second-story balustrade are constructed of galvanized iron and zinc. At the corners and angles of the main entrances and towers, the building is finished with octagonal turrets, extending the full height from the ground-level to above the roof, those on the towers being surmounted by flag-staffs. The bases of these towers are of cast iron, but the balance above is of galvanized iron.

Above the brick-work and the galvanized iron-work, the walls of the building are composed of timber framing, with glazed sashes, an upper portion of these sashes swinging on centre pivots at the sides, and capable of being opened or shut at pleasure by means of cords operated from the floor. The method of laying the floor is different from that of

Agricultural Hall, Interior View.

any previous Exhibition building, the planking being nailed to sills firmly bedded in the earth, which is also filled in between them, leaving no air-spaces beneath, and vastly decreasing the risks from fire. The roofs are covered with tin, and Louvre ventilators and skylights with glazed sashes are placed in continuous lengths over the nave, side galleries and central aisles, and in shorter lengths over the middle portions of the building. The sashes swing horizontally, and are provided with opening and closing apparatus, affording ample ventilation. The exterior finish of the building is of wood. Ample entrance- and exit-doors are provided at the main entrances, at the corner towers, and in the sides of the building. Gas is supplied throughout, for policing purposes alone, however, as the Exhibition is not opened at night. There are seven thousand burners and over thirty-two thousand lineal feet of gas-pipe of the various sizes. Water is also supplied in ample quantity through about twelve thousand eight hundred lineal feet of pipe, and a large number of fire-plugs are distributed so as to be most efficient in case of fire. The drainage system is very complete, there being about eight hundred lineal feet of thirty-inch sewer, and over fourteen thousand feet of terra-cotta pipe, varying from six to eighteen inches in diameter. These large figures give an exceedingly good idea of the vast extent of the building.

The exterior of the building is painted in an agreeable tint of buff, relieved by darker shades, with bright colors in the chamfers, the rustic and foliated work at the entrances, the caps of columns, etc., being of a green bronze picked out with gold. In the interior, the wood-work of the roofs is kalsomined in two coats of a light pearl color, and decorated with

Agricultural Hall—the Mill.

stenciling, the iron-work being painted in buff picked out with crimson, and the pendants in crimson, blue and gold. The interior side-work is painted in colors, the body-color of the columns and wood-work a light olive-green in several shades, and the decorations in crimson, blue and gold. The panel-work and flat sides of the columns are covered with decorated work. The ventilating sash with circular openings are ornamented with various emblems and figures in such a way as to produce the effect of stained or painted glass. All of the clear glass in the parts of the building exposed to the sun has been tinted with an opaque wash, producing the effect of ground glass, and very much relieving the eyes. On entering, the effect of the coloring is quite pleasing, harmonizing well with the rich display of exhibits, and fully justifying the reasons which led to its adoption.

Passing down the central aisle, we are lost in bewilderment. The construction of the building permits us to see all over it; the wealth of the world is before us, and our sight is only limited by the exhibits. Where shall we go first? What shall we do? These are the questions one hears on every side. On our left is Italy with her Roman and Florentine mosaics, her cabinets and bronzes. Next comes Norway. Oh! what beautiful silver jewelry! what handsome furs! Then Sweden, with her pottery, her rich wealth of iron manufacture, her full-size models of domestic peasant life. On our right is China, the curious carvings, the huge porcelain vases; and adjoining it, Japan. What grotesque bronzes! what lovely cabinets! As we pass on, one country after another comes before us, the exhibits of each more enticing than those just left, all demanding our fullest attention—the rich fabrics of India, the magnificent silver- and gold-ware of Russia, and the porcelain of Great Britain. We move on, passing the Egyptian department, with its enclosure modeled after an old temple of the Nile Peeping in, we ask after the Sphinx and mummies, to the evident disfavor of the gentleman in fez whom we address, and then go on past the grand façade of the Spanish section, extending almost to the roof. We are arrested for a moment by an exquisite porcelain fountain in Doulton-ware. On one side of us we are dazzled by the glass-ware of Vienna, and on the other the lovely furniture of England recalls some happy days across the waters. Finally we are under the great central roof; we sink on to one of the numerous seats and gaze around us. The Exhibition is a success; the building is a success. Resembling somewhat in interior effect that of the great Exhibition of 1851, it would have shown to better advantage perhaps with a high arched roof over this great central aisle, but that was out of the question, and on the whole it is so well adapted to its purposes, so satisfactory in every way, so utilitarian, so truly an engineer's building, and so perfectly an ornamented construction, that no one can fail to be pleased with it. Close by is the famous Elkington exhibit; across the way is the French Court with all its wealth; on our right is Germany, and over there the United States—in the front rank the rich exhibits of our American silversmiths, adding largely to the elegance of the display. Truly this centre is a lovely spot. Thanks to those who arranged that these should all group here! a fitting nucleus to the greatest Exhibition the world has ever seen. We rest and dream. The soft notes from the great Roosevelt Organ work in harmony with our thoughts, and we forget that there is much yet to be done, that we are now only taking a general glance, that it is the buildings and not the exhibits with which we have to do, and we reluctantly wend our way back towards Machinery Hall.

Machinery Hall is located five hundred and forty-two feet west of the Main Exhibition Building, with its north face on the same line. It covers an area of three hundred and sixty by fourteen hundred and two feet, having projections beyond these dimensions for doors and portals on the east, north and west sides, also an annex on the south of two hundred and eight by two hundred and ten feet, connected with the Main Hall by a passage-way ninety feet in width. It therefore presents on the north side, in connection with the Main Exhibition Building, upon one of the principal avenues within the grounds, the Avenue of the Republic, an entire frontage from east to west of three thousand eight hundred and twenty-four feet.

The boiler-houses are located as distinct buildings, on the south side of the Main Hall, east and west of the southern annex, those on the east side being the British boiler-house, or No. 1, and the Corliss boiler-house, or No. 2; those on the west side, the machine-shop and boiler-house No. 3 in one building, and boiler-house No. 4.

In designing Machinery Hall, its width was limited by the maximum distance that it was thought advisable to convey steam from the various boiler-houses, and in arranging the cross sections of the building a certain amount of low roof was desired, with stiffened tie-beams, for the purpose of hanging shafting, while the balance could be made higher, so as to improve the effect and afford facilities for light and ventilation. It was therefore arranged in five spans, the centre and two outer spans being sixty feet, with a height from floor to tie-beam in clear of twenty feet, and to ridge of thirty-three feet, and the two intermediate spans ninety feet, with a height to square of forty feet, and to ridge of nearly fifty-nine feet. These avenues extend the whole length of the building, and the exterior finish at the east and west ends is designed to harmonize with their cross section, low towers or belfries, having a height to apex of roof of eighty-one feet, being placed at the ends of the ninety-feet spans. The southern

Massachusetts Building.

annex consists of three spans—a centre span of ninety feet, and two side spans of sixty feet—the heights and outlines corresponding with those of the same dimensions in the main part of the building. The centre span of ninety feet continues on across the main portion of the building, intersecting with the longitudinal avenues of ninety feet, and forming a transept, at the northern end of which the face of the building is finished with a tower and wings similar to those at the east and west ends.

The building furnishes three principal entrances, those on the east, west and north. The projections at these entrances on the lower or main floor provide offices, retiring-rooms and restaurants, while on the upper floors, offices and galleries, the latter being favorite resorts for visitors from which to view the animated scene below. The governing dimension or unit of span in the direction of the length of the building is sixteen feet.

The entire floor-area covered by the Machinery Hall and annex is five hundred and eighty-eight thousand four hundred and forty square feet, or twelve and eighty-two hundredths

Entrance to the Art Gallery—Opening Day.

acres, and the galleries and office-floors in the upper stories increase this total to fourteen acres. The annex, which is designed especially for the exhibition of hydraulic machinery in motion, and forms one of the greatest features of the display, contains in its central portion an open tank, the top of which is level with the floor, covering an area of sixty feet by one hundred and fifty-six feet, and having a depth of ten feet. This tank is filled with water, and at the south end there is a waterfall of thirty-five feet in height and forty feet in width, supplied from the tank by pumps on exhibition. There is also a pit at the south end for trial of Turbine wheels. During the hours that the waterfall is in operation and the numerous pumps along the sides of the tank are raising water and pouring it back again in as many different streams, some large, some small, some from fire nozzles up to the ridge of the roof, returning in spray, this annex forms one of the coolest resorts for a hot summer's day that can be found at the Exhibition, and one may sit and listen to the roar of the waters, watching with untiring pleasure the ever-varying scene before him.

The foundations of the building are of good rubble masonry, and are covered with a base course of granite, returning at all door-openings through the entire thickness of the walls. On this the exterior walls are built of Trenton brown-stone laid in broken range work to a height of five feet above the floor, and pointed with dark mortar. All doors are provided with heavy stone stills, level with top of flooring, and extending through the whole thickness of walls. Foundation-piers are provided for all interior columns and finished off with granite capstones. The tank in the annex is built with stone side-walls, and lined inside with brick-work laid in cement, the bottom being covered with cement concrete.

The frame-work of the building, unlike that of the Main Exhibition Building, is constructed entirely in solid timber, excepting only certain members of the roof-trusses, more particularly the tension members, which are of wrought iron. All columns, caps, sills, rafters, cornices, sashes, scroll-work, etc., are of white pine. The purlieus, framing of Louvre ventilators and roof-sheathing are of spruce. The flooring is of yellow pine, the main flooring being laid on sills bedded in the earth in the same way as in the Main Building. The outer masonry walls are covered on top by a timber sill, securely held to the masonry by anchor-bolts, and into it are mortised the main posts with intermediate posts. The system of forming the interior columns consists in having a solid square timber of ten by ten inches section in the centre, surrounded by four pieces of four by eight inches section on each face, the whole well bolted together and acting as one column. Where the low roof joins to the column, three of these side pieces stop, the fourth only, on the inside face next to the ninety feet spans, being continued up to the top of the column. Stiffening trusses are framed in from column to column at the level of the low roof, and above these, up to the roof of the ninety feet span, intermediate framing is introduced, the same as in the outer walls. This intermediate framing is in all cases filled in with glass sash, a lower part fixed and an upper part movable for ventilation, being hung on centre pivots at the sides, and movable by means of cords operated from the floor below. The lower roofs are framed of timber, and the upper ones of iron, with timber rafters on the French triangular system, all roofs having Louvre ventilators in continuous lengths as in the Main Building. The system of ventilation is exceedingly perfect, giving a pleasant temperature within the building during the hottest days of the season. Many of the

cxxx HISTORY OF THE

Vestibule of the Rotunda. The Art Gallery—Western Pavilion. Entrance to the Western Pavilion.

lower sash on the exterior walls are made to open on hinges like French casement-windows. The roof-covering is of tin.

Gas is supplied for policing purposes only, there being about five thousand lights and over sixteen thousand feet of pipe. The water-service system is very complete, there being over ten thousand five hundred feet of pipe, six hundred and twenty-four feet of which is a special ten-inch main running from the bottom of the lake north of Machinery Hall to supply the boilers of the fourteen hundred and two horse-power Corliss engine. There are thirty-four fire-plugs exterior to the building, and forty-eight in the interior.

The building is painted on the exterior of a pearl tint, relieved by different shades, and by a dark maroon color on the chamfers. The interior is very plainly painted, as would become a building devoted to the purposes for which this is, the sides and columns of light shades of umber and white lead, and the roof kalsomined in two coats of light pearl color, the iron work, rods, struts, etc., being painted dark blue. The effect, although not by any means elaborate, is remarkably good, and has been much admired.

The boiler-houses are all of the same character of construction as Machinery Hall, differing of course somewhat in the details, especially in the sub-structure, sunken areas being required for placing the boilers and for fuel; but they present the same exterior appearance. They are provided with facilities for unloading coal directly from the cars, and have interior platforms so that visitors can have ready access. The Corliss boiler-house is perhaps the most interesting to the general visitor. Arranged around three sides of the interior—the fourth side next to Machinery Hall being the entrance with visitors' platform—are twenty Corliss upright boilers of seventy horse-power each. An underground connection by an eighteen-inch pipe three hundred and twenty feet long, passing through a tunnel, carries the steam to the great engine which has been fully described under "Mechanics and Science." Two huge brick chimneys, quite ornamental in construction, connect with the flues from the boiler furnaces and provide the necessary draught. As one leans over the railing around the visitors' platform and looks down into the area below, it seems difficult to imagine that the quiet attendants who so leisurely pile the coal into the furnaces and try the various gauge-cocks, are the active agents in whose control is the generation of that mighty power, steam, which drives all of the shafting and gives motion to the numerous machines executing such varied and complicated work. And yet so it is. Neglect on their part and all would stop; the great wheels would remain silent, the busy hum would cease, and the curious machinery would lose its life. Strolling back into the Machinery Hall, we are startled to find that what was in our thoughts has really occurred: the great Corliss engine has stopped; all activity has come to an end. It cannot be that anything is wrong. No! a moment's reflection assures us that it is only the hour of noon. Machines, like men, require repose. Work them continuously and they become technically "tired," and will soon fail if not properly cared for. The mid-day rest is essential also to the attendants and must be provided for. The vast crowds seek the open air, some to the restaurants, and others to employ their time elsewhere until the hour is over. We are drawn to the northern part of the building by the sound of singing, which we discover to proceed from a party of colored attendants belonging to the great tobacco factory of Richmond, Virginia, who, during the silence of the machinery, entertain the visitors with plantation melodies, rendered in that peculiar and attractive manner, as can only be done by natives to the soil. After enjoying this for a time, we move on, passing out at a door on the north side, and wander along the banks of the charming lake which has been so judiciously placed and gives such life to the landscape. The beautiful fountain in the centre sparkles and dances in the sun, keeping time to the music of the chimes, as the sounds of "Angels ever bright and fair," "Home, sweet Home," etc., float out from the eastern tower and gladden the ear. We pass a statue of Elias Howe, the great sewing-machine inventor, and move on, stopping a moment in Fountain Avenue to study Colonel Lienard's relief-plan of the city of Paris. When we saw it a few weeks ago it looked very neat and pretty, the ground having been graded to represent the undulations of the city, and some of the principal buildings sufficiently correct

The Swedish School-house.

in outline to be recognized, although there was a striking similarity in the various blocks of houses in the different streets, which does not appear in Paris itself; and we observed what we never knew before, that all the bridges over the Seine were of exactly the same pattern. Neglect to cover the ground with asphalt or cement before laying out the city, however, has resulted in allowing the weeds to grow up, which although at first resembling trees in the streets, have now grown so far beyond the proper proportions for such purpose as to produce a very curious effect, a city gone to seed—Paris after a reign of the Communists—and to require the speedy advent of a care-taker to restore the model to its original condition.

A few steps more and we reach the Government Building. This was constructed and paid for from a fund specially provided by the Government to furnish a place for the display of exhibits from the several Executive departments of the United States. It was desired by the Government "that from the Executive departments in which there might be specimens suitable for the purpose intended, there should appear such articles and materials as would, when presented in a collective exhibition, illustrate the functions and administrative faculties of the Government in time of peace, and its resources as a war power, and thereby serve to demonstrate the nature of our institutions and their adaptation to the wants of the people." For the purpose of securing this, the President appointed a Board composed of a representative from each of the Executive Departments of the Government, except the Department of State and the Attorney-General's Department, but including the Department of Agriculture and the Smithsonian Institution; and this Board was charged with the duty of perfecting the collective exhibit which we see before us. All took great interest in the matter, and their efforts have resulted in a display alike creditable to the country and attractive to the visitor. The building while of simple construction is quite effective in appearance and very creditable to its architect. It is designed in plan in the form of a Latin cross, covering an area of one hundred and two thousand eight hundred and forty square feet, and is constructed of wood and plaster, with a roof similar to that on Machinery Hall, and a low octagonal tower at the intersection of the nave and transept. Its extreme length is four hundred and eighty feet, and width three hundred and forty feet. The transept is one hundred feet wide, divided into a centre span of sixty feet, and two side spans of twenty feet each. The nave is one hundred and eighty feet wide, and consists of seven spans, one of sixty feet in the centre, and three of twenty feet on each side. The short arm of the building opposite to the nave has a width of two hundred and twenty feet, being increased over that of the nave by two spans of twenty feet each. The roofs are supported by columns twenty feet apart, giving an entirely unobstructed open area through the whole building. The total floor-surface available for exhibition purposes is eighty-five thousand eight hundred square feet, divided up for the several departments as follows:—

Post-Office Department,	6,000 square feet.
Agricultural Department,	6,000 "
Interior Department,	20,600 "
Smithsonian Institute and Food Fishes,	26,600 "
Navy Department,	10,400 "
War Department,	11,200 "
Treasury Department,	5,000 "
Total,	85,800 "

On entering the building there is found on one side the Post-Office Department, where are specimens of all the paraphernalia and apparatus for carrying and distributing the mails, maps showing the different mail-routes, all the different kinds of stamps, envelopes, and everything pertaining to mail service. Exterior to the building is one of the cars in complete working order, as built for the fast mail now in operation on our main trunk railroad lines, and performing such rapid and effective duty.

On the other side, in the Agricultural Department, are collected samples of all the varieties of grain grown in the United States, stuffed specimens of domestic fowls, sample-bottles of the different native wines, etc.; and on the walls are charts showing the areas in which are grown the various products of the soil—cotton, corn, wheat, etc.—with the comparative amounts in each locality, represented by intensity of color,—giving so much instruction at a glance.

As we advance, we come to the Department of the Interior, where a large space is

The Chinese Court—Main Building.

occupied with models for the Patent Office, also various appliances for educational purposes, exhibits giving information in reference to the Indian Territory and the Indians, their education, occupations, etc.; and finally in the far corner we come to that collection so absorbing to the antiquarian, which shows to the visitor the few evidences that have been discovered in reference to the most remarkable and ancient inhabitants known to have existed on this continent—the dwellers of the cave-cities.

Across the aisle from the Interior Department is the exhibit of the Smithsonian Institution and the splendid collection of food fishes, so interesting and instructive.

In the front of the building, on one side, is the Navy Department, where are seen models of various vessels—the "Constitution," hospital-ships, school-ships, etc.—life-size figures of soldiers and sailors in old-time costumes, specimens of the various signals used at sea, the different weapons of marine warfare, and numerous other objects of naval interest.

In the War Department the complete machinery for the manufacture of arms, cartridges, etc., is shown, and a very interesting collection of fire-arms from early days to the present time.

In the Engineering Department are to be found many models of fortifications and other works, including a very fine one of General Newton's Hell Gate improvements; also maps and statistics of the Coast Survey.

In the Treasury Department the exhibit of the Light-House Board is particularly noticeable; and on these hot summer days it is a great relief to look across the aisle at the immense refrigerators, through the glass sides of which may be seen fruit, fish. etc., frozen solid, and appearing so refreshingly cool.

In passing from the building, let us pause at the Trois Frères Restaurant for refreshment. We obtain it, but are glad to get away. What is furnished is good, but the prices are exorbitant, and the waiters insolent. The building itself, especially the rear view on the lake, is a disgrace to the Exhibition, a blot on the landscape, and should never have been allowed an existence.

Around the Government Building are grouped several other small buildings connected with the Government exhibit, the most important being the Post Hospital of the Medical Department, one wing of which is fitted up with twenty-four beds, bedding and other furniture as for actual service. In the remainder of this building, and in sheds and tents adjoining, are exhibited a complete series of the medical supplies used in the army, comprising medicines, medical and surgical instruments, hospital stores, clothing and furniture, meteorological instruments, etc., all the various blank forms and record-books of the Medical Department, a set of the publications of the Surgeon-General's office, selected medical, surgical, anatomical, and microscopical specimens, models of barrack hospitals, railroad-cars for transportation of sick and wounded, and hospital steamships and steamboats. There is also a selection of full-sized ambulances and medicine wagons.

The Signal Service make an interesting exhibit, to the west of the Government Building, showing a full telegraphic train of wagons with outfit complete, telegraphic tower, international and cautionary signal outfits, and a full assortment of barometers, thermometers, anemometers, etc., etc.

From here we stroll up Fountain Avenue and on towards Horticultural Hall, past lovely parterres of flowers, exquisite sunken beds in masses of color, clusters of shrubbery and roses, and groups of sub-tropical plants—the whole Lansdowne plateau between Belmont Avenue and the Hall being laid out in a way to equal even European gardens in its beauty and effect. As we approach the Hall we see to the north a sort of tent, or frame-work covered with canvas, around which a crowd is gathered, and, like all Americans, not being able to resist the desire to satisfy our curiosity, we are drawn towards it. Reaching the doorway we involuntarily give an exclamation of delight at the beautiful sight which bursts upon our view. The interior is arranged with winding walks, and mounds planted out with an assortment of English Rhododendrons, covered with lovely flowers, large in size, and varying in color from the deepest purple, crimson, pink, and cherry, to pure white,—

The Russian Educational Department—Main Building.

a perfect feast to the eye, and something that one can never forget. These plants were brought from England early in the Spring, with great care, and have surpassed all expectation in the magnificence of their bloom, enabling those who have not already seen them abroad to form some idea of the effect that acres of such plants produce at the great English country seats. With proper care, shading by groups of trees, and protecting from our hot suns and severe winters, these plants will do equally well in this country, contrary to the usually received ideas on this subject, as has been amply proved by some of our best horticulturists and amateurs. Too often they are badly treated, and result only in disappointment. The American Rhododendron is the name by which it is known abroad. These handsome varieties have a common stock in the wild rhododendron of our Alleghenies, and if treated as that which grows in the clearings on the borders of the forests, they will amply repay cultivation.

Horticultural Hall is well located on a terrace, and is an extremely ornate and commodious building, constructed in the Moorish style from appropriations made by the city of Philadelphia, to whose munificence is due this permanent attraction to the Park. The building is three hundred and eighty-three feet long, one hundred and ninety-three feet wide, and seventy-two feet high to the top of the lantern. The central portion is a conservatory of two hundred and thirty by eighty feet in area, and fifty-five feet in height, surmounted by a lantern one hundred and seventy feet long, twenty feet wide, and fourteen feet high. A gallery five feet wide extends entirely around the interior of this conservatory at a height of twenty feet from the floor. The forcing-houses, four in number, are on the north and south sides of this principal room, each covering a space of one hundred by thirty feet, and having an entrance vestibule to the conservatory, thirty feet square, dividing the two houses on each side. Similar vestibules exist at the centres of the east and west ends of the building, on either side of which are restaurants, reception-rooms, etc. Ornamental stairways from the vestibules lead to the internal galleries of the conservatory and also to four external galleries over the forcing-houses, each one hundred by ten feet. The latter connect with large platforms over the roofs of the vestibules and other ground-floor rooms, which give a superficial area for promenade purposes of eighteen hundred square yards. Flights of blue marble steps lead to the terrace around the building and to the entrances, open kiosques twenty feet in diameter being placed at the east and west ends. The basement is of fire-proof construction, and here are located the kitchens, store-rooms, heating apparatus, coal-houses, ash-pits, etc. The area for exhibition purposes amounts to one hundred and twenty-two thousand five hundred square feet. The framework of the building is of iron, the conservatory walls up to the gallery floor being of ornamental brick arches, an excellent specimen of work of this kind, supported on iron columns; and the forcing-houses are covered by curved roofs of iron and glass. The filling-in and finishing material of the building, where glass is not used, is principally of wood, which is very much to be regretted, as the damp atmosphere necessarily required in a conservatory will cause it to decay very rapidly, and soon demand its replacement with a more permanent material. The collection of plants within the conservatory is quite as good as could be expected, considering the extremely short time at command in which to make it, and the great difficulty of obtaining such large and capable conservatory plants as would here be required to produce an effect. Miss Foley's beautiful fountain in the centre adds very much to the *tout ensemble*. The impression produced by the exterior of the building as a whole is quite satisfactory, and the design of the ornamentation and coloring very pleasing. The grounds contiguous to the Horticultural Building and comprised within the jurisdiction of the Bureau of Horticulture, comprise over forty acres, and the manner in which they have been laid out does the Chief and his committee great credit.

Just south of the Horticultural Hall is an exhibit by a Cuban, of Cacti and kindred plants, very nicely arranged, presenting many quite curious varieties, and giving to the

The Women's Pavilion.

visitor an excellent idea of the landscape effect of these plants in their native climes. We glance over at the Philadelphia City Building, and then wend our way down the path along the edge of the Lansdowne Ravine, passing a small building belonging to the Bible Society, then quite a picturesque restaurant erected by the Milk Dairy Association, where pure milk and, without exception, the best ice cream at the Exhibition may be obtained, and soon reach Agricultural Avenue. In front of us is the building of the Brazilian Empire, and beyond, that of the German Government. Between the latter and Belmont Avenue is a little Moorish villa, and at the door stands an attendant, a Moor from Tangier, possessing fine features, thoroughly oriental in style, and dressed in turban and flowing yellow costume. Entering, we find the little building arranged and furnished throughout with hangings, divans, carpets, and furniture, all genuine and in complete keeping with the house a veritable Moorish villa on a small scale.

Retracing our way, we see the building of the Centennial Medical Department, a most useful and essential agency, where are treated all cases among the many visitors requiring prompt attention. Of all the beautiful features in these unrivaled Exhibition grounds, the Lansdowne Ravine carries off the palm. Its shady walks, winding in and out between the magnificent forest trees and among the undergrowth; the little babbling brook, as it leaps from stone to stone on its way towards the river, and its secluded and romantic aspect, all unite in inviting the visitor away from the crowds about him, to a contemplative stroll. From the pretty music pavilion on the other side of the ravine float the sweet strains of the Government Marine Band as it performs its afternoon programme. As we saunter down the walks into the deepest and darkest part of the ravine, we suddenly come upon a hunter's camp, perfect in all its details, and we are transported for a moment far away from the busy scenes around us to the distant western country, to the aboriginal forests. Here is a phase of American life that must be new and interesting not only to most of our foreign friends, but also to many of our own people from the eastern States, where civilization is rapidly crowding out all traces of colonial life.

The ravine is crossed by two bridges—one near its upper end, consisting of two spans of a braced arch carriage-bridge, erected as an exhibit by the King Bridge Company of Cleveland, Ohio, and used as a means of communication on foot over the ravine; and the other a more pretentious structure, across the lower part, on the boundary-line of the Exhibition grounds, built by the Centennial Board of Finance, partly as a footway for visitors to the Exhibition, and partly as a carriage-bridge for park purposes outside of the grounds, connecting the two portions of the Lansdowne Ravine which have been separated by the arrangement of the Exhibition area. The latter bridge consists of twelve spans, of which the three centre spans are eighty feet, the two intermediate ones sixty feet, and the seven end spans twenty feet each. The total length of superstructure, including spaces over piers and abutments, amounts to five hundred and fifteen feet, and there are approach walls at the north and south ends of forty-five and one hundred and twenty-five feet respectively, making an extreme length of hand-railing on bridge and approaches of six hundred and eighty-five feet. The width of roadway is sixty feet, and there are two outside footways of ten feet each, making a total width of bridge of eighty feet. The boundary-fence of the grounds extends over the bridge, including within its jurisdiction one footwalk and twenty feet of the roadway for Exhibition use. The distance from centre to centre of consecutive trusses in the same span is fifteen feet, and the projection of footwalks beyond the trusses seven and a half feet. The spans of sixty and eighty feet consist of single inter-

Belmont Avenue.

section, deck, Pratt trusses, with timber upper chords and posts, and wrought-iron lower chords and other tension members, vertical diagonal bracing being introduced between each of the posts in the trusses, and upper lateral bracing between the upper chords of the two outer trusses only, and continued to the abutments at the ends. Masonry foundations are used throughout, timber trestles being erected on the piers. These are neatly framed with combination-posts, the pieces firmly bolted and mortised together, forming a stiff, rigid system, and vertical diagonal bracing is placed between each of the posts. Wind-ties connect the lower chords of the truss-spans with the trestles, those on the outer ends of the sixty-feet spans being firmly bolted to the masonry. The foundations are carefully laid with the best quality of stone of good size and shape, and the masonry above ground is of rock range work, pointed with dark mortar, the coping and cap-stones being of light sand-stone, hammer-dressed, with sloping top and draft on the faces. The bearings at top and bottom of posts and on trestles, and the finish to hand-rail over abutments, are of cast iron. The lumber used in the structure is nearly all of white pine, that in the trestles, columns, truss-posts, upper chords, and lateral struts being dressed. The flooring on roadway consists of two thicknesses of two-inch plank, the lower layer being of white pine laid diagonally, and the upper of white oak laid at right angles to the line of travel. The foot-walks are also covered with two thicknesses of flooring, the lower layer being two-inch white pine, and the upper, one and a quarter inch yellow pine, tongued and grooved and laid longitudinally to the structure. The curbs are of white oak. The bridge has been very neatly painted in shades of buff, relieved with Indian red in the chamfers and on the ornamental parts, and presents quite a handsome appearance. The height of the floor above the ground at the centre of the structure is sixty-eight feet, and a very fine view up the Schuylkill River may be obtained from this point.

Returning along the avenue on the south side of the ravine, we first pass, with averted head, a building devoted to an exhibit exclusively of burial-caskets, very fine, no doubt, but hardly attractive to the pleasure-seeking crowd; then a very creditable structure erected by the Singer Sewing-Machine Company for its special exhibit, where is kept a book in which some one of all those registering their names may have the good fortune, at the close of the Exhibition, to draw a machine. We pause at an octagonal building containing the Pennsylvania Educational exhibit, where is a most interesting collection of apparatus, furniture and all appliances as used in the various grades of public schools in the State; also specimens from the schools of design, institutions for the blind, etc.

We are warned, however, by the fog-horn stationed near the Government Building, that the hour of closing draws near, and we must take our departure for the day. Foremost among the many excellent arrangements for the convenience of visitors to the Exhibition may be mentioned the numerous cheap routes to and fro, connecting with the very heart of the city. There are five lines of street-cars from the front on Elm Avenue, two steam railways—the Pennsylvania, and the Philadelphia and Reading—and finally the little pleasure paddle-boats on the Schuylkill. Our choice for this evening is the latter, and we turn down into Lansdowne Ravine, past the Sudreau restaurant, which, as we glance up to it, looks so invitingly cool and pleasant under the canopy over its flat-deck roof, that we inwardly determine to test its *cuisine* on the morrow. Passing down the lovely glen, we emerge into the open park beyond the Exhibition boundary, and soon reach the river. We take the opportunity to pause for a moment to examine the two large Worthington steam-pumps in a building near by, belonging to the water service of the Exhibition, and so efficiently performing their duty, and then go on to the little steamboat-landing. As the long summer day has not yet drawn to a

Interior of Women's Pavilion.

close, we decide to take the steamer up the river to the Falls, about two miles above, and so enjoy the round trip. Philadelphia may well be pardoned for boasting of her magnificent park. What city in the world has one so lovely, possessing such great natural advantages, and with so beautiful a river winding through it? We glide on, under the old bridge, past the wooded hills sloping down to the water's edge and making such exquisite reflections, and as we glance back we see the various buildings of the Exhibition in the distance, rising gradually out from among the masses of foliage, calling to mind Lewis's charming picture of "1876," as it now hangs in the Art Gallery. A few days ago one lovely afternoon we came this same way, but for how different a purpose! There are sad as well as joyful pictures in every phase of life. Then we landed at Laurel Hill, the "Home of the Dead," that we see on our right, with its white monuments peeping out from among the trees, and we joined a quiet little procession to the chapel. A young Englishman, a member of the British Commission to the Exhibition, was being carried to his last resting-place, far away from kindred and friends—no! not friends, for during his short stay of barely six months he endeared himself to a large circle, and made many warm friends, who mourned his untimely death, and sorrowfully paid him those last tributes of respect which are due from the living. There on a sunny slope, overlooking one of the loveliest scenes in this fair land, they laid him to await the Resurrection morn.

Another day, and after entering the Main Building we pass directly through to the Art Gallery on the north side, erected on the most commanding portion of the great Lansdowne Plateau, one hundred and sixteen feet above the river, and intended as a permanent memorial of this the Centennial year. Bearing this point in view, it is but natural that one should expect the building to represent in itself, in its design and construction, the progress that the nation has made in Engineering and Architecture during the past one hundred years. In this respect it is a disappointment. There is a want of harmony in the proportions; the dome should have been larger and higher, and a simpler and bolder treatment throughout, with less of commonplace ornamentation, would undoubtedly have produced a better result.

Passing up the steps of the approach, we see on our right and left the great bronze horses, so mistakenly brought from Vienna several years since, after rejection from the Grand Opera House of that city, and here so singularly given a chief place in our World's Fair embellishments. Further on and near the building are two handsome bronzes—on the right, Wolf's "Dead Lioness," and on the left, Mead's group of "The Navy," for the Lincoln monument at Springfield, Illinois. The building covers an area of about an acre and a half, and is intended to be fire-proof. Its general outline in plan is a rectangle of three hundred and sixty-five by two hundred feet, with a pavilion of forty-five feet square at each corner, and a central projection ninety-five feet in width on the south, extending ten feet beyond the general face of the building.

The style of architecture is the Renaissance. The central portion of the southern front, seventy-two feet in height, contains three colossal arched main-entrance doorways, and is connected on each side by groined arch arcades ninety feet long and forty-five feet high, with the corner pavilions, which are sixty feet in height. In the rear of these arcades are open courts, ninety by thirty-six feet, paved, ornamented with fountains and plants, and intended for the display of statuary. The main cornice of the front is surmounted with a balustrade having emblematic figures at the corners and candelabras at intermediate points. Over the centre of the structure an octagonal dome, on a square base, rises to a height of one hundred and thirty-four feet, crowned with a colossal figure of Columbia, the top of which is one hundred and fifty-seven and a half feet above the ground. On the four corners of the base are figures typical of the four quarters of the globe. The corner pavilions have large windows; the walls of the east and west sides are relieved by niches intended for statues, and the rear or north front is designed in the same general character

New Jersey State Building.

as the south front, except that the arcades are omitted, and in their place are walls pierced with windows.

The main entrance vestibule on the south is a hall of eighty-two by sixty feet, and fifty-three feet high, lighted by windows opening into the courts, and by transoms over the entrance-doors. Beyond, under the dome, is the grand central hall, eighty-three feet square, with a height of seventy-nine feet to the uppermost point of the ceiling. East and west of this are the two main picture-galleries, each ninety-eight feet long by eighty-four feet wide, and thirty-five feet high, connected with the central hall by three arched doorways, and forming with it a grand area of two hundred and eighty-seven by eighty-four feet, capable of containing eight thousand people. Beyond these are smaller picture-galleries of twenty-eight by eighty-nine feet, running north and south between the corner pavilions. A corridor of fourteen feet in width extends along the whole length of the north side of the central hall and main galleries, opening on its outer line into a series of private rooms intended for studios, etc., and having a second story of rooms above. The central hall is lighted from the dome, the main picture-galleries from the roof, and the smaller rooms by side windows.

The foundations of the building are of rubble masonry, the exterior walls of granite, backed with brick, and the inner walls of brick. On piers of masonry in the basement are iron columns supporting wrought-iron beams, which carry the flooring of brick arches and tile. Where a second story occurs, the floors are laid in the same way. Over the four corner pavilions and the small rooms on the north side of the building, the roofs are carried by light boiler-plate girders; while over the main vestibule, wrought-iron open trusses are used, the covering being of tin on sheathing-boards and wooden purlines. Wrought-iron trusses are also employed over the picture-galleries, supporting roof-lights of three-eighth inch rolled glass. The arrangement of the dome is somewhat unique. Had it rested on the main walls, it would have been of the same size as the grand central hall below; but in order to reduce it, four trusses parallel with these walls, and situated at a distance of eight and a half feet from them, are used as supports. The frame-work of the dome consists of sixteen, built, wrought-iron ribs, resting at their lower ends on these trusses, and meeting at the crown in a heavy wrought-iron ring, the whole forming in plan a square figure with the corners cut off. Horizontal tie-rods connect the opposite ribs together, and wind-bracing is introduced above. Horizontal struts at four points stiffen the ribs laterally, and assist to carry the iron frame-work for the glass. The false dome on the interior is constructed with a light frame-work of wrought iron, footing on the supporting trusses of the main dome, and having at the crown a ring twelve and a half feet in diameter.

The interior finish is of plaster, the heavy cornices, ceilings, mouldings, etc., being worked on light iron frame-work and wire-netting, and the ceilings of the picture-galleries are of glass in wooden frames, suspended from the roof-trusses. Steam-heating is provided from boilers in the basement.

We cannot stop to examine the pictures. The crowd is too great, and it would take days to study them properly. We can therefore only glance at them hurriedly, and move on to the next building, The Art Annex, north of Memorial Hall. The latter building having been found entirely too small for the vast collection being sent to the Exhibition, it became necessary at the last moment to erect this additional edifice, which was hurriedly constructed—a plain brick building without any pretension, consisting of a number of rooms, opening one into the other, and furnishing the requisite wall-space. Two corridors, each twenty feet in width, cross each other at right angles at the centre of the building. Passing in at the south entrance, we find ourselves in a large gallery, one hundred by fifty-four feet, devoted to statuary, paintings, and mosaics from Italy. Moving on through this, we enter the north and south corridor, on both sides of which are, first, a series of three, and then one of four galleries, those of a series being arranged one beyond the other, each room forty feet square. This brings us into the east and west corridor at the centre of the building. North of this the arrangement is symmetrical with that on the south, except that at the extreme north end

Main Building—Central Avenue looking West.

the corridor passes through to the entrance, with a gallery on each side, instead of all being thrown into one large room, as at the south end. The various galleries communicate with each other at their corners—an excellent arrangement, securing the greatest possible amount of useful wall-surface. The floor-area of the building amounts to sixty-four thousand two hundred square feet, and the available space for paintings to sixty thousand square feet.

Making our exit at the eastern doorway, we find ourselves in close proximity to three buildings connected with the French Government exhibit—one being devoted to Stained Glass, another to Lines of Art, and the third and largest to the exhibit of the Department of Public Works. The latter building is interesting as being entirely fire-proof, and constructed of iron, brick, tiles, and glass, all of which were brought from France for the purpose. The collection in it is one of great value and of special interest to the engineer, consisting of beautiful models of famous works erected under direction of the Government, a more detailed description of which will be found under the head of "Mechanics and Science."

We are now so close to the Vienna Bakery that we cannot resist the temptation to try some of Gaff, Fleischmann & Co.'s world-renowned bread, with one of those delicious cups of "Chocolat à la Crême" which so delight the taste. Thus refreshed, after glancing in at the large windows and observing the process of bread-making, we move on, and taking a look at the police barracks and fire-patrol buildings, so useful if not otherwise interesting, we are attracted by a neat structure having the appearance of a railroad station, and which we find to be an exhibit of the Empire Transportation Company. Exterior to the building is a section of railway track, on which is a beautiful and complete working model of a locomotive, one-fourth full size, drawing a train of model freight cars. In the interior are seen a most interesting set of models, very fully illustrating the freight shipping business of this Company; propellers and grain-elevators of the Lake Transportation; models of petroleum oil-wells in working order, with all the adjuncts; oil pipe lines, showing the method of loading the oil on cars; models of shipping piers and depots on the large rivers, and many other matters of great interest.

Near by is the Bankers' exhibit, and further on is the Photographic Building, quite a neat structure, well adapted for its purpose, and covering an area of two hundred and fifty-eight by one hundred and seven feet. The walls are crowded with admirable displays of photographs from almost all parts of the civilized world, and some of the English landscapes are perfectly exquisite, far surpassing the most extravagant hopes of the photographer of fifteen years ago. Moving through this building, and continuing on the avenue past the front of Memorial Hall, we arrive at the Carriage Annex, a building three hundred and forty-six by two hundred and thirty-one feet, constructed of timber-framing, covered with corrugated iron, in which is made an exceedingly handsome display of carriages from many of the prominent builders of the world. Here are found the famous London drags, Philadelphia phætons, beautiful carriages from San Francisco, sleighs from Russia, also Pullman palace-cars, and in one part of the building, household appliances, cooking-ranges, etc.

Just to the rear of the Carriage Annex is the Swedish School-house, one of the most thoroughly national buildings on the grounds. It covers an area of about sixty by thirty-six feet, and is twenty-five feet high, being designed exactly after the school-houses used in

Sweden, although more neatly finished on the exterior. It is constructed of white pine logs, laid horizontally, with the curved faces visible, and having a bold roof, carried on massive brackets formed by the projecting ends of the logs. The material was all framed and brought

Main Building—Japanese Court.

from Sweden ready for erection. We are met at the entrance and conducted through the building by a most genial gentleman, Mr. C. J. Meyerberg, the Swedish Commissioner for the Educational Department, and one of the first Government school-inspectors of Stockholm. Nothing receives so much attention in Sweden as the subject of education, and every inhabitant of the country is placed on an equal footing in this respect, it being not only free but obligatory, the Government paying for the entire cost. No difference is made between the children of the peasant and those of the nobleman; each may acquire precisely the same education. If the workman is too poor to clothe his child for school, the Government does it. If he refuses to send him, and prefers to keep him at work, he is summoned first before his clergyman, and if that fails, then before the Board of Education. If obedience is still declined, the Government has the power to put the parents into the work-house, and take the children and educate them.

The building which we see before us represents a country school-house intended for primary classes, and capable of accommodating about fifty children. Its dimensions, light and ventilation are all regulated by law, the school-room being forty by twenty-two feet, and twelve feet high, giving two hundred and eleven and two-tenths cubic feet of air, and seventeen and six-tenths square feet of floor-space, to each scholar, the area of the windows being such as to allow three and six-tenths square feet per child. Two rooms on the ground-floor next to the main room are provided for the schoolmaster as a dwelling, and the upper story gives a sleeping-room and store-room. Outside is a space of ground for a garden, where he may practically instruct the scholars in horticulture.

Every appliance is provided in the school-room that will facilitate the teacher in imparting instruction, as well as add to the comfort and health of the children. The science of object-teaching appears to have been well considered. We see here cubes and other geometrical forms; bundles of sticks for counting; an abacus, or instrument for performing calculations by means of balls on wires; maps of various districts of Sweden, giving the mountain-ranges, the political divisions, the water-surface, etc.; and illustrations showing the occupations of the inhabitants in the different parts of the country, such as lumbering, mining, hunting and fishing. There are also good collections of minerals, fine specimens of pressed plants, insects, stuffed animals, shells, etc.; a cabinet-organ, and good serviceable and comfortable desks and stools. Everything in the school-room is characterized by extreme cheapness, with good quality and solid usefulness, and the brightness and attractiveness of the room is in marked contrast with the bare school-rooms of our own country. How much more likely is a child rendered willing to go to school and to study if everything is made pleasant and cheerful around him, instead of being dull and gloomy.

The morning hours are occupied in study, and the afternoon, for boys in practical instruction in carpentry, cabinet-work, drawing, boot-making, and other trades, while for girls, in sewing, knitting or drawing. Every one in Sweden learns at least to read and write. No one can be confirmed without so doing, and all must remain at school until that time, the minimum age being twelve years. No one can be married, give evidence or become a soldier unless confirmed. The conscription takes place in Sweden between the ages of twenty and twenty-five, and all must necessarily have received at least an elementary education.

The lowest class of school is the infant, the teachers generally being females; the course of instruction comprising reading, writing, arithmetic, history, singing and religion. The latter is not taught if the child does not belong to the established faith, unless it is particularly so requested. Then follows the primary school, as here seen, where are taught in addition, natural history, physics, geography, grammar, drawing, geometry, chemistry, gymnastics and drilling. The teachers of the primary schools must have been at least three years at the Normal school, and have obtained from it a certificate. Next come the higher primary schools, where the same instruction is continued, only of a higher grade, and then the grammar or high schools, of which there are two kinds. In one are taught German, French and English, mathematics and the natural sciences; while in the other, instruction is given also in Latin, Greek and Hebrew, thus preparing the student for either of the two universities, Upsala or Lund. In addition to all of these, there are seven higher technical schools, also special schools of navigation, agriculture and forestry. There are also national high schools, where persons of from twenty to forty years of age may be taught to

Ohio State Building.

be good citizens, instructed in the laws, municipal institutions, and general administration of the country; may be taught surveying, book-keeping, etc., and where once a week a sort of court of common council is held in order to train the people to administrative duties. It is to these schools, greatly aided by private contributions, that Sweden largely owes her high position of independence and truthfulness of character.

Having, however, spent too long a time already in this interesting spot, we take leave of our kind friend, and stepping across the avenue, find ourselves in front of the Japanese Bazaar. It is a long, low, wooden building, strictly national in its character and construction, built by Japanese workmen with materials brought from the "Kingdom of the Rising Sun." The north front is open with overhanging eaves, and here are counters on which are displayed the numerous goods for sale. The roof is covered with heavy corrugated earthen tiles, and the sides are enclosed with sliding wooden shutters and paper screens. Exquisite designs in woodwork and carvings adorn and beautify the building, and the ceilings, walls and floors are painted in tile patterns. The grounds adjoining are laid out as closely as possible in accordance with the rules of Japanese landscape-gardening. Two large catalpa trees with their long hanging beans stand in the foreground, and lend their aid to the effect, being decidedly in keeping with the scene, although natives to the soil. A little fountain occupies a place

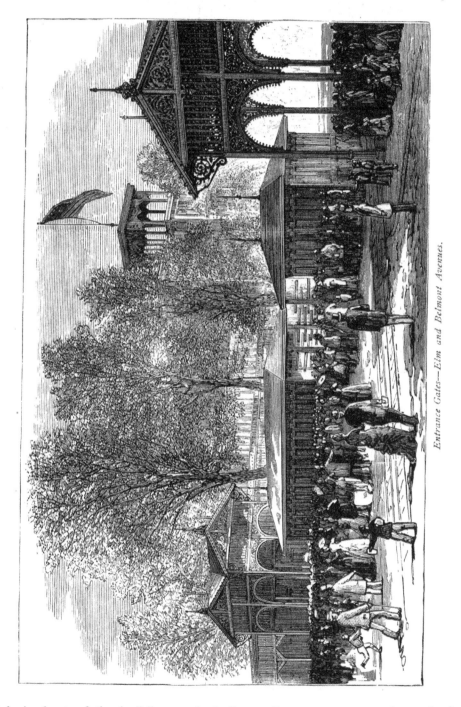

Entrance Gates—Elm and Belmont Avenues.

immediately in front of the building, and winding walks among grassy slopes lead down to the main avenue. A number of gigantic cranes in antique bronze are placed around under the trees, lifting their heads way up almost to the leaves, and one sees also on the ground some very curious bronze pigs, exceedingly ludicrous, and without the least particle of beauty, more like infant hippopotamuses than anything else. Lawn vases, urns and other adornments of strange design aid in giving a foreign air to the surroundings; and on the east side of the building is a most interesting garden, so quaint and so evidently entirely Japanese, that as we wander up and down the regular walks and look at the beds of lilies and strange flowers, it takes very little effort to imagine one's self transported to that far-off country in the Pacific Ocean. Here is a little square pit sunk in one corner of the garden for the cultivation of aquatic plants, and there are curious dwarf-trees, like the figures on the old vases at home, planted in pots and stood on odd-looking benches. Under a bamboo awning are certain plants which we presume could not be exposed to the strong rays of the sun. The counters in the front part of the building are crowded with articles of porcelain, bamboo and lacquered ware, which are being disposed of in large quantities, the courteous Japanese attendants attracting a large share of attention on their own personal account, many of our country friends having evidently never seen a "Jap" before.

Between us and the Avenue of the Republic is the Department of Public Comfort, a building erected for the comfort and convenience of visitors, having a frontage of two hundred and seventy-five feet, with a depth of one hundred feet, and containing restaurants and reception-rooms, halls and baggage-rooms, telegraph-rooms where messages may be forwarded to all parts of the world, and rooms for the United States Centennial Commission.

Next to this building, on the west, is the Judges' Hall, fronting directly on the centre of the large open space between the Main Building and Machinery Hall. It is a neat, plain building, with rather a pleasing outline, having an arched roof over the centre portion, showing the construction lines, and ornamented with two belfries. It is designed for the meetings and discussions of the Judges of the Exhibition, and for all business connected with the giving of awards. It contains a large central hall having a gallery, and surrounded with two stories of small committee-rooms for the different groups of Judges. It was here in the latter part of June that the brilliant and interesting wedding-ceremony of the marriage of the daughter of the Swedish Commissioner-General to the Norwegian Commissioner took place—a pleasant and novel incident, of the kind seldom recorded in the history of exhibitions, and adding a charm to the memories of the year.

In the corner next to Belmont Avenue is the Pennsylvania Railroad Ticket Office, and directly across the way, the establishment of the world-renowned Cook Tourist Agency, a little many-sided building, in the rear of which is pitched a Palestine encampment, illustrative of the manner in which the "Cookies," as a noted traveller calls them, are taken care of when journeying through that country. In the interior of the building is a bazaar for the sale of ornaments in olive-wood, and various other oriental articles of bijoutry, and in one little room is a genuine mummy, claiming to be a princess of the royal blood, who departed this life some two thousand years ago, not exhibiting any special beauty, however, at the present time, nor showing stronger credentials than the thousand and one other mummies that have been exhumed on the banks of the Nile.

Nearly opposite to Cook's office is the building of the Centennial Photographic Association, under whose exclusive direction all the photographs of the Exhibition are taken. It is very conveniently arranged for its purposes, having facilities for doing a large amount

INTERNATIONAL EXHIBITION, 1876.

Horticultural Hall and Grounds.

of work, and to one interested in the details of the art, this department of the Exhibition is well worth a visit.

We will close the labors of the day by one of the really most delightful pleasures that the Exhibition affords, especially towards evening, when the heated hours are past, and the tired body needs a little rest before starting for home. We will take an excursion on the narrow-gauge railway. This feature is something entirely new, has never been introduced before at an exhibition, and has proved a most signal success. Nothing gives the visitor a better idea of the topography and extent of the grounds, and fixes more satisfactorily in his mind the locations of the various buildings, than a series of trips on this line. Entering at the station in front of the Department of Public Comfort and paying our five cents fare, a train draws up and we are soon off. The cars are open and airy, being merely platform-cars with seats, and roof supported by posts, admitting an unobstructed view on all sides. We turn the corner into Belmont Avenue at a rapid rate, the summer breeze fanning our cheeks, and the little "Emma," a diminutive locomotive, an infant among engines, as it were, puffing and blowing, but coming up nobly to the work. On we go, past the lake with its fountain glistening in the setting sun, and we draw up in a few moments at the station near the Women's Pavilion, having only just time, before we stop, to glance at the lovely grounds and flowers in the direction of Horticultural Hall. Then on we start again, passing the picturesque New Jersey Building, and suddenly swinging on a curve to our right, around the Southern Restaurant, we skirt the upper part of Belmont Ravine, and come out in front of the Agricultural Building, where we make another station. Near by is the American Restaurant, and under the trees we see the little tables, all crowded with hungry and thirsty occupants. Then by a series of graceful curves we leave all this in the distance, and reach the land of wind-mills. Mounting a steep grade, and curving to our left, we approach the north side of the Agricultural Building. The grade now descends rapidly; our speed is increased until one almost holds his breath, and sweeping round the corner of this building we find that we have doubled on ourselves, and are back again at the Southern Restaurant, the four tracks all running together for a short distance. Here, however, we take a new departure, and darting across Belmont Avenue in the rear of the Government Building, we go through, as yet to us, unexplored ground. On our right are the State Buildings—Ohio, Indiana, Illinois, etc.—one after the other, until the British Government Buildings loom into sight. As we turn and look back towards the Main Building, we are treated to one of the most beautiful sights it has ever been our fortune to witness. We are on a slightly rising slope, and the whole extent of the Main Building and Machinery Hall, nearly four thousand feet in length, comes into view. The Main Building is one blaze of light, of flaming fire, from end to end, owing to the reflections on the glass of the rays from the departing sun. It is a grand illumination. In the foreground the fountain has ceased to play, and the now quiet lake, a bright gem in its green setting, reflects every line and flash. The dome of Memorial Hall looks up over the trees, and the lesser buildings are grouped at various points. Restless, happy crowds are flitting from point to point, and the whole looks like fairy-land, an incantation scene, something that we wish would never pass away.

But our train moves on, and sweeping through a village of buildings, we take another great circle, and turning past the west front of Machinery Hall, are back into the Avenue of the Republic, soon reaching our starting-point. All in fifteen minutes! is it possible? Round the world in miniature in fifteen minutes? It is so exhilarating and so enjoyable that we must do it again; and thus it is that nearly every evening during our stay do we close with this dessert, as it were, of the day's feast, until every feature of the grounds becomes thoroughly impressed upon the mind, never to be effaced.

On our next visit we pass through the Belmont entrance directly to the Women's Pavilion, the site of women's work. Yes, all by women—the money furnished for the building, and all of the exhibits made exclusively by women. And much more than this has been done by them. Where would the Exhibition have been to-day, and its international features, if it had not been for Women?

Early in 1873, at the suggestion of the Citizens' Committee of the Board of Finance, it was decided to enlist the efforts of women in the cause of the Centennial Exhibition, and on the 16th of February of that year, thirteen patriotic ladies, citizens of Philadelphia, were named and invited to convene for discussion of the subject. They were met by the President of the Centennial Board of Finance and several other gentlemen, and after the objects of the meeting had been stated, these gentlemen withdrew, leaving matters entirely in the hands of the women to manage for themselves. The ladies came together necessarily without any very clear ideas of what was to be their work, and unaccustomed to business management. Their first move was to elect Mrs. E. D. Gillespie as President, as they insisted that nothing could be done without her in charge. She accepted the honor conferred upon her, but, to use her own words, "she felt the position a novel one: she had never before presided, and she did not know what to do. She thinks she suggested to some one to adjourn." So the ladies accordingly adjourned to meet the Monday following, at which time it was agreed that the main object was to arrange some plan for raising public enthusiasm. It was decided to commence work in the city and to take it by wards, endeavoring to enlist the services of all who could give time to the cause, and as thirteen women had been chosen to represent the thirteen original States, it was suggested that a chairman be appointed for each ward, with a sub-committee of thirty-six to represent the present number of States in the Union, and that these women should solicit subscriptions to Centennial stock. So the work began, and by the 9th of June of the same year the ladies had already collected fifty thousand dollars in stock subscriptions.

In the meantime the Executive Committee of the Commission had sanctioned the appointment of the women, and the Commission itself had passed complimentary resolutions, not containing much, it is true, but sufficient to give official recognizance to the organization.

Although the great mass of its work was naturally in Philadelphia, especially during the earlier periods of its existence, yet the Committee soon decided that it would not confine itself to local work, but would branch out all over the country. Communications were addressed to the Commissioners of the different States and to the Governors of each State and Territory, asking for the names of suitable ladies to represent their districts, who would be willing to work for the cause. From many of these no answers were received, and from others came replies that were worth nothing, merely expressing great interest in the

Centennial, and containing promises to write when the proper person was found. The disappointments were great, but by continued efforts the ladies were finally rewarded in uniting to the organization the women of thirty-one States.

In starting the work in Philadelphia, each lady took under her charge two or three wards, and distributed an appeal written by the President. It was not always possible to form sub-committees of thirty-six, but as soon as a sufficient number of capable and willing ladies had been found, the work was started. Stock subscriptions were obtained through personal solicitation from door to door, a large part of the collections being made in this way. These women, many of whom knew nothing of book-keeping nor of business matters, soon learned to bring in their accounts showing the stock subscribed, the number of shares, the name of the subscriber, the amount of money, and the name of the lady collecting, all in systematic

Connecticut State Building.

shape. In addition there were proceeds from tea-parties, loan exhibitions and other entertainments, the success of these being due to hard and continuous labor by women alone. In the short space of three months the subscriptions added to the Treasury of the Board of Finance by the women amounted to forty-two thousand and sixty dollars, sums being obtained in many cases from those who would have been reached in no other way, and in all cases after the wealthy and business class had contributed, the women being "patient gleaners in a field that had already yielded a rich harvest."

If prosperity had continued, there is no doubt but that the women of this country would have collected at least a million of dollars, but with the autumn of 1873 came the panic, and the work practically sank to nothing. Application was made to be allowed to collect money in smaller sums than the ten-dollar stock subscriptions, but this could not be granted, as the Board of Finance had no power to allow it.

Then came the news from Washington that the International feature of the Exhibition was in peril, and fourteen women, representing that number of States, went to Washington, with Mrs. Gillespie at the head, including also Mrs. Goddard, the grand-daughter of General Cass, and Mrs. Etting, the grand-daughter of Roger B. Taney. These patriotic ladies worked nobly, using every effort in their power to the sustaining of the international feature of the Exhibition. Letters from women all over the country, especially from the Southern States, were printed and distributed among the members of Congress. An audience was granted by the Committee on Appropriations, and there is no doubt but that the favorable report of this Committee for the retention of the international feature was largely owing to the labors of these women.

Then in the midst of this came a time when it was judged essential to petition the Councils of Philadelphia for an additional appropriation of one million of dollars. To obtain this it seemed necessary to give evidence that the request was approved of by a large number of citizens, and the women were the only ones that could be called upon to obtain their names. The President was telegraphed to return immediately on important business. She was met by one of the prominent gentlemen of the Board of Finance, who said: "Mrs. Gillespie, we want signatures of citizens to a petition to go before Councils asking for another million. We think that the fact that Philadelphia has given another million will operate on the minds of many. The petition must be laid before them in two days. We have no authority or organization to collect these signatures, and *thee has.*" The chairmen of the ward committees were telegraphed, the work was cheerfully started and vigorously pushed from door to door, and on the day appointed the petition was returned with eighty-two thousand signatures, and the desired appropriation was granted.

Mrs. Gillespie then returned to Washington to her duties there, came home, and again went back. It was at this time that the meeting with the Appropriation Committee took place. One gentleman connected with Congressional matters, who was violently opposed to the Exhibition, said to Mrs. Gillespie, with eyes flashing, "I don't like a female lobby!" But said Mrs. Gillespie, "Major, we have not lobbied: we merely came here to interview several of the Senators." Then said the Major, furiously, "The most effective lobbying is done on the other end of the avenue. I can count on my fingers the names of the gentlemen you have won over."

Prominent among the pleasurable incidents in connection with the Centennial were the entertainments organized by the ladies for the purpose of adding to their collections. Late one afternoon in the autumn of 1873, during a meeting of the ward chairmen of the Ladies' Executive Committee, a gentleman from Gloucester, New Jersey, came in and said, "I called to ask whether you knew that the Centennial Anniversary of the Boston Tea-Party will take place on the 16th of December of this year." The ladies had not thought of it. He then said that he "had come to suggest that there should be a tea-party in each ward of the city in commemoration of it." This, however, did not seem feasible, but taking the benefit of his suggestion, the President thought she could manage to have a "big" tea-party in the Academy of Music. So it was finally arranged, and the tea-party was given on the evening of December 17th, and repeated the night after, the tea being served by the ladies and their aids in Martha Washington costume, and special tea-cups, made for the purpose, provided, which were sold as mementoes. Three thousand of these cups were ordered, and the President was frightened and thought she would never be able to pay for them, until reassured by a gentleman friend devoted to the cause, who promised to take the responsibility. So far, however, from losing on them, more had to be supplied, and ten thousand were disposed of. The price paid for them was one dollar and fifty cents per dozen, and they were sold at twenty-five cents apiece. After this, tea-parties became the rage. The tea-cups used at a party in Cincinnati were painted by the ladies with their own fair hands.

On January 26th, 1874, was held in the Academy of Music the Washington Assembly, a

Interior of the New England Kitchen.

superb affair. Then in June there was a Fête Champêtre in the Park, a splendid success in every respect except financially, as crowds of people came who did not pay. The next year stock subscriptions came in more rapidly. On the twenty-second and twenty-third of February, 1875, were given two International Assemblies, where the ladies wore costumes representing different nationalities. By these two entertainments the sum of fourteen thousand dollars was realized after all expenses were paid. On the same dates of the year following, the Carnival of Authors netted over eight thousand dollars.

In other localities the same course was pursued. The work was opened in Rhode Island with a Martha Washington Tea-Party at which was cleared thirty-six hundred dollars. Afterwards the one hundred and third anniversary of the Burning of the "Gaspey" was celebrated by a clam-bake, enlivened with music and the actual burning of a rigged vessel representing the "Gaspey," the evening closing with fireworks. Everything was arranged and managed entirely by the ladies, except the rigging and operating of the boat. Ten thousand people were on the grounds, and between two and three thousand dollars were gained for the cause.

During all this time, in addition to the assistance which the ladies were pledged to give to the Board of Finance, one project lay very near their hearts—an exhibit of women's work, separate and distinct from all others, by which those who obtain such a scanty subsistence by the labor of their needle would have an opportunity of seeing what could be done by their sex in other and higher branches of industry. They would then discover that some women had gone far ahead of others, and the more timid would be encouraged to that perseverance which is almost always sure to bring success. The first intention was to have a separate space in the Main Building, but the proportions of the general exhibition increased so rapidly as soon to make it evident that it would be impossible to afford the ladies the area they required. Steam-power was wanted by the women of Massachusetts for the female operatives of Lowell, the Educational Committee desired to have a Kindergarten, and so many applications came pouring in for all sorts of exhibits of women's work, that nothing but a large space would satisfy them. The ladies had made up their minds to have plenty of room for a complete exhibit, and it was in danger of not being obtained.

Then it was that the proposition was made for a separate structure, to be paid for from the contributions raised by themselves, and such a building, it was found, could be erected for the sum of thirty thousand dollars. After proper consideration it was decided upon, and application was made to the ladies of the different States for assistance. The first answer came from Florida with forty-seven dollars and twenty-five cents. Then Rhode Island pledged herself for three thousand dollars, after which Philadelphia gave five thousand dollars which she had already raised, and trusted to Providence for more. Massachusetts came forward with five thousand, and Trenton and Camden each with one thousand dollars. A noble countrywoman abroad, on hearing of the project, sent a check for five thousand dollars, and subscriptions continued to be given in until the money was raised. While all this was going on, the ladies still continued their efforts to obtain stock subscriptions, and succeeded in disposing of over a hundred shares.

America should be proud of her women. From the days of the Revolution until now they have always come nobly forward when occasion required, and their work for the Centennial is their crowning glory. They have obtained subscriptions to ninety-six hundred shares of stock, and have collected one hundred and seven thousand three hundred and sixty-three dollars and twenty-eight cents additional, of which they gave to the Board of Finance, as a free gift, eight thousand four hundred and forty-eight dollars and eleven cents, and from the balance paid for their building and all of their expenses, including the running expenses during the Exhibition. From this fund also came the cost of the grand Chorus for the opening and closing days, and the price of four thousand three hundred and seventy dollars for the Centennial March by Wagner. Besides the amount given above, six thousand dollars was collected from the sale of commemorative medals.

The labors that have given these results have been voluntary—for love of country—and full credit should be awarded for the patriotism which induced them. It may be proper to state that the only pecuniary allowance made by the Board of Finance to the women's organization has been fifty dollars per month for a Secretary, and since the month of June, 1874, a salary of seventy-five dollars per month to the President—nothing beyond this. Only one thing would have perfected the women's work: they should have had a woman for architect, and this could have been done if it had been thought of in time.

Pennsylvania State Building.

The plan of building is not as well adapted to its purpose as it might be, but the ladies have endeavored to do their best with it. One-fourth of the area is devoted to foreign countries, and among the articles here displayed is some of the handiwork of Queen Victoria and also of her daughters. One-third of the building is assigned to works of Art, and the Schools of Design of New York, Boston, Lowell, Cincinnati, and Pittsburgh have their exhibits here. Mrs. Wormley's microscopic engravings and the wood-carvings of the women of Cincinnati are especially worthy of attention, attracting large crowds. Many lady artists refused to have their work classified with that of their own sex, preferring to put it in the general Art Department, but still there is much here that is creditable. We find in this building a weekly newspaper, called "The New Century for Women," its entire make-up taking place here, printed and published exclusively by women, and giving a full and exact account of the exhibits of the Women's Department. Here also is "The National Cookery Book," compiled from receipts contributed by the women of the whole country. In "The Bureau of Charities" are the statistics of a great number of the charitable institutions of the world; and a large album from the Empress Augusta, of Germany, contains the pictures of such institutions under her charge in the city of Berlin. The Pharmaceutical Exhibit of the Women's Medical College in Philadelphia makes a fine display, showing great care in its preparation and scientific ability of the highest order.

In an annex near the Women's Pavilion is the Kindergarten—a genuine Froebel Kindergarten. The building was erected by the women, the contributions coming largely from the organization in Pawtucket, Rhode Island, and the expenses of the school are borne by a few ladies who have collected funds for the purpose. It is under the charge of a most able teacher, Miss Ruth R. Burritt, and the pupils are from the Nursery of the Northern Home for Friendless Children, a Philadelphia institution. They were placed in training with her for several months prior to the opening, and now certainly do full justice to her exertions. Crowds of visitors throng around the building to see the exemplification of Froebel's method of nature, and to ask questions in reference to it, making it difficult for one to obtain a favorable place of observation. We arrive in time for the opening exercises, and as we enter, the little orphans, dressed in pink and blue, so innocent, bright and happy, are standing in a circle, singing their morning hymn. This over, they are asked, one by one, for a little story, each telling what he or she had done that morning or the day before, or perhaps one of them making up a little narrative; exercise being thus given to their memories in an interesting way, at the same time teaching them to put their thoughts into shape and to express themselves properly. Then came some songs—"Happy every morning," "Little Mamie loves to wander from one child to another," etc.—followed by a vigorous march, also, to a "Happy, merry song," the children winding in and out among each other in regular figures, until all come opposite to their little tables. Each table has its upper surface laid out in regular squares by lines an inch apart running in opposite directions, and there is a comfortable seat provided. The squares formed by the lines are the units of measure for the child in all its work. To-day is what is called "Clay-Model day." Every day there is something different to do; sometimes it is working in colored papers, folding them into shapes, or weaving mats, and gaining a knowledge of colors; another time blocks are used—cubes, oblongs, cylinders, etc.—teaching the solid forms; or again it is something else. Nothing is made tiresome or monotonous. There is always variety, and never too long a time at one thing. It is a point never to make the child weary. Clay-modeling is a favorite occupation, and it is wonderful what the little children will do. Each one takes its seat, and a restless little boy, full of nervous energy, is allowed to work it off by giving him the distribution of the tools. A piece of oil-cloth is placed in front of each, and on it a small lump of moist clay. It is a refined way of playing mud-pies. First they make balls, then from these cubes, working the sides flat alternately, and learning the law of opposites. Afterwards each child is allowed to exercise his own invention and to model what he chooses. Some make birds' nests with eggs, some apples, and one bright little fellow that we watch makes a little baby in its bath-tub. Many of these things are rude, it is true, but all are inventive, all original, and some show considerable ability. No copying is allowed or thought of. After this is over, the things are put neatly away, hands are washed, and the children are marched again, first in a circle, each half passing in opposite directions and chaining into the other; then comes an arm- and foot-exercise, all done to singing, and after that rubber balls are brought out and held and tossed to song.

Main Building—Spanish Court.

Then the children march back to their seats and have a little lunch. Occasion is taken to teach them politeness, and to wait on and help one another. This over, the tiny napkins are folded carefully away, and some little play-songs follow, giving gymnastic exercise in the most pleasant manner, the songs being acted out. We have the "Jolly little Chickadees," "This is right and this is left," etc.; and when the happy play is over, they return to the tables to have the metallic rings. These are about an inch or more in diameter, some whole, some half and some quarter segments. One little boy takes them and lays a certain number of each kind, one at a time, before each child, placing them, by direction of the teacher, on the intersections of certain of the cross lines with which the table is covered. The teacher meanwhile takes occasion to explain all about the rings, how they are made, and where the iron to make them comes from. The children are then directed to place the rings in position—first a whole ring, then a half segment at the bottom, then one at the top, another at the right side, a fourth at the left, and so on until a certain figure is formed. Each child then gives his own idea of what this figure is like, these ideas for different children being quite varied. One says a fountain, and another a brush or a fan. Afterwards they are allowed to make their own figures, and it is curious to watch how they study out and work up such original designs. Now putting these away and taking up a maple-leaf, the teacher explains all about the tree from which it was plucked, shows the form of the leaf, impressing it upon their minds, tells its use and everything in reference to it. And afterwards, when the child walks through the woods, it picks out the maple-leaf, knows the tree from all the others, and is led to study nature as a delightful pastime.

Then come the closing exercises with song, and the little good-byes are said, and the courtesies and bows made, and all depart, bright and cheerful, not tired nor worn out, ready for their play, and anxious for the morrow to come with its Kindergarten again.

By the Froebel Kindergarten system the child is taught, unknown to itself, habits of order, attention, application, cheerful obedience, careful manipulation, and a knowledge of geometrical and natural forms and figures; and when the time comes for higher studies, he will be found far in advance of those who have not had these preliminary advantages.

Grouped near the Women's Pavilion are a number of very interesting structures. On the north is the New Jersey Building, one of the most picturesque and characteristic of the State edifices, and farther on is the Southern Restaurant. Near by is the Kansas Building, for which an appropriation of ten thousand dollars was granted by that State, the first of all to select a site in the Exhibition grounds for such a purpose. The interior is quite unique in its decorations. Around the sides are hung sheaves of wheat, rye and barley; under the dome is a fine bronze fountain from the ladies of Topeka, and above it a full-size model of the Independence bell, formed entirely of agricultural products of the State. Wheat-stalks are on exhibition from five to six and a half feet high, with heads, some of them, six inches long, and corn is shown up to seventeen and a half feet high, with ears twelve to fifteen inches in length, and eight to ten feet above the ground, there being from seven to thirteen ears to the stalk. One wing of the building is appropriated to Colorado, whose exhibit is confined exclusively to wild animals and birds native to that State. Near by is a very small building which serves to give Virginia at least a representation, and next to it is that old-time structure, attracting attention from every one, the "New England Log-House"—a little low building, characteristic in its appearance of the early colonial days, with a rustic portico in front, over which is a quaint sentence, "Ye Olden Time," and on one side is nailed a horse-shoe to scare away the witches. On entering we find ourselves in a room of one hundred years ago. Ancient dames in flowered gowns are spinning and performing other domestic duties. In the open fire-place is a spit with a turkey slowly turning and roasting for a Thanksgiving dinner. There are shelves of old crockery, plain-fashioned furniture of that time, and on one of the tables an old clasped Bible. Herbs and other stores of the careful housewife are hanging from the rafters, and in the room adjoining is a canopied bed with a patchwork quilt, and alongside a little old cradle. All are veritable relics of a farmer's home of a century ago, and it is to Miss Southwick and her able corps of assistants from New England that we are indebted for this picture of our forefathers' life. We must complete the realization by partaking of some of the New England dishes so deftly prepared by the good ladies, such food as the old Puritans grew and waxed strong on.

Main Building—Egyptian Court.

In striking contrast to this is the next building, an Algerian booth, devoted to the sale of trinkets. The attendants are dressed in national costume, true to character, but whether genuine natives themselves or not, is a doubtful question.

Passing up the adjacent avenue, we soon reach Agricultural Hall, with its Gothic gables and huge green roof, not very prepossessing externally, its design, however, exemplifying a somewhat novel adaptation of the materials used, wood and glass, to a construction of large size, giving economy in cost, capability of rapid erection, and at the same time producing a fine interior effect. The building consists of a nave eight hundred and twenty-seven feet in length by one hundred feet in width, crossed by three transepts, the central one having a breadth of one hundred feet, and the two end ones seventy feet each. These avenues are formed of shallow Howe-truss arches, springing vertically from the ground and laid together at the top in Gothic form, the height from floor to point of arch in the nave and central transept being seventy-five feet, and in the end transepts seventy feet. The courts enclosed between, and the four spaces at the corners of the building, are covered with roofs of ordinary construction,—the whole comprising a rectangular area of about seven acres.

The exhibits are very profuse, nearly all of the foreign countries being well represented. Huge stacks of wine-bottles from Spain, Portugal and other wine-growing countries rise up before us. Brazil sends an immense cotton trophy. In the United States section are the great mowing-, reaping- and binding-machines, drills, thrashing-machines, and all those farm appliances for which we are becoming so famous, attracting the notice of a large number of agriculturists from abroad. At the upper end of the nave an old wind-mill looms up, and near it is a large collection of stuffed animals and some skeleton specimens of the extinct fauna of another age. The tobacco exhibit is exceedingly perfect, and just beyond it we find a most complete display of the processes of India-rubber manufacture. On one side is the plant, next the raw material, and then the different methods of working, and the resulting products ready for the market, all exhibited in regular order.

Leaving by one of the eastern doors, we enter directly into the Pomological Annex, now being used for a poultry show. Such a chirping and chattering never saluted one's ears before. Not only fowls, ducks, geese, etc., of all kinds are shown, but all varieties of pigeons, most lovely cooing doves, rare birds of beautiful plumage, canaries, fancy rabbits, etc., making a most attractive and unique display. Several hours might well be spent here with profit and pleasure if we had but the time at our command.

Near by is the Wagon Annex, filled with farm vehicles of all descriptions, and on the other side is the Brewer's building, where all the operations of manufacturing that favorite beverage may be observed. Then on the hill to the east of us are the numerous wind-mills for pumping water and performing other duties, and down below them, nearer the Belmont Ravine, we come to the butter- and cheese-factory, where we find ourselves among churns of all kinds, the industrious exhibitors actively engaged in transforming cream into butter, or pretending to do so, the same butter probably doing service for several weeks, each one trying to convince you that his churn is the best, and that if you propose starting a large dairy, it is the only one you should purchase. In another room are cheese-presses, and in yet another, long rows of cheeses drying and seasoning to be ready for the market. Cheese-making in the United States has become a very large and growing business, all of the more celebrated European varieties being successfully imitated, and we are surprised to learn of the great quantities annually exported to Great Britain and the West Indies.

Farther on is the Tea and Coffee Press Building, where are exhibited all the different sorts of apparatus as used in our large hotels for making tea and coffee, and where one may step in with his pocket-lunch, and sitting down at a small table, or standing by the counter, order a nice hot cup of coffee or tea to take with it; or if he wishes a more extensive repast, he can step into the two-story car of the single-rail elevated railway near by, and in an instant, almost, be transferred over the ravine, through the tops of the trees, to the other side, where the German Restaurant will provide all he can possibly desire.

View of the International Exhibition from Belmont.

Another day we determine to visit the various State and National Buildings, or such of them as we have not already seen, and after entering the grounds, we pass up Belmont Avenue to its farther end, beyond the Government Building, and turn into what is called State Avenue. Just at its entrance is located the Centennial Fire Patrol, where fire-engines are kept in constant readiness for any necessity, the splendid horses harnessed and the attendants on hand, to move at a moment's notice, forming a model exhibit of the Fire Service as now employed in all of our great cities. Next in order on our right is the Ohio Building, a substantial cottage-like edifice, constructed of Ohio sandstone, and showing samples from the various quarries in the State, of this well-known material. The roof with its alternate squares of tin, painted in colors of marked contrasting shades, produces rather a curious effect, not to be commended. Adjoining this building is that of Indiana, which evidently was never favored with the services of an architect in the preparation of its design, the sky-line of the gable front being beyond all criticism in ugliness. Then next is Illinois, a plain, neat, frame building, of Gothic outlines, painted white and having an open portico around it, a fair specimen of a comfortable residence in a Western village, but nothing more. Wisconsin follows, also a simple structure, not pretty, merely useful, succeeded by the Michigan Building, quite handsome in contrast with the others, having porches and balconies, and ornamented with scroll-work. For fear of exhibiting too much good taste in the same neighborhood, New Hampshire steps in next, with a frame structure of what might be designated the "no-style" order of architecture—certainly never intended as an example for the improvement of our people in æsthetics. We proceed on to Connecticut, which presents a good, substantial, rural-like building, somewhat after the old Colonial style, having the motto "Qui Transtulit Sustinet" over the entrance-porch; and see beyond it, Massachusetts, with an edifice of considerably greater pretensions than most of the others, designed in good taste, commodious and well built. Then comes little Delaware with a small sea-shore cottage, and Maryland, her sister State, adjoins on the other side with a neat and unassuming structure. Next is Tennessee, followed by Iowa and Missouri—all small buildings of no special attraction.

We wander in and out of these various houses, more for the purpose of saying that we have seen them than from any particular interest that the mass of them incite. A few contain very interesting exhibits of local productions from the States they represent, but generally they appear to serve merely as a rendezvous or headquarters for the people of the State, books being kept in a conspicuous position for the registering of names and for reference. Many curious incidents are told of old-time friends and relations meeting at the Exhibition who had not seen or heard of each other for years, each one supposed by the other to have departed this life long, long ago. The State buildings have contributed not a little towards this bringing together of those originally from the same section of country. The last building in this row is Rhode Island. Here we have the work of an architect without doubt—a little gem in its way, about twenty by forty feet area, with an addition to the rear, and an open porch in front. The construction is in solid timber, the frame-work showing on the outside, and the roof is of slate. The architects have done themselves credit. Across the avenue from Rhode Island is Mississippi, a pretty, rustic structure, built of native wood still covered with bark, the whole decorated with the hanging moss so profuse in the Southern States, and producing quite a picturesque effect. Retracing our steps, we first refresh ourselves at the Hungarian Wine Pavilion, and passing a restaurant nearly opposite the Missouri Building, soon reach the California Building, a heavy structure of no beauty, containing in the interior, however, some interesting exhibits. Still farther back, opposite to Delaware, is the New York Building, designed in the Italian style, with porticoes and a sloping roof with gablets, the whole surmounted by a sort of tower or cupola, and reminding one forcibly of the frame residences so much in vogue several years ago in the neighborhood of the Empire City.

Turning down an avenue to our right, and passing in front of the New York Building, we see, a little beyond, something that strikes the eye at once—a group of three structures in that picturesque, half-timbered, sixteenth-century style of Old England, so expressive of the home-

View of the International Exhibition from George's Hill.

liness of the English character, that we know at once they can be no other than the buildings of the British Commission. The principal one, covering an area of almost five thousand square feet, is called St. George's House; the others are the Barrack's and Workmen's Quarters. The architect of these buildings was Mr. Thomas Harris, of London. The walls are of half-timber work, with lath and rough-cast plaster between, the base being a plinth of red brick, coped with stone. Lofty chimney-stacks, also of red brick, well grouped together, tower above the roofs, which are covered with red plain tiles, tile-ridging, hips, and finials, sent from England for the purpose by Messrs. Eastwood & Co. The windows are glazed in lead quarries, the opening casements being of wrought iron. A kettle-drum is being given this afternoon by the British Commissioners, and we will avail ourselves of a polite invitation to visit the interior. The walls of the various apartments are hung with exquisite designs of English papers, provided with paneled dados, and the wood-work is stained dark and varnished. In the hall and verandas are Minton encaustic tiles, and on the floors of various rooms are rich, illuminated Indian carpets, subdued in style, however, to correspond with the "lovely" furniture which so enchants every one. Beautiful ornamental vases, elegant table-plate, etc., from Elkington, of London, and damascened works of art, adorn the rooms, while English water-color paintings and well-selected engravings are on the walls. Open grates, with mantles of artistic tile-patterns, are in every room, enhancing the domestic and home-like aspect—just such a house, we say, as one would like in the country, with the addition of a few more porticoes to suit our climate. The artistic manner in which the house is fitted up, calling forth the admiration of all who see it, is due, we are informed, to the excellent taste of Mr. Henry Cooper, who came specially from London to attend to it, and whose praise is in every mouth. After passing from room to room, conversing with our friends and partaking of the hospitalities of the occasion, we bid adieu and pass on our way.

In the rear of the Barracks is the Japanese Dwelling, that curious structure erected by those peculiar workmen whose methods of work were to us so novel. It is related that during its construction, one of our own people loaned to the Japanese some wheel-barrows for the purpose of removing the earth from the foundations. They tried them for a while, but finding difficulty in wheeling them according to our custom, finally gave up the attempt and carried them, fore and aft, like hand-barrows, to the great amusement of the bystanders. We notice the odd-looking tiled roof, the entrance-door with its gabled projection, the arrangement of the walls of the building so that they may be opened and closed laterally by sliding panels, giving an airy house in fine weather, or a close one during storms, and then proceed to examine the interior, under the guidance of one of the residents, by whom we are shown the various rooms and also some beautiful fans, vases and other Japanese handiwork.

Across the avenue from the Japanese Dwelling are the buildings of the Spanish Government, one of them used as headquarters for the Spanish soldiers brought to this country, and the other containing a most interesting collection of exhibits, notably that of the Spanish War Department, consisting of beautifully constructed models of fortresses, barracks, etc., specimens of mountain artillery, fire-arms, models of field artillery and pontoon trains, specimens of the celebrated Toledo sword-blades, etc. An exceedingly large proportion of the exhibit is on the subject of education, and although Spain is far behind most other countries in this respect, yet she shows a most commendable desire for advancement. Here are seen specimens of pupils' work, desks and other school-furniture, text-books, scientific and philosophical instruments, engineering and architectural models, maps, also decorations, mosaic, inlaid work, a fine collection of woods, and many other things worthy of close study. Near by, West Virginia is represented by a very neat building, two stories in height, which, in addition to being the headquarters for the State Commissioners and visitors, contains quite a large exhibit of minerals, coals, ores, agricultural products, etc. The next building is that of Arkansas, an appropriate octagonal structure of timber and glass, designed as an exhibition building, and containing a large display of the agricultural and mineral productions of the State. Just east of Arkansas, and close to the narrow-gauge railway, is an edifice that appears modeled after an old Grecian temple, but which upon close examination proves to be the Canadian Log-House, erected by the Canadian Commission, and constituting an exhibit of the timber of that country. The heavy vertical logs distributed symmetrically, and supporting the roof of the portico around the structure, strongly resemble Doric columns, while the arrangement of planking, boards and lath, and the construction of the roof with its cupola or ventilator, all

United States Government Pavilion.

show a considerable amount of ingenuity and taste. Close by, at the junction of Fountain Avenue with the Avenue of the Republic, is the fountain erected by contributions from the numerous societies of the Catholic Total Abstinence Union of America. It consists of a circular platform, from which four arms project out at right angles with each other, each arm terminating in a smaller circular platform. In the centre is a mass of rock-work of marble, sixteen feet in height, crowned by a statue of Moses smiting the rock. From this the water descends out of numerous fissures into a basin forty feet in diameter. On each of the four smaller circular platforms is a drinking-fountain twelve feet in height, surmounted by a statue, the four statues being Father Mathew, Charles Carroll, Archbishop John Carroll, and Commodore John Barry. The work is as yet only partially completed.

The day is drawing to a close, and we descend the massive flight of steps from the Fountain Plateau to the Avenue of the Republic, stopping a moment, as we pass out, to glance at the Vermont Building, a small plain structure, and to obtain a cup of warm coffee at the Turkish Café adjacent, an ornamental octagonal building with a heavy projecting roof, and painted in an attractive oriental style. Taking a seat at a small marble table, we are handed a beverage that might be considered enjoyable perhaps to one accustomed to Turkish manners and customs, but to our own taste proves anything but agreeable. At the Jerusalem and Bethlehem Bazaars near by, we find for sale a great variety of trinkets, rosaries, etc., and articles made from olive wood, all of which are evidently genuine, although this cannot be said for the wares sold at many of the other booths. Wandering on towards the exit-gates, we pass the Pennsylvania State Building facing the lake, quite a pretentious Gothic structure of two stories, with tower, and containing the usual reception-rooms and offices observed in all of these State buildings.

There is still a portion of the grounds that we have not seen, and taking another day for this purpose, we pass out to the western end of Machinery Hall, first entering a building erected by an enterprising manufacturing firm for the exhibition of stoves, ranges, heaters, etc. Another, near by, is used by an opposition firm for the same purpose, and a little to the west of this is a small building painted in divers colors, like Joseph's coat, which proves to be a patent paint exhibit. Still farther on is the Saw-Mill Annex to the Machinery Hall, a substantial, open, shed-like structure, covering an area of two hundred and seventy-six by eighty feet, and containing a large and interesting display of steam saw-machines, gang-saws, etc., principally for wood, but including also several very excellent stone-cutting machines, where the practical use of the black diamond is fully exemplified. A boiler-house close by supplies all the steam required for running the machinery of this building. Crossing the narrow-gauge railway we come to a large glass-ware exhibit, in a one-story frame building, in which the process of the manufacture of various articles in glass is shown on quite an extensive scale. The house is always crowded with curious visitors, making it difficult for us to observe the work as closely as we would like, but it is marvelous to see with what dexterity the material is fashioned into articles of ornament and utility. Purchasing a tiny glass slipper as a souvenir, we move on, glancing at a saw-mill near by, and then following the line of the narrow-gauge railroad past its engine-house, until we reach an exhibit of the Pennsylvania Railroad Company, consisting of a section of railroad double track, laid in complete shape, with ballast, ties, steel rails, etc., according to the standard rules and regulations of that Company. Just across the narrow-gauge road this same Company exhibits an interesting relic in the shape of an old locomotive, the "John Bull," and an attendant train of cars. The engine was constructed by Messrs. George and Robert Stephenson, at Newcastle-upon-Tyne, England, in 1831, for the Camden and Amboy Railroad Company, and had its first public trial near Bordentown, on November 12th of the same year. In 1833 it was put into active service, continuing in use until 1866. The train consists of two odd-looking passenger-cars, being of the identical ones formerly drawn by the "John Bull," and built about the year 1850, the whole train presenting quite an old-time appearance. The engine and cars were lately repaired and put into working condition, and were actually run from the shops of the Company, near Jersey City, to the Exhibition grounds, as we now see them, making the journey of nearly ninety miles at an average rate of two minutes and thirty-five seconds per mile.

In this same locality is a building containing a complete working model illustrating the

New York State Pavilion.

Krohnke Silver Reduction Process as used in Chili for the reduction of the rich ores of the sulphurets and antimoniates of silver. This model, which is kept in operation as continuously as practicable, is made to one-sixth full size, and was originally constructed for the Valparaiso Exhibition, where it was shown last year. The process of working is divided into three parts or sections, each having a separate building. The first comprises the crushing-machinery, by which the ore is broken up and pulverized. It is passed through a double set of crushing-rolls and then into pulverizing-mills, from which it is carried off by a constant stream of water into settling-pits. From these it is shoveled out and thoroughly dried, when it is ready for chemical treatment, amalgamation, etc., which takes place in the next building. The retorting and smelting are carried on in the third building—the silver mass, after the mercury is volatilized, being melted down and cast into bricks. The model is exceedingly complete, and makes a very interesting exhibit to the metallurgist, the process being one of great thoroughness, not leaving behind, it is said, over one ounce of silver per ton of ore treated, and sometimes giving even higher results than shown by assay. On account of the great cost of the plant, however, compared with capacity of working, this does not appear to be an available method except for very rich ores.

North of this exhibit, near Machinery Hall, is a structure containing a display of the various printing-presses manufactured by the Campbell Press Company, of Brooklyn, N. Y., together with specimens of type-printing, it is stated, from the date of the invention, and a complete printing-office, modeled after those of 1776, in actual operation.

Returning towards the rear of Machinery Hall, we pass some gas-machines of various kinds and iron pipe exhibits, look for a moment at some hoisting machinery, then at pneumatic tubes, busy transmitting messages from one end to the other, and stop to examine a gunpowder pile-driver, near by, in active operation. Entering a building containing a large exhibit of special iron-castings, lamp-posts, hydrants, stop-valves, etc., not, however, of any special interest, we soon move on, entering and passing through Annex No. 3 of Machinery Hall, into an area devoted to machinery for brick-making, rock-drilling, artesian-well boring, etc., and in which, next to Machinery Hall, is located a building of the State of Nevada, containing a quartz-mill in full working, separating gold from the rock, according to the most approved method. Having satisfied our curiosity on all of these exhibits, we go on through Annex No. 2 and through the Hydraulic Annex, past the various boiler-houses, to the Shoe and Leather Building, a neat structure of about one hundred and sixty feet in width by three hundred and fourteen feet in length, devoted to exhibits of all kinds connected with leather and its manufacture into the numerous articles of the trade. We are struck, on entering, with the tasteful interior decoration of the roof in red, white and blue bunting, and the exhibits prove of considerably more interest than we had expected they would be. Near to the rear entrance-door is a heavy tanned hide, which we find to be that of the great elephant "Empress," which died a short time since at the Zoological Gardens. The material is of great use for polishing purposes. We see here excellent trunks, fine harness and saddles, all sorts of saddlers' furnishing goods, boots and shoes, including some exceedingly curious and handsome varieties from Russia, India-rubber and other fabrics, all kinds of leather, morocco and sheepskin, and a large amount of leather and shoe machinery. It is wonderful how far machinery has been applied to the making of boots and shoes, so reducing their cost, and giving to New England the supremacy of the world in that manufacture. Here is a machine for sewing soles to boots and shoes, that will sew nine hundred pairs per day. It is almost impossible to believe it, and yet it is said on good authority that thirty-five million pairs are annually sewed on these machines in the United States. Near by is a riveting-machine, which will rivet on three hundred pairs of soles per day, and around us are numerous machines for trimming, heel-burnishing, pegging, etc. We are lost in wonder. Truly, Yankee invention is equal to everything.

We make our exit by the eastern door of the building, and passing through the exhibit of the New England Granite Company, consisting of various specimens of stone, monuments, etc., out in the open air, we find ourselves back again at the plaza of the Bartholdi Fountain, whence we started many days ago to explore the Exhibition. We have now been over the whole ground. We are through—we have seen everything. Have we? We hear some one say, Yes. No! we reply, we have not. We might take you again and again, on this pleasure-trip, through the various buildings, out on the avenues, down in the ravines and into the shady nooks, and you will find many new things, many exhibits, that must have been there before, and yet which you passed over and either did not see or have forgotten. It would take the full six months of the Exhibition, and perhaps more, to see all thoroughly. The body tires, the feet wear out, but the enjoyment of the eyes—never! We are different beings now from what we were before we came here. We have advanced in our ideas fifty years. We go

New Hampshire and Connecticut State Pavilions.

home with new inspirations and with enlarged capacities, ready to do our share in the advancement of our country in the Arts and Manufactures during the new century that has just dawned for it.

While all these happy days are passing, one department connected with the Exhibition has been hard at work, that of the Judges, flitting from one building to another, studying this exhibit and that one, making comparisons, testing and experimenting, so as to be fully prepared to give fair and impartial reports. Their work having been accomplished, the evening of September 27th is set apart by the Centennial Commission for the announcement of the awards. The ceremony takes place in Judges' Hall, in the presence of a brilliant audience of invited guests. On the stage are the distinguished officers of the Centennial Commission, the

Board of Finance, the Board of Judges, the various Foreign Commissioners, and many others. The exercises are opened with prayer, and after an appropriate anthem is rendered by the Temple Quartette, of Boston, Hon. D. J. Morrell, Chairman of the Executive Committee of the Commission, gives an address. The orchestra then strikes up a medley of national airs, and the Director-General follows with some remarks in reference to the Exhibition, its appropriateness, the benefits that will result from it, the profound impression produced by the high standing and qualifications of the gentlemen connected with the various Commissions, their close attention to their duties, and the great degree in which the Exhibition is indebted to them for its success. He also refers to the eminent body of men, both foreign and American, selected as Judges, the delicate and difficult task they were called upon to perform, and the good will, earnestness and zeal with which they accepted the charge and carried out their work. After another interlude of vocal music by the Temple Quartette, the President of the Centennial Commission moved forward to the front of the stand and explained the system of awards, the departure made from that usual at previous exhibitions, and the advantages derived from the change. He dwelt upon the obligations due to the tens of thousands of exhibitors, many of whom, not only from the United States, but also from other countries, were here to testify their good will in this our fraternal year; also on the many purely governmental exhibits, and the friendly interest shown by many sovereigns, tending to perpetuate international friendship; and in conclusion he stated that the awards would be announced to the several countries in alphabetical order, giving no precedence to one over the other, and that if any were warmer friends than others, he trusted they were those with whom we had sometimes quarreled. He then called forward in alphabetical order the Chief Commissioners of the various governments, and delivered to them copies of the awards made to the exhibitors from their several countries. As the list of nations was called, beginning with the Argentine Confederation, and as the representative of each, respectively, stepped forward to receive the roll containing the list of names for his country, he was received with enthusiastic applause. This portion of the exercises took considerable time, after which the evening closed with music.

The total number of exhibitors amounted to twenty-six thousand nine hundred and eighty-six, of which eight thousand five hundred and twenty-five were from the United States. There were thirteen thousand one hundred and forty-eight medals awarded, being a little over forty-eight per cent. of the number of exhibitors, and five thousand one hundred and thirty-four of these awards were to this country.

Let us take a glance at some of the results which may be deduced from the Exhibition, more particularly in reference to our own country, its capabilities and development. First, in reference to that great industry, the Iron Manufacture. The exhibit of minerals is very large, and one fact is brought forth above all others, in that the United States give evidence of the possession of great mineral wealth. The Smithsonian Institute is represented by a magnificent collection; very many of the States have on exhibition the natural productions of their respective territories, well selected and arranged, and individual manufacturers also furnish numerous specimens. Immense coal exhibits show the presence of the required fuel to reduce these ores, and the display of finished iron and steel gives proof of the complete ability and metallurgical knowledge possessed by those connected with the manfacture, and necessary for the production of the best results. The large dimensions and thorough finish of the manufactured articles are evidence of the strength and perfection of the machinery as well as the skill of the men employed in their production. Immense iron-ore deposits exist all over the country. The amount of ore smelted in the year 1875 was about four million three hundred and seventy thousand tons, of which about one million tons came from Lake Superior, three hundred and fifty thousand from Lake Champlain, about one hundred and fifty thousand from the great Cornwall ore banks of Pennsylvania, and four hundred thousand tons from little

Ohio State Pavilion.

New Jersey. Iron Mountain, in Missouri, gave two hundred and fifty thousand tons. In 1875 there were seven hundred and thirteen blast-furnaces, with a joint capacity of about five million four hundred and forty thousand tons, and the puddling-furnaces were four thousand four hundred and seventy-four in number, the total capacity of the works for production of rails and other wrought iron being about four million one hundred and ninety thousand tons. American iron manufacture has kept pace with the age, and the works appear to be fully up to those of Europe as to the latest and best details of manufacture, in fact even serving as examples for the instruction of metallurgists from abroad, nearly all of whom have evinced the deepest interest in them, and have expressed the greatest surprise at the freedom with which information has been given and access allowed to what in Europe would be considered important trade secrets. Some processes were quite novel to them, and among these may be particularly mentioned that of cold-rolling, or passing a bar of iron a number of times through rolls when cold, and reducing it about six per cent. in its section, thereby materially increasing its tenacity and hardness, giving it a highly finished surface, and adapting it directly for shafting, piston-rods, etc., without further manipulation.

Many fine exhibits are made of American steel, the various processes of blister and puddled, crucible, Siemens-Martin, and Bessemer manufacture being fully represented, and the qualities of metal produced by the different methods will bear comparison with any in the world. The Siemens-Martin method is in very successful use, having an annual production of some forty-five thousand tons. The Bessemer process is shown to be fully up to that of England in its details of operation. Indeed it even surpasses it in perfection of machinery for handling the material. The capacity of the various works is about five hundred thousand tons annually, principally iron rails, and greater than that demanded by the railroads of the country. In the figures we have given on iron and steel, we do not mean to infer that the annual production is up to the full capacity of the various works, as it is not, but only that the works have a capability equal to that amount. The perfection to which the Bessemer steel works of the United States have arrived is due to the fact that the Government afforded a heavy protective tariff on steel rails just when it was most needed. At the time Bessemer works were first commenced in this country, steel rails were selling at one hundred and fifty dollars per ton, but when these works had gone into operation, the price fell to a hundred and twenty dollars, and now, to-day, the manufacturers are able to furnish rails at forty-five dollars per ton. This shows the value of a protective tariff and the good results coming from it when properly applied. Had there been no duty on steel rails, the works never could have been started, England never would have reduced her prices, and we would have been paying to-day very nearly what we did ten years ago. Perfection in machinery for these works, owing to American invention, has contributed not a little to these results, in enabling our manufacturers to turn out a greater number of casts per day than at any other works in the world.

In regard to the exhibits of iron from foreign countries, Sweden is especially conspicuous for the number and exceedingly high standard of her specimens, and their excellent arrangement. Sweden has long been noted for her close dependence upon scientific knowledge in reference to the proper manipulation of iron, and it may be said that to her is due the success of the Bessemer process, an invention which, on first application in Great Britain to the less pure form of pig-iron, was a failure, and it was only when Swedish experts showed its practicability, and Mushet suggested Spiegel-eisen as a corrective to the impurities in the iron, that the difficulties experienced were overcome. We must not omit to mention, in this connection, the fine display made by Prussia of this Spiegel-eisen, so essential in the Bessemer manufacture, one of the few materials of which we are as yet so deficient, only a little coming from New Jersey and Connecticut, and almost all that is used has to be imported.

No one, unless particularly informed on the subject, would have supposed that the United States could make much display at the Exhibition in "Ceramic and Glass Wares," and would have been much surprised to learn that out of five hundred and ninety-two exhibitors, one hundred and ninety-nine were from this country. Such, however, is the case, and the display is an important one, not only on account of its extent, but also from the fact of its showing the existence of an abundance of excellent natural material, and the requisite industrial skill to manipulate it. The resulting wares are here in direct competition with those of the same kind from Great Britain and other European countries, and they challenge comparison without fear. Taking into consideration the vast extent of the general display, including porcelain of all kinds, hard and soft, biscuit, Parian, stone-ware, glazed and unglazed; stone china, "granite" ware and the softer cream-colored wares, faïence, majolica and Palissy wares, terracotta, tiles, etc., our own exhibits, while more of the practical and useful kind, are really very satisfactory. The industry has developed in this country with most wonderful rapidity, reflecting great credit on the ability and energy of those who have taken hold of it, most of them without previous training or knowledge, and in the face of innumerable difficulties.

The Art Gallery.

Heavy and coarse wares were manufactured in the United States as far back as the middle of the last century, and more than one hundred years ago porcelain works existed for a short time in Philadelphia. During the war of 1812, numerous potteries were started, but ceased to exist under foreign competition after peace was restored. A determined effort was made again in 1830 in reference to this industry, by establishing a porcelain manufactory in Philadelphia, but it closed in a few years, involving the founders in considerable loss. After this time a number of potteries for coarser wares, gray and yellow stone, sprung up, and they have been generally successful. About the year 1854, however, the subject of the manufacture of a higher grade of wares, such as the English "white granite," was taken up at Trenton, New Jersey, and after long labor, many efforts, and much loss, the industry was established on a firm footing in 1866, resulting in a commercial success about 1870. Other manufactories have developed in various parts of the country from this, and there are now works situated in Chicago and Philadelphia, at Greenpoint, New York, where porcelain as well as earthenware is made, and in Ohio,—Trenton, however, being the chief point of production, and rapidly becoming, as it were, the Staffordshire of America. The wares exhibited are of most excellent body and glaze, entirely free from iron spots or other impurities, showing a high quality of material and a great perfection attained in what may be said to be almost a new body in pottery wares. The glazes are of good medium hardness, well incorporated with the body, and have, it is claimed, little tendency to crackle—far less than foreign wares. They receive colors well, and although the decorations as a general rule are deficient in originality, and often copies of foreign designs, it is to be hoped that the results of the Exhibition and the efforts of our Art-schools will make a great change for the better in this respect.

In reference to wares from abroad, Great Britain comes first in importance, displaying a large range of manufacture, from objects of the finest texture down to cheap household goods. The porcelain, having a body compromised between hard paste, like that of Dresden, and soft, like old Sèvres, is compact, homogeneous and translucent, and the glaze hard and brilliant. Most excellent table-, dessert- and tea-services are shown, and large collections of decorative objects of exceedingly artistic design and execution. One variety, called "Ivory" porcelain, is very elegant, having a soft rich surface and most agreeable tone of color, some specimens being delicately perforated, showing great skill in manufacture. A large and interesting exhibit is made of ornamental stone-ware, showing its application to architectural decoration, a new and most successful use for this material. The specimens of stone-ware for sanitary and chemical purposes are very fine, and among these may be mentioned a sewer-pipe fifty-four inches in diameter, and a stone-ware jar of six hundred and twenty gallons capacity. The terra-cotta exhibit is very large, the most important object being the colossal group of "America," in Memorial Hall, reproduced from one of the corner groups of the Albert memorial in Hyde Park. A pulpit of combined red terra-cotta and stone-ware produces a very striking effect, as also a large wall fountain and a font. One should notice in these the elaborately wrought out relief-work, scarcely ever attempted so successfully. Terra-cotta is now being quite effectively employed in architectural works for decorative purposes, not only in Europe, but in the United States, and a large field is opened for its use. A Chicago firm has developed the manufacture in this country to a high degree. The display of English floor and wall tiles is very fine, most of the large manufacturers being well represented.

France has a large and interesting exhibit of porcelain and other ceramic wares, Palissy, majolica and decorative faïence, and one will never forget the exquisite terra-cotta statuettes of M. Eugene Blot & Son, illustrating fishing-life at Boulogne, so full of artistic expression, and having such force and freedom of touch. In the Memorial Hall are some very large and elaborate vases from the National Manufactory at Sèvres, all fine examples of that kind of work. Among the other European exhibitors, Sweden is worthy of particular mention for an excellent display of porcelain and pottery of various kinds, showing evidence of energy, enterprise and skill fully adequate to make her independent of other nations in this industry.

Of course the exhibit of porcelain and pottery from Japan is far beyond that of any other nation in importance, not only in the extent of its collection and its varied character, but in its general high standard of excellence and in the great superiority of its individual specimens. Taking into account the nature of the material, many of the pieces shown are really colossal in size, and not only are many curious objects of early date exhibited, but also imitations or reproductions of the ancient wares on a large scale, and so accurately as to defy detection. Vases are shown six and eight feet in height, perfectly potted, and fine examples of effective decoration. Large flat slabs of decorated porcelain are exhibited, one nearly six feet in diameter, finished and glazed on both sides, and showing no marks of points of support in the

The Japanese Pavilion.

oven, being most remarkable pieces of work. It has been stated by one fully capable of giving a reliable opinion, that the Japanese display surpasses anything that has ever been shown by a single country at any previous International Exhibition.

Concerning the exhibit of glass, that from the United States is large and important, including almost all descriptions of ware, and it is evident that before many years America will successfully compete in all branches of this industry with the countries of the world.

In the exhibition of "Chemical Products," the display is very large, showing great excellence, particularly in the collections of pharmaceutical chemicals displayed by American firms. Philadelphia, especially, having been long celebrated for her chemical manufactures. Important exhibits are made from our own country, as well as England and Germany, in mineral-oil products and those of alkali manufacture.

In textile fabrics, the manufacturers of the United States show very decidedly the vast progress they have made in the various branches of this industry, and give striking proof of their ability to cope with foreign competition in these goods. The exhibits of cotton, linen and other fabrics from abroad are not nearly so extensive as might have been expected. The collection from France is very scanty. Some goods of excellent quality are shown from Würtemburg and Elberfeld in Germany, and Hanover furnishes a most artistic display of cotton velvets and velveteens, resembling silk in appearance, and particularly noticeable for their texture and finish, and for the variety and blending of the colors. An admirable and unsurpassed display of woolen tweeds and cassimeres, heavy cheviots, flannels, woolen blankets, heavy sheetings, etc., comes from Canada, and Ireland takes the lead in linen fabrics, although the goods sent from Dresden and other noted European localities are of exceedingly high class and fully up to their well-deserved reputation. The exhibit of oil-cloths and other enameled tissues is exceptionally fine, and the display of American exhibitors unrivaled, nothing contributed by foreign exhibitors being equal to it. The raw cottons are almost all American, although some excellent specimens come from Brazil, and small samples from very many other localities.

The United States has a very large and most important exhibit of wool and silk fabrics, outrivaling that made by the cotton manufacturers, and the industry as far back as 1870 involved a capital of about one hundred million of dollars, and nearly three thousand establishments. The display of Great Britain is very fine, notwithstanding that some of the most enterprising English firms are not represented.

In carpets our own country makes a very large display of all the leading varieties, the specimens being well made and containing a good combination of colors, arranged with taste. Too many of the designs are copies of foreign goods, although some are original, but it is hoped that this defect will be remedied in time, and that the manufacturers will see the policy of employing competent designers of their own. Our carpet industry is becoming very rapidly a most important one, entering into competition most successfully with foreign importations, and it deserves every encouragement. Great Britain makes a very choice display of carpets, also France and Belgium,—the tapestry carpets of the latter being of most admirable design and color.

In jewelry, watches and silver-ware the United States makes a most excellent exhibit, and in reference to watches, has caused great consternation among the Swiss manufacturers, owing to the superior facilities which this country shows she possesses for their manufacture, and the very high standard which she has attained in their quality.

In paper, stationery, printing and book-making, the majority of the exhibits are from the United States. The great natural facilities for paper-making possessed by the country, and the ingenious adaptations of machinery to the processes—no hand-paper being now made—have added very much to the development of the industry. A very large variety of printing-presses are shown, from the old original press of Franklin, down to the large and powerful machines of the present day,—the English Walter press, and the American Bullock, Hoe, and Campbell machines.

Hardware forms a most prominent display, and the exhibits for building and household use from the United States are remarkable for variety, beauty of design and artistic finish, surpassing all those from foreign countries in these points. Locks have formed an American specialty since the day of the Exhibition of 1851, when Hobbs picked the famous Bramah Lock, and the number and variety that are now manufactured may be called legion. The combination- and time-locks for safes seem to be very much admired by our cousins from abroad. In edge-tools more than half the exhibitors are from this country, and the quality of the material is without any superior. Every one knows the world-wide reputation of the American axe, and the enormous demand for it in foreign countries. All hand-tools for carpenters' use show marked improvements, due to the inventive genius of the Yankee. In agricultural and laborers' tools, very marked advances have been made in the United States,

United States Government Building.

they being much more solidly and permanently constructed, while at the same time possessing greatly increased lightness and a freedom of working that is very desirable. Exhibitors in cutlery from this country make a most extensive and handsome display, showing great improvement both in the style and quality of their goods. There are exceptionally fine exhibits from Great Britain, Germany and Russia, and it seems to be generally admitted that while America holds the first place in table cutlery, tools, and fire- and burglar-proof safes, Great Britain has the pre-eminence in pocket and fine cutlery.

In railway plant, rolling stock, engines, etc., the number of foreign exhibits is very limited, although most of those displayed possess peculiar merit. Thus we notice important switch-locking and signaling systems from Great Britain, tires and axles from Sweden, etc.

As would naturally be expected, however, the mass of the exhibits in this department comes from America. We have the permanent way of the Pennsylvania Railroad, the rolling-stock of the Pullman Palace Car Company, the Miller platform and coupling, the various styles of springs, a most prominent series of car-wheel exhibits, the Westinghouse, Smith's and Henderson's car-brakes, etc. There is a large exhibit of American locomotives, but only one from abroad, a narrow-gauge Swedish engine. To foreigners, our railway exhibits have been most interesting, presenting the peculiar features of a system different in many respects from anything in their own countries.

The United States is largely represented in hydraulic motors, transmitters and pneumatic apparatus. There is an extensive exhibit of Turbines, generally of excellent design and workmanship, and although mostly constructed in the usual form, in some cases presenting features of novelty. The collection of shafting and belting is specially prominent, and the cold-rolled shafting of Messrs. Jones and Laughlin, of Pittsburgh, is particularly noticeable. A very large exhibit is made of pumps of the various classes, and the admirable arrangement of the Hydraulic Section of Machinery Hall enables them to be shown to the utmost advantage.

The display of machinery is greatly in excess of anything at previous exhibitions, and the United States is far ahead of any other country. Those best qualified to judge state that it is really a most remarkable exhibit, full of new ideas, refined in mechanism, and most encouraging for the future. The display of machine tools, especially, has never been equaled, either in number, quality or adaptability, and is full of novelty and progress. Great Britain shows a magnificent exhibit of steam hammers and some textile machinery from Leeds, but otherwise is exceedingly meagre. Canada, for a young country, makes an excellent exhibit. The display from France, although small, is very fine, the wood machinery of Arbey, of Paris, being specially worthy of attention. As we have devoted a large space in our "Mechanical and Scientific" department, to this branch of the Exhibition, we cannot do more than refer to it here.

A very large exhibit is made in sewing, knitting and embroidering machines, clothes-making machinery, etc., sewing-machines of course taking the first position, both in number and importance, all, or nearly all, coming from the United States. America has always occupied a very prominent position in sewing-machines at previous exhibitions, and it was only to be expected that she would in this instance make a display surpassing anything that has ever been seen before. That she has done this, every one will admit. The competition between rival firms is very great; new improvements are constantly being made, and each manufacturer endeavors in every way possible to keep a front position with the public. All sorts of machines are exhibited. There are the family machines, those for cloth, shoes, and even for boots, harness, saddles, etc., all doing most wonderful work; and to choose a machine, or to decide which is the best or most worthy of award, must be the most bewildering work that ever mortal man was entrusted with. Some of the knitting-machines are very curious, and a very novel apparatus is exhibited for darning stockings.

In electric and telegraphic apparatus some notable exhibits are made. Gray's, Edison's and Bell's Telephones may be mentioned among others as having a most brilliant future before them. The end to which they may develop, and the immense value they may prove to the world, no one knows.

The civil engineering exhibit from the United States is a very important one, although many most extensive and interesting works are not represented. Many exhibits are under the charge of the American Society of Civil Engineers, and a number of engineering works being prosecuted by the Government are represented by models in the United States Building, such as the Hell-Gate improvement in New York Harbor, the construction of breakwaters in the great lakes, models of light-houses, etc. France and Holland both make exceedingly fine displays of their Public Works.

The International Boat-Race.

In agricultural machinery the exhibits are confined, with a few unimportant exceptions, to the United States and Canada. Implements of tillage and planting, machines for thrashing, winnowing, corn-husking, and shelling, portable and stationary engines, grinding-mills, dairy fittings and appliances, etc. etc., are displayed in great profusion, showing a high degree of perfection attained by American manufacturers in this department, and attracting the attention of the agricultural world. The Exhibition will undoubtedly open a large foreign market for our people in this industry, particularly in South America and Australia.

Exterior to the regular Exhibition grounds, but under the same management, international live-stock exhibitions of great interest have been held, a large area having been enclosed and arranged with the necessary sheds, etc. Dogs, horses, cattle, sheep, and swine have had their turn in rotation, a few days being given to each, and to those particularly interested in live stock, the displays have proved quite an attraction.

At length the time of the Exhibition draws to a close. On the evening of the 9th of November the Centennial Commission gives a farewell dinner to the various Foreign Commissioners and other distinguished guests, making a company of about two hundred and fifty. The morning of the 10th dawns, and is announced by thirteen guns from George's Hill and from the steamer "Plymouth" in the harbor, simultaneously. Regret is felt by all, and yet it is not unmixed with a sensation of relief at the thought of coming rest—of a return to the quiet life of former times. But Philadelphia never will fall back to quite the old-fashioned routine. She has been thoroughly awakened and enlarged in her ideas, and will undoubtedly remain more cosmopolitan. The day sympathizes with the feelings of the people, and a slow and steady November rain pours down from the clouds, rendering utterly useless the extensive preparations that have been made in the open air for the closing ceremonies. There is no diminution in the number of visitors to the buildings—the records giving nearly one hundred and twenty-two thousand on that day—but the vast rows of temporary seats, one above the other, at the west end of the Main Building, facing the Bartholdi Fountain, look cheerless and deserted. The ceremonies must take place in the Judge's Hall, a very small building for that purpose, and to reduce the number of invited guests, admission is refused to ladies' tickets. Some few of the more adventurous, however, pass the guards, one lady claiming the right as a descendant of one of the signers, another as having had an ancestor on board the "Bon Homme Richard," etc. etc. On a platform at the upper end of the hall are the President of the United States, his Cabinet, the various Foreign Legations, the Centennial Commission and Board of Finance, etc.—the Philadelphia City Troop acting as Guard of Honor, an office they have always performed for every President of the United States when a guest of Philadelphia, since the time that they formed the Body-Guard to General Washington in the Revolutionary War.

The final ceremonies open with Wagner's Inauguration March, recalling vividly the scene of six months before; then a prayer, and after that a choral and fugue of Bach's. Addresses follow from the Chairman of the Executive Committee, from the President of the Centennial Board of Finance, the Director-General, and the President of the Centennial Commission, the intervals between each being occupied with musical selections by the orchestra and chorus. Next comes the hymn, "America"—

"My country, 'tis of thee,
Sweet land of liberty,
 Of thee I sing;
Land where my fathers died,
Land of the pilgrims' pride,
From every mountain-side
 Let freedom ring!

"Our fathers' God, to Thee,
Author of liberty,
 To Thee we sing;
Long may our land be bright
With freedom's holy light;
Protect us by Thy might,
 Great God, our King"—

during the singing of which, the original flag of the American Union, first displayed by Commodore Paul Jones on the "Bon Homme Richard," is unfurled and saluted by forty-seven guns, one for each State and Territory of the nation. The President of the United States now performs the last act of the drama, by declaring "the International Exhibition of 1876 closed," and at the same moment, by a touch of his hand on a telegraphic signal, the great Corliss engine—that pulse which has been the life of Machinery Hall for six months—is stopped, never to resume its work there again; and the audience rise up, and uniting with the grand orchestra and chorus, break forth in the Doxology—"Old Hundred"—

> "Be Thou, O God, exalted high!
> And as Thy glory fills the sky,
> So let it be on earth displayed,
> Till Thou art here as there obeyed"—

the chimes on Machinery Hall at the same time ringing out their last peal in honor of the Exhibition of 1876.

No such exhibition has ever been held before, either as to extent, number of admissions or receipts. From the 10th of May to the 10th of November, inclusive, there was a grand total of 9,910,966 visitors, of whom 8,004,274 paid admission fees, amounting to $3,813,724.49, 1,815,617 were connected with the Exhibition, and 91,075 had complimentary tickets. After the closing day, up to December 16th, there were 213,744 visitors, of whom 43,327 paid admissions amounting to $19,912; 168,900 were connected with the Exhibition, and 1517 had complimentary tickets. The total admissions, therefore, from the 10th of May to the 16th of December amounted to 10,164,489, for which the total receipts were $3,833,636.49. The largest number of visitors occurred on Pennsylvania Day, September 28th, being 274,919, and the smallest number on the 12th of May, being 12,720. There were nearly two hundred buildings on the grounds, and the narrow-gauge passenger-railway carried 3,812,794 passengers.

View in the Book Department, Main Building.

More books from CGR Publishing:
www.CGRpublishing.com

1939 New York World's Fair: The World of Tomorrow in Photographs

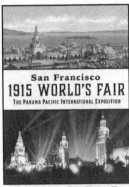
San Francisco 1915 World's Fair: The Panama-Pacific International Expo.

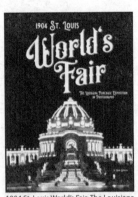
1904 St. Louis World's Fair: The Louisiana Purchase Exposition in Photographs

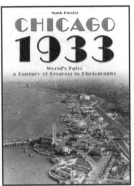
Chicago 1933 World's Fair: A Century of Progress in Photographs

19th Century New York: A Dramatic Collection of Images

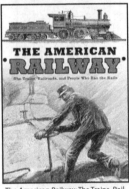
The American Railway: The Trains, Railroads, and People Who Ran the Rails

The Clock Book: A Detailed Illustrated Collection of Classic Clocks

The World's Fair of 1893 Ultra Massive Photographic Adventure Vol. 1

The World's Fair of 1893 Ultra Massive Photographic Adventure Vol. 2

The World's Fair of 1893 Ultra Massive Photographic Adventure Vol. 3

1901 Buffalo World's Fair: The Pan-American Exposition in Photographs

The White City of Color: 1893 World's Fair

Ethel the Cyborg Ninja Book 1

Ethel they Cyborg Ninja 2

How To Draw Digital by Mark Bussler

The Kaiser's Memoirs: Illustrated Enlarged Special Edition

Visit our complete catalog at:
www.CGRpublishing.com

Ultra Massive Video Game Console Guide Volume 1

Ultra Massive Video Game Console Guide Volume 2

Ultra Massive Video Game Console Guide Volume 3

Ultra Massive Sega Genesis Guide

The Illustrated History of the 1876 Centennial Exhibition Volume 1

Chicago's White City Cookbook

The Art of World War 1

How To Grow Mushrooms: A 19th Century Approach

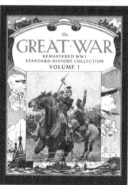
The Great War Remastered WW1 Standard History Collection Vol. 1

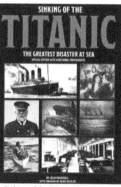
Sinking of the Titanic: The Greatest Disaster at Sea

All Hail the Vectrex Ultimate Collector's Guide

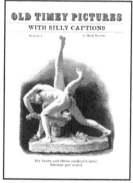
Old Timey Pictures with Silly Captions Volume 1

P.T. Barnum The Greatest Showman on Earth

Electricity at the World's Fair of 1893 Columbian Exposition

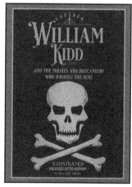
Captain William Kidd and the Pirates and Buccaneers Who Ravaged the Seas

The Aeroplane Speaks: Illustrated Historical Guide to Airplanes

THE MASTERPIECES

OF THE

CENTENNIAL INTERNATIONAL

EXHIBITION

VOLUME III

Made in the USA
Middletown, DE
07 October 2023

40248289R00309